Smart and Connected Health: Applications and Real-World Use Cases

Bahar Farahani • Farshad Firouzi •
Krishnendu Chakrabarty
Editors

Smart and Connected Health: Applications and Real-World Use Cases

Springer

Editors
Bahar Farahani
Cyberspace Research Institute
Shahid Beheshti University
Tehran, Iran

Farshad Firouzi
Arizona State University
Tempe, AZ, USA

Krishnendu Chakrabarty
Arizona State University
Tempe, AZ, USA

ISBN 978-3-032-09252-6 ISBN 978-3-032-09253-3 (eBook)
https://doi.org/10.1007/978-3-032-09253-3

This Springer imprint is published by the registered company Springer Nature Switzerland AG
The registered company address is: Gewerbestrasse 11, 6330 Cham, Switzerland

To my beloved father and mother, my first teachers, with gratitude and love.

—Bahar Farahani

To my father, Fereidoun Firouzi—always with me.
To my mother, Latifeh (Fatemeh), with deepest love.

—Farshad Firouzi

In loving memory of my father, Ramendra Narayan Chakrabarty.

—Krishnendu Chakrabarty

Preface

The rapid convergence of Artificial Intelligence (AI), the Internet of Things (IoT), edge-fog-cloud computing, blockchain, and other emerging technologies is fundamentally transforming healthcare, driving the evolution of Smart and Connected Health (SCH). This paradigm shift represents a profound transition from traditional, hospital-centric, and reactive healthcare models to a patient-centric, proactive, and data-driven model, facilitating personalized, efficient, and accessible solutions through real-time monitoring, intelligent data processing, and advanced analytics. This two-volume book serves as a comprehensive resource, offering both foundational knowledge and practical applications of these cutting-edge innovations.

Volume 1, *Smart and Connected Health: AI, IoT, and Trustworthy Technologies*, is designed to serve as both a reference guide and a textbook for students, researchers, and professionals. Covering essential topics such as machine learning, deep learning, Large Language Models (LLMs), sensor technologies, cloud computing, software architecture, security, and blockchain, this volume provides a structured introduction to the principles shaping modern healthcare systems. With a strong focus on foundational concepts and methodologies, this resource is well-suited for use in academic courses and training programs, equipping readers with the theoretical and technical knowledge necessary to develop SCH solutions.

Volume 2, *Smart and Connected Health: Applications and Real-World Use Cases*, presents real-world implementations and case studies. From AI-driven clinical decision support to smart toilets for health monitoring and AI in diabetes management, this volume showcases the transformative impact of emerging technologies in healthcare.

Whether you are a student seeking a structured learning path, a researcher exploring new frontiers, or a practitioner striving to implement and integrate emerging technologies, we hope this book provides valuable insights into the future of smart and connected health systems.

Tehran, Iran — Bahar Farahani
Tempe, AZ, USA — Farshad Firouzi
Tempe, AZ, USA — Krishnendu Chakrabarty

Contents

Electronic Health Records: Toward Digital Twins in Healthcare

Muhammet Alkan, Hester Huijsdens, Yola Jones, and Fani Deligianni

> *"The good physician treats the disease; the great physician treats the patient who has the disease."*
>
> *—William Osler*

Abstract The pivotal shift from traditional paper-based records to sophisticated electronic health records (EHRs) enabled systematic collection and analysis of patient data through descriptive statistics, providing insight into patterns and trends across patient populations. This evolution continued toward predictive analytics, allowing healthcare providers to anticipate patient outcomes and potential complications before they occur. This progression from basic digital recordkeeping to sophisticated predictive modeling and digital twins reflects healthcare's broader evolution toward more integrated, patient-centered approaches that combine data-driven insights with personalized care delivery. This chapter explores the evolution and significance of healthcare information systems, beginning with an examination of the implementation of EHR in the UK and the USA. It provides a comprehensive overview of the International Classification of Diseases (ICD) system, tracing its development from ICD-9 to ICD-10. Central to this discussion is the MIMIC-III database, a landmark achievement in healthcare data sharing and arguably the most comprehensive critical care database freely available to researchers worldwide. MIMIC-III has democratized access to high-quality healthcare data, enabling unprecedented opportunities for research and analysis. The chapter examines its structure, clinical outcome analysis capabilities, and practical applications through case studies, with a particular focus on mortality and length of stay metrics, vital signs extraction, and ICD coding. Through detailed entity-relationship diagrams and practical examples, the text illustrates MIMIC's complex data structure and demonstrates how different querying approaches can lead to subtly different results, emphasizing the critical importance of understanding the database's architecture for

---

M. Alkan · H. Huijsdens · Y. Jones · F. Deligianni (✉)
School of Computing Science at University of Glasgow, Glasgow, Scotland, UK
e-mail: fani.deligianni@glasgow.ac.uk

B. Farahani et al. (eds.), *Smart and Connected Health: Applications and Real-World Use Cases*, https://doi.org/10.1007/978-3-032-09253-3_1

accurate data extraction. The chapter concludes by discussing the progression from descriptive analytics to digital twins in healthcare, highlighting the transformation toward more sophisticated and personalized healthcare information systems.

1 Electronic Health Records

1.1 Electronic Health Records

Electronic health records (EHRs) are digital versions of a patient's health information. EHRs exhibit a diverse composition, encompassing both structured and unstructured data types, such as medical images, electrocardiograms (ECGs), laboratory results, narrative and diagnostic reports, medical interventions, and hospitalizations.

Unlike clinical trials, which follow up patients at regular intervals (e.g., 6 months, 1 year, 5 years), electronic health records capture data at irregular intervals by recording the dates of each event. For example, someone may be hospitalized for 3 days, receive extensive medical follow-up during these 3 days, and then not be seen by the medical system for the following year (see, e.g., Fig. 1).

1.2 The Transition to Electronic Health Records in the UK and the USA

The shift from paper-based to electronic health records (EHRs) has profoundly influenced healthcare data management, enhanced the quality of clinical decision-making, and opened new avenues for biomedical research. It also signaled a transition from independent records to integrated databases.

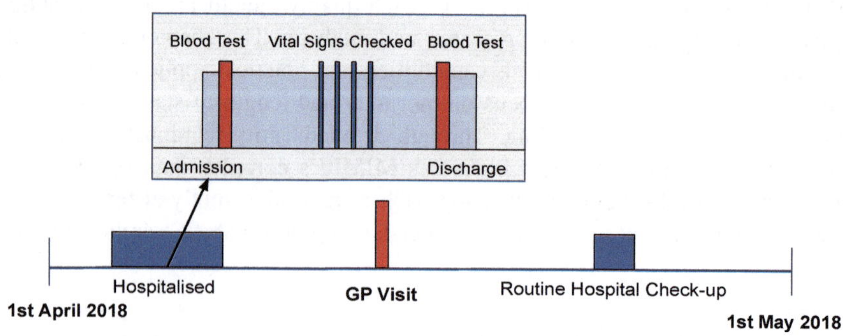

Fig. 1 Patient's timeline

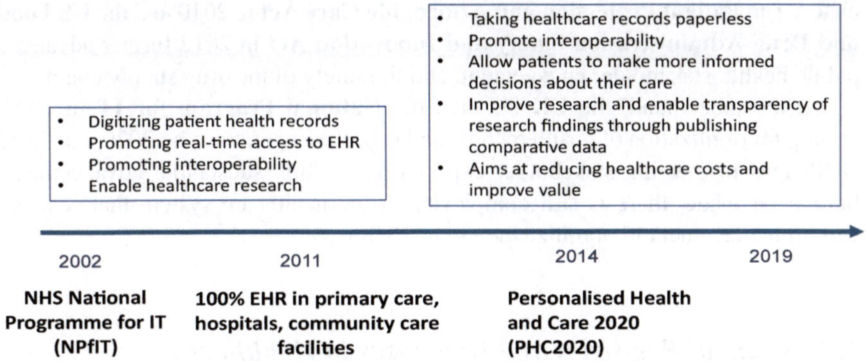

Fig. 2 The transition to EHRs in the UK

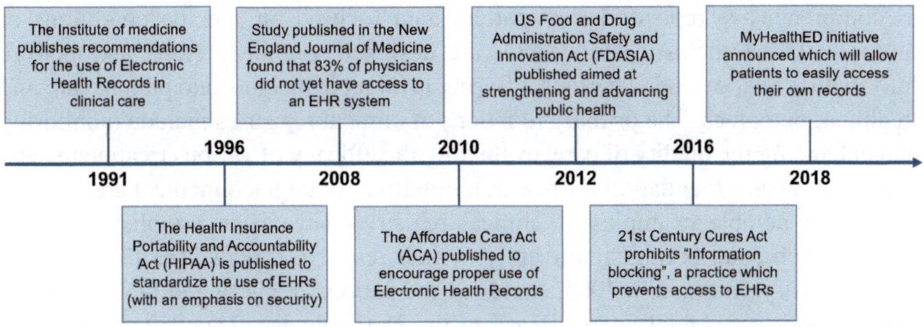

Fig. 3 The transition to EHRs in the USA

An overview of the adoption of EHRs in the US and UK healthcare systems is shown in Figs. 2 and 3, respectively. This transition took over the period of two decades and it highlights the challenges and requirements for a successful integration. The US and the UK healthcare systems are organized very differently [25]. The UK has one of the largest public sector systems. On the other hand, the USA has the largest private-sector system and one of the largest healthcare expenditures in the world.

The Health Insurance Portability and Accountability Act is the first of the US laws to affect IT in healthcare [15]. It was enacted in August of 1996. Part of this sets the rules and requirements for the privacy and security of information. This information was referred to as protected health information. Rules were set in place for what and how it could be accessed and shared. The purpose was to encourage the use of electronic data interchange by explaining how data could safely be shared between parties. Part of it was to define standards for corporations to abide by and thus help build interoperability.

Despite HIPAA, by 2006, over 80% of US physicians lacked access to EHRs, leading to efforts to establish a nationwide health information network and the extension of protected health information (PHI) responsibility to business associates. The **Patient Protection and Affordable Care Act** in 2010 and the **US Food and Drug Administration Safety and Innovation Act** in 2012 further advanced public health, stakeholder engagement, and the safety of the drug supply chain.

On the other hand, the UK initiated the **National Program for IT** in 2002, aiming for digitization of health records and paperless operations by 2020, but faced challenges in meeting stakeholder expectations. While substantial advancements have been made, there is still complexity in the healthcare system that requires careful management to optimize the use of EHRs [6].

1.3 Clinical Registries and Databases in Healthcare

Clinical registries play a crucial role in modern healthcare systems by serving as comprehensive repositories of patient data focused on specific conditions or diseases. These sophisticated systems are designed to collect, store, and analyze information over extended periods, offering invaluable insights into patient care and treatment outcomes. The primary functions of clinical registries include monitoring and enhancing the quality of care, evaluating the efficacy of various treatments, and providing a solid foundation for research initiatives in targeted medical areas.

Several notable examples of clinical registries have been established across Europe, each contributing significantly to their respective fields. The Dutch CONCOR Registry, initiated in 2005, stands as the national registry and DNA bank for patients with congenital heart disease in the Netherlands [23]. This registry has been instrumental in advancing understanding and treatment of congenital heart conditions in the country. Similarly, SWEDCON, the Swedish National Registry on Congenital Heart Disease, exemplifies another comprehensive approach to data collection and analysis in this specialized field [2]. By consolidating information from across Sweden, SWEDCON has become an essential tool for researchers and clinicians alike, driving improvements in patient care and outcomes. In Belgium, the BELCODAC (BELgian COngenital Heart Disease Database combining Administrative and Clinical data) represents a more recent development [13]. Established in 2020, this registry takes an innovative approach by merging administrative and clinical data, providing a more holistic view of patient experiences and treatment efficacy.

While clinical registries focus on specific conditions or diseases, clinical databases cast a wider net, encompassing health information from a broader range of sources. These databases are not limited to particular conditions but instead offer a comprehensive overview of patient health data. A prime example of this broader approach is the hospital database, which typically contains a wealth of information on various medical conditions, treatments, and patient outcomes within a single healthcare institution or network.

The distinction between clinical registries and databases lies in their scope and focus. Registries are tailored to specific conditions, allowing for deep, targeted analysis and research. Databases, on the other hand, offer a more expansive view of health information, facilitating broader studies and analyses across multiple conditions and patient populations. Both clinical registries and databases are invaluable tools in the healthcare landscape. They not only support day-to-day patient care but also drive medical research, inform policy decisions, and contribute to the overall advancement of medical knowledge and practice. As healthcare continues to evolve, these data collection and analysis systems will undoubtedly play an increasingly vital role in shaping the future of patient care and medical research.

2 International Classification of Diseases System

2.1 The World Health Organization and the International Classification of Diseases System

The International Classification of Diseases (ICD) system, maintained by the World Health Organization (WHO) [7], plays a crucial role in global healthcare. This standardized coding system enables the collection of reliable data for statistical and machine learning applications, facilitating comparisons across different locations and time periods [4, 16]. The WHO, representing 194 member states, strives to promote optimal health standards for all individuals, regardless of their socioeconomic status, race, gender, religion, or political beliefs. At its core, the organization upholds the principle that access to affordable and comprehensive healthcare is a fundamental human right.

Health is influenced by various factors, including biomedical, genetic, behavioral, socioeconomic, and environmental determinants. To measure health and well-being effectively, common metrics are essential. These include objective measures like life expectancy and mortality rates, as well as subjective assessments of an individual's perceived well-being. The ICD system serves as the international standard diagnostic tool for epidemiology, health management, and clinical purposes. It translates diagnoses and health problems into alphanumeric codes, enabling systematic storage, retrieval, and analysis of data. This standardization allows for meaningful comparisons of morbidity and mortality data across hospitals, nations, and time periods. Developed collaboratively between the WHO and international centers, the ICD system aims to group medical terms reported by physicians, medical examiners, and death certificates for statistical purposes. Its broad acceptance and official agreement for use have made it an invaluable tool for international health reporting.

The ICD system benefits various stakeholders in the healthcare sector. Clinicians, including physicians, nurses, and health workers, use it to support patient care decisions. Administrators, such as health information managers, policymakers, and insurers, rely on it for resource allocation and program management. Researchers and epidemiologists utilize the data for population health studies and the development of AI-based decision support systems.

2.2 Evolution of the ICD System

The International Classification of Diseases (ICD) system has a rich history spanning nearly four centuries (Fig. 4). Its roots can be traced back to seventeenth-century England, where epidemiologist John Graunt recognized the need to organize mortality data systematically [3]. His pioneering work, the London Bills of Mortality, marked the first statistical study of disease, though it was limited by inconsistent nomenclature and poor data quality.

The modern era of disease classification began in the late nineteenth century with the efforts of William Farr, often regarded as the first medical statistician. Farr's 1855 report on the nomenclature and statistical classification of diseases laid the groundwork for a more comprehensive approach, emphasizing the importance of classifying diseases based on the organic systems they affect.

In 1893, the International Statistical Institute adopted the first edition of the International List of Causes of Death, marking the birth of the modern ICD system. This classification, based on the work of Jacques Bertillon, distinguished between general diseases and those localized to specific organs or anatomical sites. The system was widely adopted by American and European countries, with an agreement to update it approximately every decade.

Throughout the twentieth century, the ICD system underwent significant transformations [12, 7]. The sixth revision in 1948 marked a pivotal moment as the World Health Organization (WHO) adopted it as its official classification system.

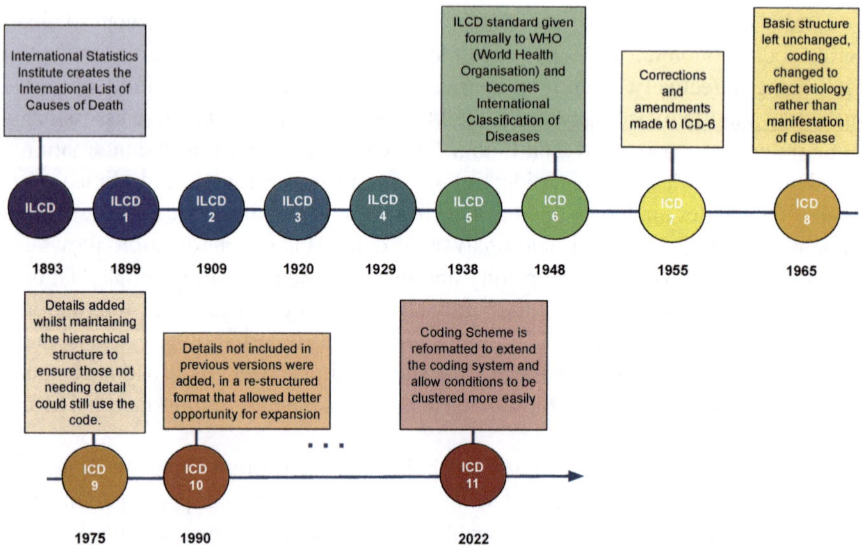

Fig. 4 Evolution of the international classification of diseases system

This version expanded beyond mortality to include morbidity, recognizing the importance of nonfatal but chronic conditions.

Subsequent revisions brought further refinements. The ninth revision introduced a dual classification system reflecting both etiology and manifestation, while the tenth revision implemented an alphanumeric coding structure for greater flexibility and expansion.

The latest iteration, ICD-11, adopted in 2019 and effective from 2022, represents a significant leap forward [10]. Designed for the digital era, it aims to improve diagnostic accuracy, enhance the coding of healthcare quality and safety, and better account for socioeconomic factors influencing health. ICD-11 also simplifies diagnostic descriptions, particularly in mental health.

The ICD system's evolution reflects the ongoing effort to create a comprehensive, flexible, and internationally accepted classification of diseases. Its widespread adoption has made it an indispensable tool for comparing health statistics across different countries and time periods, playing a crucial role in global health management and research.

2.3 *Understanding ICD-9 and ICD-10 Coding Systems*

2.3.1 ICD-9 Coding System

The ICD-9 coding system consists of a five-digit structure: the first three digits reflect a category, while the last two digits indicate the cause or location. Categories encompass a wide range of classifications, including epidemic diseases, constitutional or general diseases, local diseases arranged by system (e.g., circulatory, respiratory), developmental diseases, and injuries. The World Health Organization mandates a minimum of three-character categories for international reporting and comparison, with the fourth digit filled with "X" when no further subdivision information is available.

ICD-9 categories cover a broad spectrum of medical conditions, starting from epidemic diseases and progressing through various systemic disorders such as neoplasms, endocrine and metabolic diseases, blood disorders, mental disorders, and diseases of specific organ systems. The classification also includes developmental diseases, injuries, and poisoning. The last two categories use letters as the first digit and offer supplemental classifications.

2.3.2 From ICD-9 to ICD-10

The International Classification of Diseases (ICD) has evolved significantly from its ninth revision (ICD-9) to its tenth revision (ICD-10), resulting in substantial changes in disease classification and coding, as it is shown in Figs. 5 and 6. ICD-10 represents a major advancement over ICD-9, offering greater specificity and

ICD-9

Codes	Description
001-139	Infectious and parasitic diseases
140-239	Neoplasms
240-279	Endocrine, nutritional and metabolic diseases and immunity disorders
280-289	Diseases of the blood and blood forming organs
290-319	Mental disorders
320-389	Diseases of the nervous system and sense organs
390-459	Diseases of the circulatory system
460-519	Diseases of the respiratory system
520-579	Diseases of the digestive system
580-629	Diseases of the genitourinary system
630-679	Complications of pregnancy, childbirth and the puerperium
680-709	Diseases of the skin and subcutaneous tissue
710-739	Diseases of the musculoskeletal system and connective tissue
740-759	Congenital anomalies
760-779	Certain conditions originating in the perinatal period
780-799	Symptoms, signs and ill-defined conditions
800-999	Injury and poisoning
V01-V91	Supplementary classification: classification of factors influencing health status and contact with health services
E000-E999	Supplementary classification: classification of external causes of injury and poisoning

ICD-10

Codes	Description
A00-B99	Certain infectious and parasitic diseases
C00-D49	Neoplasms
D50-D89	Diseases of the blood and blood-forming organs and certain disorders involving the immune mechanism
E00-E90	Endocrine, nutritional and metabolic diseases
F01-F99	Mental and behavioural disorders
G00-G99	Diseases of the nervous system
H00-H59	Diseases of the eye and adnexa
H60-H95	Diseases of the ear and mastoid process
I00-I99	Diseases of the circulatory system
J00-J99	Diseases of the respiratory system
K00-K94	Diseases of the digestive system
L00-L99	Diseases of the skin and subcutaneous tissue
M00-M99	Diseases of the musculoskeletal system and connective tissue
N00-N99	Diseases of the genitourinary system
O00-O99	Pregnancy, childbirth and the puerperium
P00-P96	Certain conditions originating in the perinatal period
Q00-Q99	Congenital malformations, deformations and chromosomal abnormalities
R00-R99	Symptoms, signs and abnormal clinical and laboratory findings, not elsewhere classified
S00-T88	Injury, poisoning and certain other consequences of external causes
V00-Y98	External causes of morbidity
Z00-Z99	Factors influencing health status and contact with health services

Fig. 5 ICD-09 and ICD-10 chapters

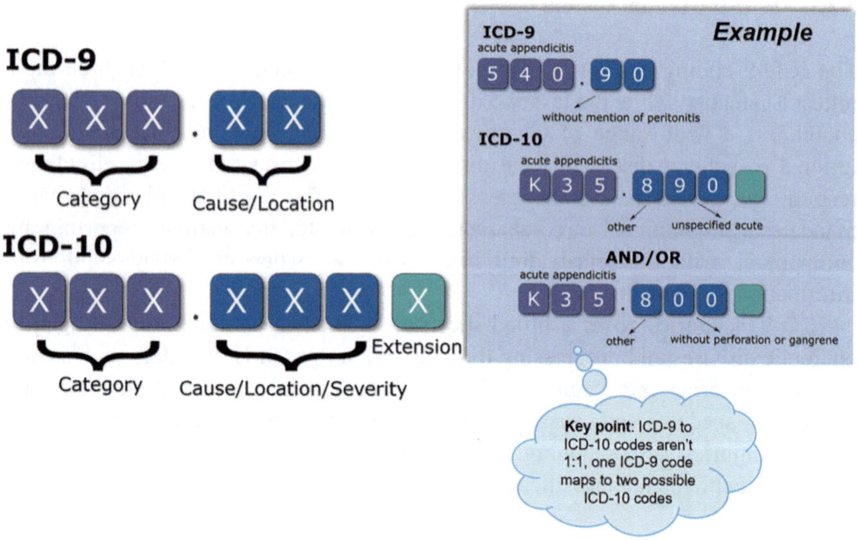

Fig. 6 ICD-9 vs. ICD-10

flexibility in disease classification [14]. While ICD-9 primarily used numeric codes with occasional use of letters E and V, ICD-10 adopts an alphanumeric system. The first character in ICD-10 is always alphabetic, followed by two numeric digits for the category. Subsequent characters can be either alphabetic or numeric, allowing for more detailed classification.

The structural differences between the two systems are significant as is also shown in Fig. 5. ICD-10 contains 21 chapters compared to ICD-9's 17, providing more comprehensive coverage of health conditions. The expanded structure of ICD-10 allows for more precise coding of diseases, injuries, and health status factors. For instance, ICD-10 separates sense organ disorders from nervous system disorders and reorganizes injury classifications by anatomical site.

One crucial aspect to understand is that there is not always a one-to-one mapping between ICD-9 and ICD-10 codes (Fig. 6). A single ICD-9 code might correspond to multiple ICD-10 codes, or vice versa, depending on the specificity required. This lack of direct correlation can pose challenges when transitioning between the two systems or when analyzing historical data. ICD-10 introduces several special signs and symbols to enhance coding precision. The dagger (†) and asterisk (*) system is particularly noteworthy. The dagger code represents the underlying disease process, while the asterisk code indicates its manifestation in a particular organ or site. This dual coding allows for a more comprehensive description of complex medical conditions. Other important symbols in ICD-10 include parentheses for supplementary information, square brackets for synonyms or alternative phrases, and colons to indicate the need for additional digits to complete a code. The terms "Not Otherwise Specified" (NOS) and "Not Elsewhere Classified" (NEC) are used to handle cases where information is incomplete or a specific code is unavailable.

The transition to ICD-10 has significant implications for healthcare information systems. It allows for more detailed patient records, improved data analysis, and better tracking of health trends. However, it also requires substantial changes in coding practices and software systems.

Looking ahead, ICD-11 has been introduced to address some limitations of ICD-10. It offers improved functionality in digital environments, incorporates more detailed morbidity information, and provides better support for multilingual applications. These ongoing developments underscore the dynamic nature of disease classification systems and their critical role in modern healthcare management.

In conclusion, understanding the differences between ICD-9 and ICD-10, as well as the special features of ICD-10, is crucial for healthcare professionals, researchers, and information systems specialists. The increased specificity and flexibility of ICD-10 offer enhanced capabilities for disease classification and health data analysis, paving the way for more precise and effective healthcare management and research. ICD-10 was mandated for use in UK healthcare system in 1995. ICD-9 is still used in publicly available EHR databases such as MIMIC-III.

3 Datasets for Research and MIMIC

3.1 MIMIC Critical Care Dataset: The Impact

The MIMIC (Medical Information Mart for Intensive Care) dataset stands as a cornerstone in healthcare research, offering a comprehensive collection of de-identified electronic health records from hospital and intensive care unit settings [9]. Developed by the Laboratory for Computational Physiology at MIT, MIMIC has evolved through several iterations. This dataset's significance lies in its volume, quality, and representation of real-world hospital data spanning over a decade. It includes a wide range of clinical information such as lab tests, medical diagnoses, vital signs, and medications, providing researchers with a rich source of complex physiological signals and clinical data.

The creation of MIMIC was driven by the need to generate new knowledge from the vast amount of data captured daily in intensive care units. By making this de-identified dataset freely accessible under a data use agreement, MIMIC has opened up unprecedented opportunities for both academic and industrial research in healthcare and related fields [18, 19, 9]. MIMIC's impact on research has been substantial and far-reaching. Since its initial release, the dataset has garnered thousands of citations, with an exponential growth in its use across various disciplines. While its primary impact has been in medicine and critical care, MIMIC has also significantly influenced research in computer science and in particular in machine learning and natural language processing. The dataset's comprehensive nature has enabled sophisticated research in deep learning models, clinical decision support systems, and big data analytics in healthcare. It has become a valuable resource for developing and validating new algorithms and technologies in medical informatics.

Moreover, MIMIC has set a standard for de-identifying free-text and clinical information, serving as a model for other clinical databases. Its public availability has facilitated reproducibility in clinical studies, a crucial aspect of advancing medical research.

In conclusion, the MIMIC dataset represents a significant advancement in healthcare data accessibility and has become an indispensable tool for researchers across multiple disciplines. Its continued growth and impact underscore its importance in driving innovation in healthcare analytics and clinical research.

3.2 An Overview of MIMIC-III: Data Linkage

Designing a database schema to encode the complexity of EHR information presents challenges cite [11]. The database must be accessible to doctors and other healthcare providers simultaneously and in a unified manner. Interoperability is crucial, involving enhanced quality, efficiency, and effectiveness of the healthcare

MIMIC-III Tables

Patient Tracking:
- **admissions**
- **icustays**
- **patients**
- callout
- transfers

ICU Data:
- **chartevents**
- inputevents_cv
- inputevents_mv
- datetimeevents
- **outputevents**
- procedureevents_mv

Hospital Data:
- caregivers
- cptevents
- **diagnoses_icd**
- drgcodes
- **labevents**
- microbiologyevents
- noteevents
- prescriptions
- **procedures_icd**
- services

Dictionary Tables:
- d_cpt
- d_icd_procedures
- **d_items**
- d_icd_diagnoses
- **d_labitems**

Fig. 7 MIMIC-III as a relational database: tables

system. Information should be provided in the appropriate format when needed, eliminating unnecessary duplication.

The choice of database and its schema architecture influences the effective management of medical data, flexibility, scalability, query performance, and interoperability. Nonproprietary standardized models are essential for building electronic health record systems that meet interoperability requirements. MIMIC-III serves as an excellent example in this regard (Fig. 7).

During ICU stays, various signals are monitored, including vital signs, waveforms, alarms, fluids, medications, and progression reports from doctors. Hospital data includes billing details, International Classification of Diseases codes related to pathology and symptoms, patient demographics, and other notes concerning medical images and discharge summaries.

A key step in building the MIMIC-III database was the de-identification process, ensuring compliance with the Health Insurance Portability and Accountability Act. All fields related to patient identification were removed, including names, telephone numbers, and addresses. Dates were shifted into the future by a random offset for each patient, preserving intervals while protecting privacy. This shift maintained seasonality, time of day, and day of the week. Special measures were taken to protect individuals with rare age characteristics, and text fields like diagnostic reports and physician notes were carefully processed to remove any protected health information.

In conclusion, MIMIC-III links data across hospitals, ICU units, and death registries. It provides a wealth of information, including lab examinations, medications, International Classification of Diseases coding, and vital signs. As the only freely available database of its kind without major usage restrictions, MIMIC-III represents a valuable resource for healthcare research and analysis.

3.3 MIMIC-III as a Relational Database

This section explores the structure of MIMIC-III as a relational database, with the aim of providing an understanding of the database schema, its basic tables (Fig. 7), and navigation methods. The MIMIC-III database contains over 53,000 distinct hospital admissions for patients aged 60 years and above, admitted to critical care between 2001 and 2012. It also includes data on more than 7800 neonates from 2001 to 2008.

To effectively extract information from this electronic health record database, it's crucial to understand both the database schema and the data sources. Most queries will involve joining information between the basic tables holding data for patient admissions in the hospital and ICU stays.

The MIMIC-III is a relational database that comprises 26 tables, reflecting the inherent hospital information sources. Its structure closely represents the raw data, with only minor adjustments made based on user feedback. These tables can be categorized into four groups: patient tracking, ICU data, hospital data, and dictionary tables. The subject ID refers to a unique patient, the hospital admission ID to a unique hospital admission, and the ICU stays ID to a unique ICU admission.

The core tables of MIMIC-III are the Patients table, the Admissions table, and the ICU stays table. These tables form the foundation for extracting various summary statistics. The Patients table contains a subject ID identifier linking it to the Admissions and ICU stays tables. It includes information such as date of birth (shifted for patients over 89 years old to protect privacy) and different versions of date of death. The Admissions table records every unique hospitalization for each patient, including admission and discharge times, type of admission, and diagnosis information. The ICU stays table documents every unique ICU stay, with details like care unit, ward locations, and length of stay.

Other significant tables include the Transfers table, which provides information on patient movements within the hospital, and the Callout table, which records ICU discharge clearance and actual discharge times. The Chartevents table is crucial for ICU data, containing all chart event observations for patients. The Outputevents table focuses on patient output measurements.

When working with ICU data, particularly chart events data, it's essential to consider the Dictionary tables. These provide definitions for identifiers, allowing researchers to cross-reference codes against their respective definitions.

In the hospital data category, frequently used tables include the Labevents table, containing laboratory test results. It's worth noting that in cases of disagreement between measurements, lab events should be considered the ground truth. The "procedures_icd" and "diagnoses_icd" tables contain procedure and diagnosis information coded using the International Classification of Diseases (ICD-9) system.

Researchers are advised to develop database views and transform data as needed rather than combining tables within the MIMIC data model. This approach maintains the integrity of the original data sources while allowing for flexible analysis.

3.4 *MIMIC-III: Descriptive Statistics*

Descriptive statistics serve as a powerful tool in retrospective studies, allowing researchers to overview historic data and explain trends. This section explores how to formulate simple yet effective queries using the MIMIC-III database. It's important to note that extracting patient data can yield different estimations depending on which table identifier is used, underscoring the necessity of thoroughly understanding the database schema and its data encoding methods.

Descriptive statistics in MIMIC-III typically involve common queries that provide insights into the distribution of data within the database. These queries, while relatively simple, offer powerful ways to interrogate the database and generate intuitive summary visualizations. They often focus on patient characteristics, intensive care unit utilization, and patient outcomes such as mortality.

When estimating the number of patients, multiple approaches can lead to similar but not identical results. For instance, one can examine the number of distinct patients across care units, unique hospital admissions, or unique admissions to ICUs. Each method yields slightly different results, as some patients may have multiple admissions or transfers between units.

A fundamental query in electronic health record databases involves examining age distribution across different departments. In MIMIC-III, this is particularly useful for understanding age distribution across intensive care units. By combining the Patients and ICU stays tables, researchers can compute age by subtracting the time of ICU admission from a patient's date of birth. The first_careunit identifier in the ICU stays table provides information about different care units.

Similarly, gender distribution across critical care units can be extracted by combining the Patients and ICU stays tables based on the subject_id identifier. This query collects information about the ICU each patient was admitted to, along with gender, date of birth, and hospital admission time. These data points allow for age computation and filtering of patients older than 16 years, focusing the analysis on the adult population.

Visualizing these statistics can provide insights into which conditions that affect males and females differently across various critical care units. For instance, observations might reveal higher numbers of men in intensive care units related to cardiac problems and trauma surgery.

Another valuable descriptive statistic involves extracting distinct patient hospitalizations (Fig. 8). This query combines the Patients and ICU stays tables to obtain each patient's unique identifier and corresponding ICU admission information. It's worth noting that the ICU stay identifier groups all ICU admissions within 24 hours of each other, meaning a patient transferred between ICU types might have the same ICU stays ID.

Researchers might also be interested in the number of unique hospital admissions and their distribution across intensive care units (Fig. 9). This query combines the Patients and ICU stays tables based on each patient's unique subject identifier,

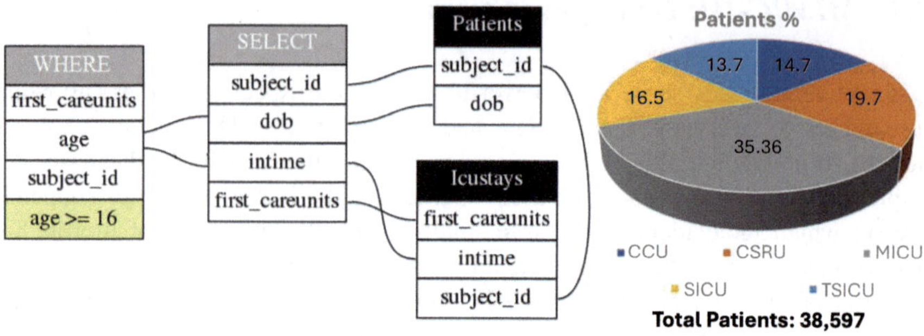

Fig. 8 MIMIC-III: distinct patients' hospitalizations

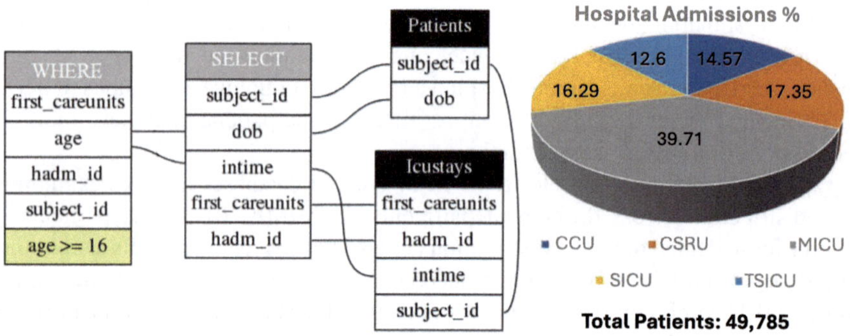

Fig. 9 MIMIC-III: hospital admissions

collecting information about patient identifiers, dates of birth, and unique hospital stay identifiers.

To identify distinct ICU admissions, a similar process is followed, but using the unique ICU stay identifier instead of the hospitalization identifier (Fig. 10). This distinction is important because the MIMIC-III database was constructed by linking hospital and ICU information, resulting in unrelated identifiers between the two systems.

Comparing the results of these queries highlights the differences between unique patient admissions, unique hospital admissions, and unique ICU stays. Generally, the number of distinct patients across all care units is slightly less than the number of hospitalizations, which in turn is less than the number of ICU admissions across different units.

Descriptive analytics can provide valuable insights about patient numbers and resource utilization, revealing trends and anomalies over time. For example, studies based on descriptive statistics have shown steady increases in chronic illnesses, like diabetes mellitus and hypertension, as well as an increasing proportion of elderly patients and women in hospital populations.

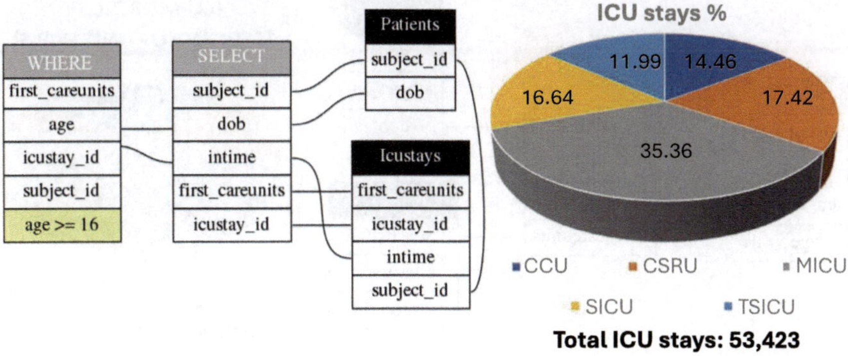

Fig. 10 MIMIC-III: distinct ICU admissions

These analytics can also shed light on how emerging technologies impact patient outcomes, length of stay, associated risks of complications, and resource consumption. They can reveal trends such as declining hospital mortality from acute coronary syndromes or the increasing age of admitted patients over time.

While descriptive analytics offer substantial information about historic data and can explain trends, they are limited to retrospective studies. They cannot predict future outcomes or prevent diseases and high mortality rates. Nevertheless, they remain a crucial tool in understanding the complex landscape of critical care and hospital utilization patterns.

3.5 MIMIC-III and Clinical Outcomes: Mortality and Length of Stay

3.5.1 Mortality as a Clinical Outcome

Mortality is the primary outcome measure employed in prediction studies across critical care and much of medicine. As a reliable indicator (surrogate) of illness severity, clinicians and researchers use this signal to identify trends and correlations between clinical data and patient outcomes. However, defining mortality is more complex than it might initially appear, particularly when dealing with retrospective data.

In the MIMIC database, complications arise from situations such as patients whose consecutive hospitalizations ultimately result in mortality. While these are considered as distinct hospital admissions, the underlying cause is the same. Mortality data is also prone to selection bias depending on the information source. For instance, hospital mortality records might not include deaths occurring after discharge to home care, potentially biasing any trained models if there's a routine process for discharging patients with severe illnesses to home care.

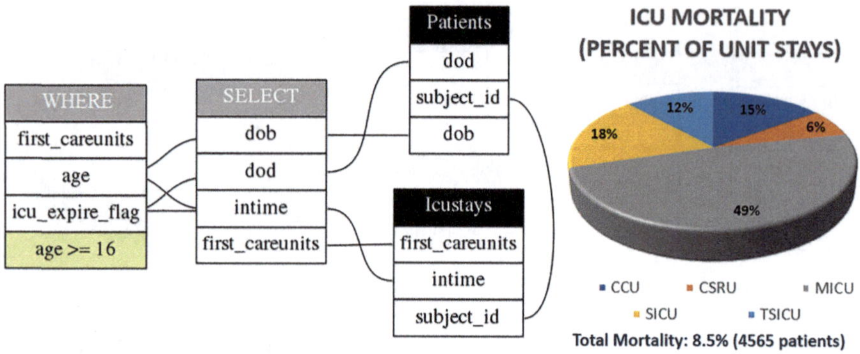

Fig. 11 MIMIC-III: ICU mortality as a clinical outcome

Researchers have some control over how mortality is defined as an outcome. The definition of mortality as death during the first 30 days of admission provides a more robust signal for immediate physiological abnormalities associated with the patient's admission. Conversely, 1-year mortality emphasizes chronic illnesses and parallel conditions.

To extract mortality numbers for adult ICU patients in MIMIC (Fig. 11), we combine the Patients and "ICU stays" tables using the "subject_id." We extract the date of death, date of birth, ICU transfer time, and the type of ICU first admitted to. A patient is considered to have died in the ICU if their death was registered while in the ICU or within 6 hours before admission or after leaving. This query focuses on patients older than 60 years. Visualizing these results shows that total ICU mortality is about 8.5%, corresponding to approximately 4,565 patients. The medical ICU accounts for 49% of these deaths, despite having only about 39% of total ICU patients, indicating a higher mortality rate in this unit compared to others.

For hospital mortality, we use the hospital_expire_flag in the Admissions table (Fig. 12). This query combines the ICU stays, Patients, and Admissions tables, filtering for patients older than 16. The results show an increase in total mortality to 11.5% when considering hospital-wide deaths, with 50% occurring in the medical ICU.

3.5.2 Length of Stay as a Clinical Outcome

Length of stay is another crucial metric, relating to both patient outcomes and resource utilization. To extract this data, we combine the Patients, ICU stays, and Admissions tables, focusing on patients older than 60. Length of stay is computed by subtracting the discharge time from the admission time, using the unique hospital admission identifier to ensure each admission corresponds to only one ICU stay (Fig. 13).

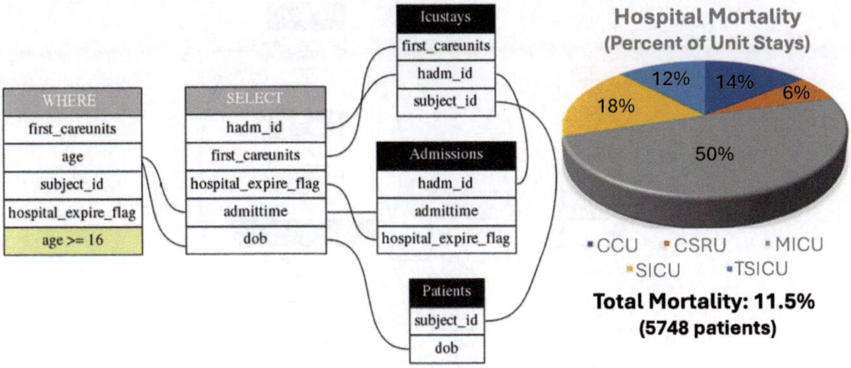

Fig. 12 MIMIC-III: hospital mortality as a clinical outcome

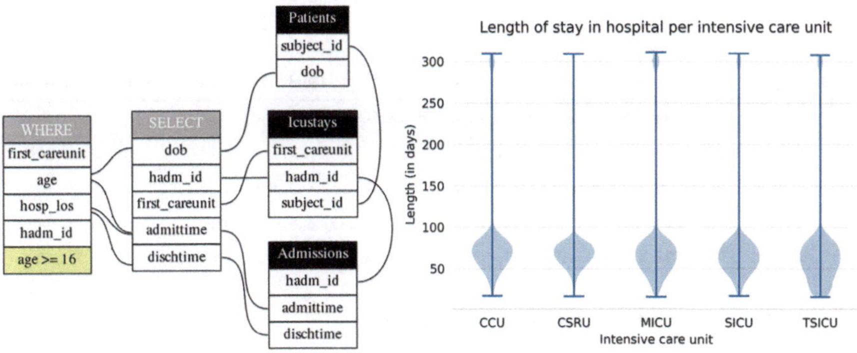

Fig. 13 MIMIC-III: hospitalization length as a clinical outcome

In conclusion, while mortality and length of stay are the most common outcomes used in predictive models, their seemingly simple definitions can be deceptive. Small differences in estimation methods can lead to significantly different conclusions. The MIMIC-III database provides a valuable resource for understanding how to form queries and extract this critical data, offering researchers a robust platform for developing and refining predictive models in critical care settings.

3.6 MIMIC-III: Exploring Patient Data—Vital Signs Extraction for a Single Patient

Understanding vital signs and medication data for individual patients admitted to intensive care units is crucial, as it forms the foundation for extracting clinical variables across multiple patients. These variables can include lab exams, signs and waveforms, as well as doctor reports and prescriptions.

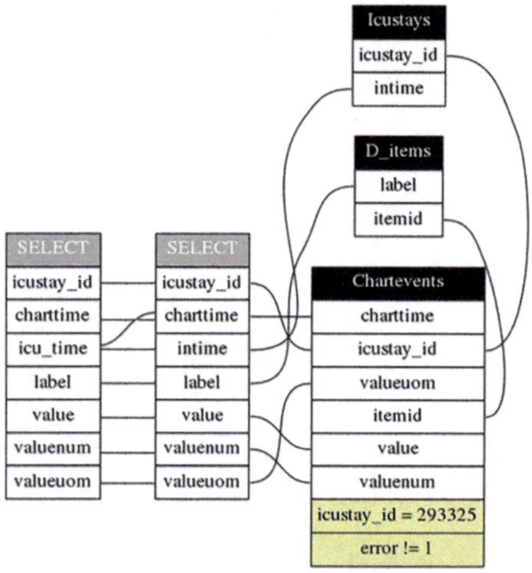

	icustay_id	charttime	icu_time	label	value	valuenum	valueuom
0	293325	2120-10-29 10:30:00	-1 days +23:57:10	Inspired O2 Fraction	70	70.0	None
1	293325	2120-10-29 10:30:00	-1 days +23:57:10	O2 Flow	10	10.0	L/min
2	293325	2120-10-29 10:30:00	-1 days +23:57:10	O2 Delivery Device(s)	Aerosol-cool	NaN	None
3	293325	2120-10-29 10:32:00	-1 days +23:59:10	Language	English	1.0	None
4	293325	2120-10-29 10:32:00	-1 days +23:59:10	Marital Status	Single	1.0	None

Fig. 14 MIMIC-III: charted events for a single patient

To begin exploring patient data, we start by selecting a random ICU stay identifier and its corresponding subject identifier. This allows us to focus on data related to a specific patient and ICU admission. By combining the Admissions, ICU stays, and Patients tables, we can create a comprehensive view of the patient's stay, including their unique subject identifier, hospital identifier, admission type, diagnosis, ICU stay identifier, first and last care units, and ICU admission time (Fig. 14).

The Chartevents table is a key source of data obtained from patients in the ICU. It contains information from both CareVue and MetaVision ICU databases, displaying routine vital signs and additional care-relevant information such as ventilator settings, laboratory values, code status, and mental status. To extract all charted events for a single patient, we combine the Chartevents table with ICU stays and the Dictionary items table. This allows us to retrieve the time of charted events, the time elapsed since ICU admission, event labels, values, and measurement units.

Similarly, the Outputevents table provides crucial data on body fluids and other measurements, such as those from Foley catheters. We use a similar process to extract this data, joining the ICU stays table with the Outputevents table and the Dictionary table for item verification (Fig. 15).

	icustay_id	charttime	icutime	label	value	valueuom
0	293325	2120-10-29 10:45:00	0 days 00:12:10	Foley	50.0	mL
1	293325	2120-10-29 11:30:00	0 days 00:57:10	Foley	100.0	mL
2	293325	2120-10-29 12:00:00	0 days 01:27:10	Foley	35.0	mL
3	293325	2120-10-29 13:00:00	0 days 02:27:10	Foley	50.0	mL
4	293325	2120-10-29 14:00:00	0 days 03:27:10	Foley	40.0	mL

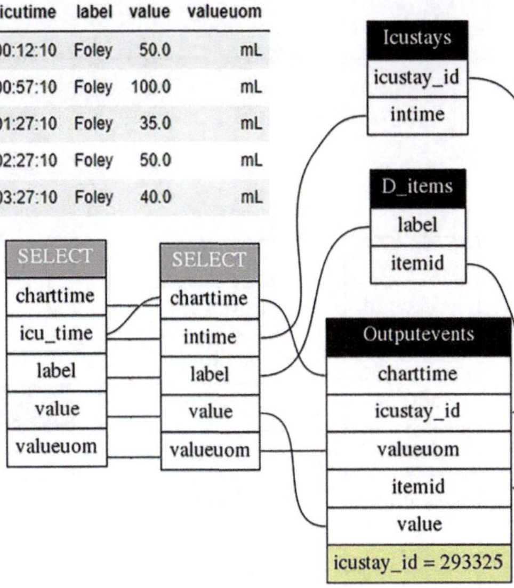

Fig. 15 MIMIC-III: output events of ICU stay for a single patient

Medication intake during an ICU stay is recorded in the Inputevents_mv table. By joining this table with ICU stays and Dictionary items, we can collect all input events for a single patient's ICU stay, including start and end times, labels, amounts, measurement units, and administration frequency (Fig. 16).

Lab events, which are part of the hospital database rather than the ICU stays database, are summarized in the Labevents table (Fig. 17). These events might include measurements such as red blood cell count or magnesium levels. It is important to note that when there is a discrepancy between Labevents and Chartevents, the Labevents table should be considered the ground truth.

Visualizing this extracted data provides a comprehensive view of the patient's condition over time. We can plot various measurements against the time since ICU admission, typically in hours. This visualization might include the Glasgow Coma Scale (which measures consciousness level), lab event results, medication administration (both frequent and sporadic), and vital signs such as heart rate, O2 saturation, respiratory rate, and output volume. An example is shown in Fig. 18.

These visualizations allow clinicians to understand a patient's progress in the intensive care unit over time. They provide a holistic view of the patient's condition, combining various data points to create a comprehensive picture of the patient's health status and treatment.

In summary, extracting data from MIMIC-III for a single patient in a single ICU stay involves examining charted events (primarily vital signs), output events (secretions and urine), input events (administered medications), and lab events (laboratory tests). This detailed exploration of individual patient datasets the stage for broader analyses across multiple patients, enabling researchers and clinicians to

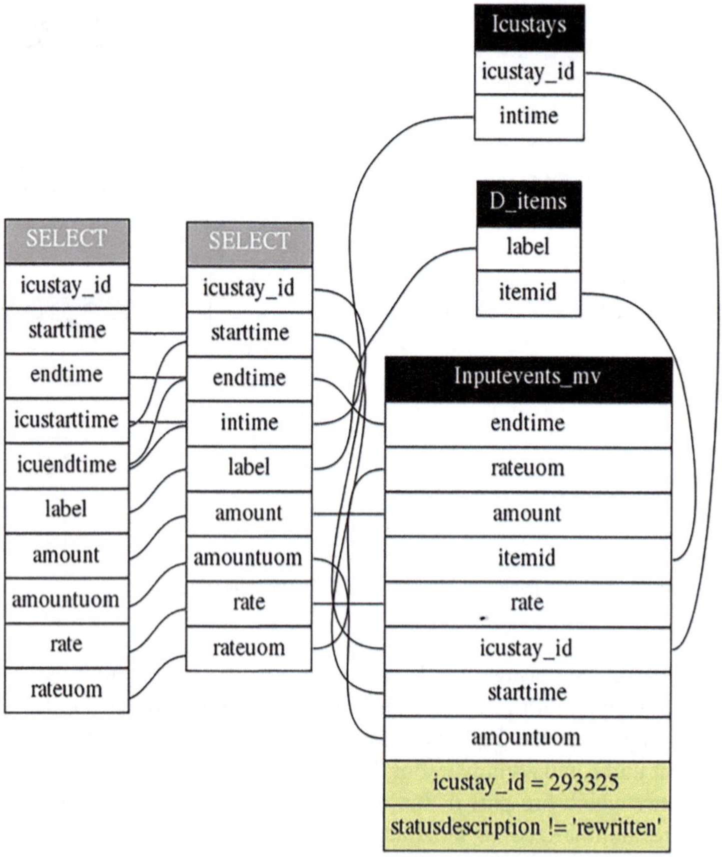

Fig. 16 MIMIC-III: medication intakes during an ICU stay for a single patient

identify patterns, trends, and potential areas for improving patient care in intensive care settings.

3.7 *MIMIC-III and ICD Codes: ICD-9 in MIMIC-III*

The International Classification of Diseases, Ninth Revision (ICD-9) serves as the disease classification system used in the MIMIC-III database. This section demonstrates how ICD-9 codes can be utilized to extract summary statistics from the MIMIC-III database, such as the number and distribution of patients across age groups diagnosed with specific diseases.

In MIMIC-III, ICD codes are stored in the diagnoses_icd table, which can be linked to the Patients and Admissions tables via the subject_id. To extract

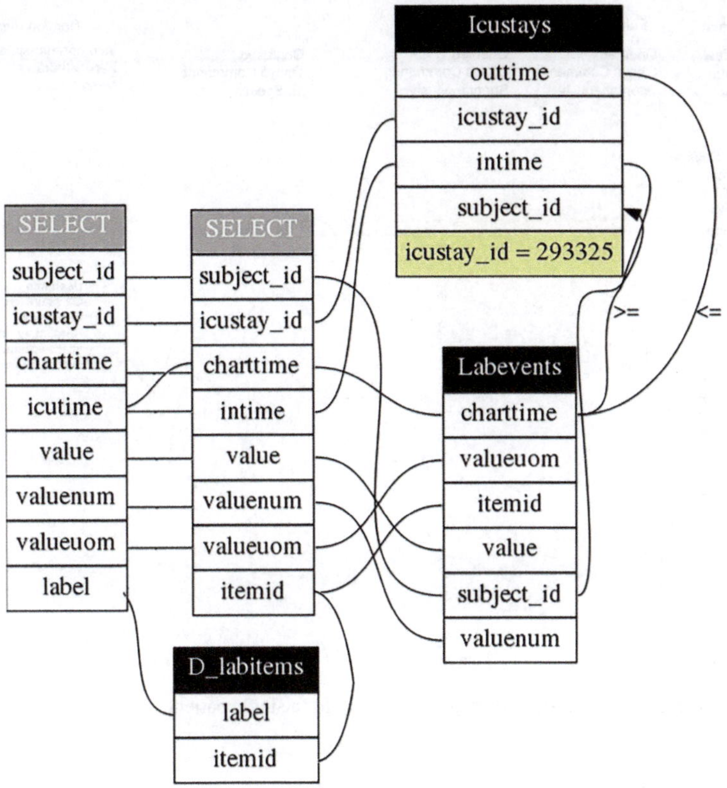

Fig. 17 MIMIC-III: lab events during an ICU stay for a single patient

meaningful data, such as patients diagnosed with hypertension who are at least 30 years old, we need to join these tables (Fig. 19). This allows us to find the age of patients based on their date of birth and admission date, filter for age, and then identify patients assigned specific ICD-9 codes related to hypertension. Visualizing these results can provide valuable insights. For instance, a histogram of age versus the number of patients with hypertension typically shows a peak between 60 and 80 years old, aligning with clinical expectations.

Another useful query involves extracting the distribution of ICD-9 codes across intensive care units in the MIMIC-III database. This process involves sorting ICD-9 codes by prevalence, selecting the top five codes, and focusing on patients older than 60 years admitted to an ICU. By combining the Patients, Admissions, ICU stays, and Diagnoses tables, we can extract information such as patient age, primary diagnosis, ICD-9 code description, and hospital admission identifier (Fig. 20).

Visualizing these results across different ICU types reveals interesting patterns (Fig. 21). For example, in coronary care units and cardiac surgery recovery units, diseases related to the circulatory system are most prevalent. Medical ICUs show a

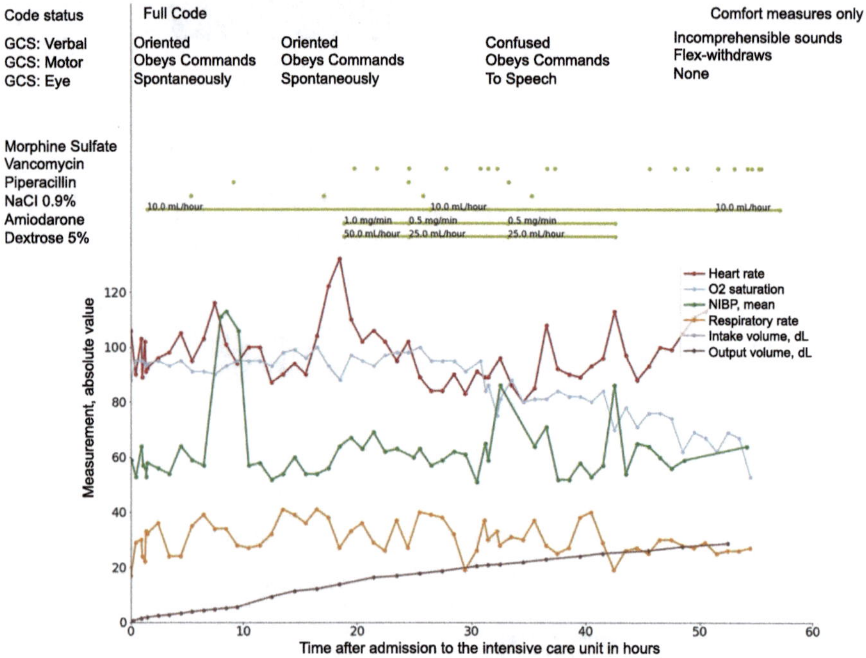

Fig. 18 MIMIC-III: vital signs during an ICU stay for a single patient

mix of circulatory and pulmonary diseases, while trauma surgery ICUs, as expected, feature more trauma-related diagnoses. Surgical and medical ICUs tend to display a more diverse profile of ICD codes.

It's important to note that a single hospital admission can be assigned multiple ICD-9 codes, reflecting the fact that a patient may suffer from several diseases simultaneously upon admission. When querying, attention should be paid to the priority of each ICD-9 code to focus on primary diagnoses.

In summary, understanding the structure of ICD-9 codes and their implementation in MIMIC-III allows researchers to extract valuable information about disease distribution across age groups and ICU types. This knowledge forms a crucial foundation for more advanced analyses and can provide important insights into patient populations and care patterns in intensive care settings.

3.8 *Clinical Concepts in MIMIC-III*

Clinical concepts in MIMIC-III represent definitions and models that provide information about the severity of illness, based on prespecified sets of clinical variables provided by field experts. These concepts were developed well before

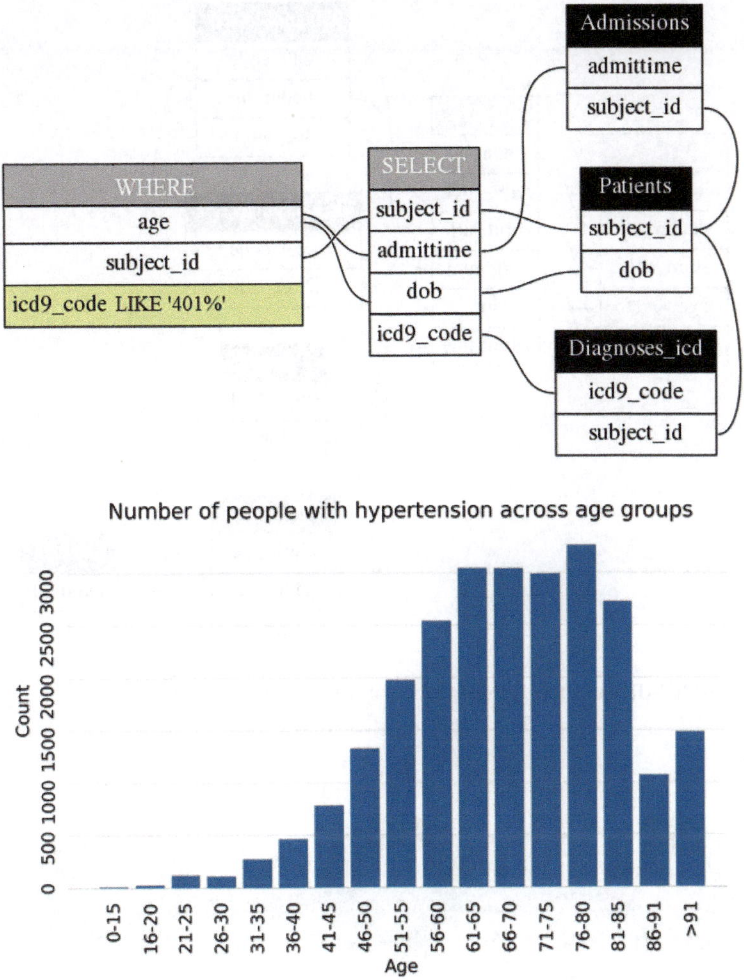

Fig. 19 MIMIC-III: ICD-9 code queries in MIMIC-III

the success of machine learning models to quantify the severity of illness or organ failure, predict specific patient events, and allow for timely intervention and treatment to improve outcomes and provide better care administration.

The development of clinical concepts based on electronic health record databases is challenging and resource-intensive, requiring close collaboration between data scientists and domain experts. MIMIC provides an ideal substrate for building clinical concepts collaboratively, bridging the gap between those familiar with the database and clinical workflows and data scientists or statisticians. These models are precursors to machine learning models, with some exploiting techniques such as regression to refine initial expert intuition.

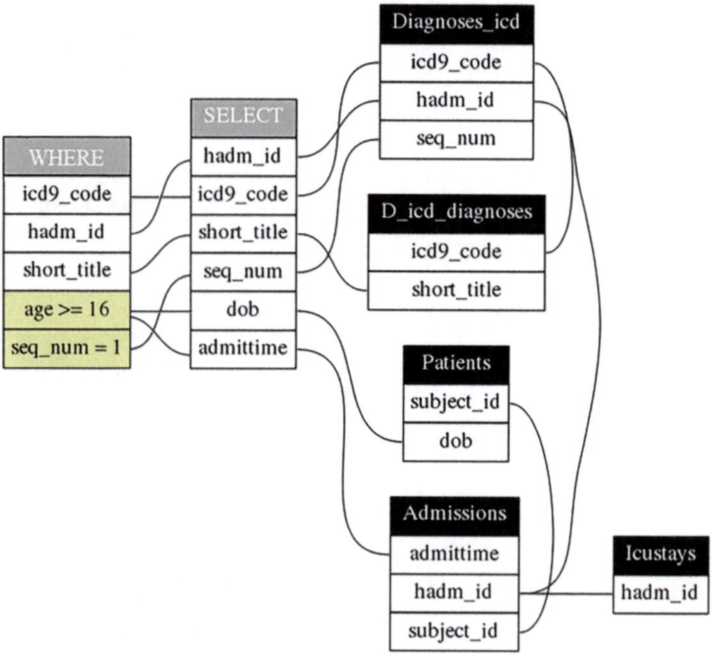

Fig. 20 MIMIC-III: extracting the distribution of ICD-9 codes across ICUs

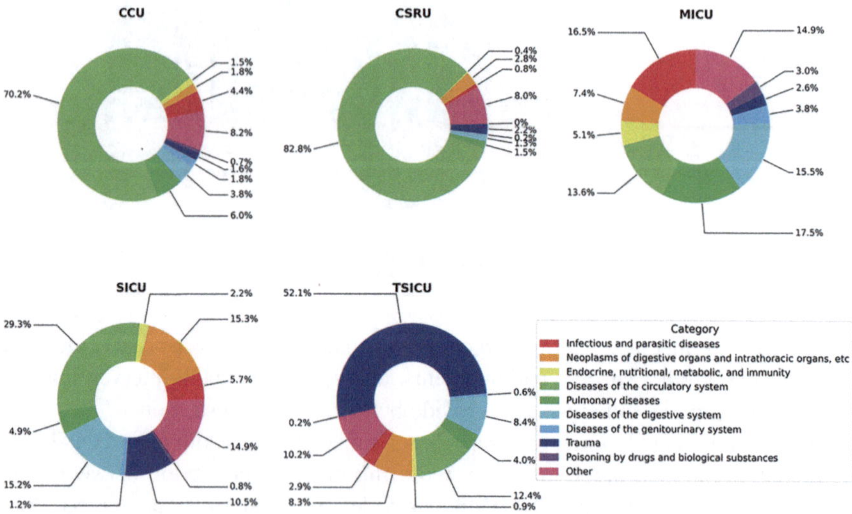

Fig. 21 MIMIC-III: the distribution of ICD-9 codes across ICUs

Concepts in MIMIC include severity of illness scores, organ dysfunction scores, timing of treatment, definition of sepsis, and comorbidities. The inclusion of these concepts in the MIMIC repository helps remove barriers for data scientists unfamiliar with the clinical environment.

Severity of illness scores, such as the Acute Physiology Score (APS) and its variations, have been developed to assess patient acuity, particularly at the time of ICU admission. These scores examine disturbances in normal bodily functioning and typically comprise at least 10 independent components calculated using data from the first 24 hours of a patient's stay. The APS, also known as APACHE (acute physiology, age, and chronic health evaluation), has evolved through several versions, incorporating more sophisticated statistical techniques and expanding its applications to include risk stratification, mortality prediction, and benchmarking hospital performance.

The Simplified Acute Physiology Score (SAPS) aimed to streamline the APS by reducing the number of required physiological parameters. Its subsequent versions incorporated data-driven feature selection and expanded the geographical scope of the data used in its development. Organ dysfunction scores, another important concept in MIMIC, assess multiple organ systems or focus on specific organs. Examples include the Sequential Organ Failure Assessment (SOFA), Logistic Organ Dysfunction System (LODS), Model for End-Stage Liver Disease (MELD), and various kidney disease assessment tools. These scores are used for prediction and prognosis, following a similar development pathway to illness severity scores.

The timing of treatment and sepsis detection are crucial concepts implemented in MIMIC. Sepsis, a life-threatening organ dysfunction caused by a dysregulated host response to infection, is a major concern in ICUs. MIMIC-III provides case selection for sepsis based on ICD-9 codes, enabling the development of early warning systems using high-resolution physiological data and machine learning algorithms.

Comorbidities, referring to chronic conditions that patients have prior to ICU admission, are another important concept in MIMIC. The database defines 29 categories based on ICD-9 codes, recognizing the impact of these conditions on a patient's probability of surviving significant illness.

While these clinical concepts provide valuable tools for population studies and quality of care assessments, their predictive power for individual patient outcomes is limited. This limitation stems partly from variant disease definitions and patient population heterogeneity. Advanced machine learning methods promise to deliver better warning systems and potentially outperform these traditional clinical concepts. In summary, concepts in MIMIC-III implement well-known scores of illness severity and organ dysfunction, providing an ideal substrate for comparing their performance and enabling reproducibility. However, these concepts are primarily based on expert knowledge, and their prognostic power is limited. With the availability of big data, there is hope that state-of-the-art machine learning approaches will provide personalized outcome predictions, potentially outperforming the clinical concepts developed over the past few decades. This evolution in clinical prediction models represents an exciting frontier in critical care research and practice.

3.9 *Flowchart of Patient Inclusion: Relation of Catheterization to Mortality—an Example Study*

This section discusses a clinical study that utilizes the MIMIC database to estimate the relationship between mortality and an intervention in intensive care unit patients. The study serves as an example of a complex selection pipeline for patients and clinical variables, followed by the construction of a model that enables statistical comparisons between control and intervention groups.

Indwelling arterial catheters are widely used in intensive care units for continuous hemodynamic monitoring and blood gas analysis. While these catheters provide crucial information about oxygen delivery to tissues and organs, they also carry risks such as bloodstream infections and vascular complications. A study published in CHEST in 2015 aimed to test the hypothesis of whether catheterization was associated with mortality in the ICU, prompted by concerns about the high rate of bloodstream infections associated with central venous catheters and, to a lesser extent, indwelling arterial catheters [8].

The study's purpose was to examine the association between indwelling arterial catheters and outcomes in a large cohort of hemodynamically stable intensive care patients with respiratory failure undergoing mechanical ventilation. This cohort was extracted from the MIMIC database using specific selection criteria based on clinical expertise.

The patient inclusion flowchart began with all patients in MIMIC-III older than 16 years with administration records (Fig. 22). The study focused on patients requiring mechanical ventilation within 24 hours of ICU admission. To eliminate confounding factors, several exclusion criteria were applied. Patients with sepsis were excluded due to the life-threatening nature of this condition. Those who were administered vasopressor drugs were also excluded, as these drugs can cause further complications by constricting blood vessels. Patients with indwelling arterial catheterization placement prior to ICU admission and those in the cardiac surgery recovery unit were also removed from the study cohort. The final cohort was split into two groups: patients who received catheterization and those who did not, with the latter serving as the control group for comparison. This division allowed for hypothesis testing on the significance of the relationship between mortality and catheterization.

The primary outcome of the study was 28-day mortality, identifying which patients died within 28 days of ICU admission. Secondary outcomes included length of stay in the ICU and hospital, duration of mechanical ventilation, and the average number of arterial and venous blood gas measurements performed per day.

The study defined the presence of an indwelling arterial catheter as placement of the invasive catheter at any point after the initiation of mechanical ventilation. The developed model included variables reflecting patient demographics, comorbidities based on ICD-9 codes, vital signs, and laboratory results. The measurements used were those immediately preceding the onset of mechanical ventilation.

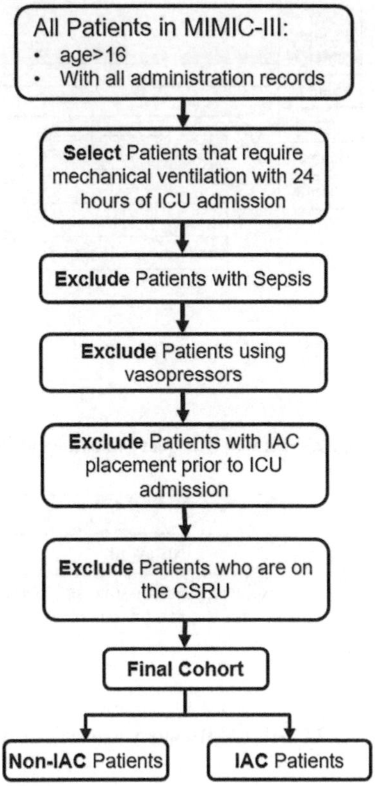

Fig. 22 MIMIC-III: flowchart of patient inclusion and exclusion criteria

The model estimated a propensity score, representing the likelihood of catheter placement. A genetic algorithm was employed to select the most relevant clinical variables and optimize the performance of the propensity score model. The propensity score matching process paired each patient from the catheterization group with patients from the non-catheterization group, allowing for an estimation of the intervention's impact.

The study's findings revealed no significant differences in 28-day mortality between the two groups. However, patients with indwelling arterial catheterization had a lower likelihood of discharge from the ICU within 28 days.

In summary, this example demonstrates how the MIMIC database can be used to conduct complex clinical studies. It highlights the importance of careful patient selection criteria to focus on relevant patients while eliminating confounding parameters. The study also illustrates the definition of primary and secondary outcomes and the development of a propensity model to examine differences between intervention and control groups. This approach provides a framework for using large-scale electronic health record databases to investigate clinical interventions and their outcomes in critical care settings.

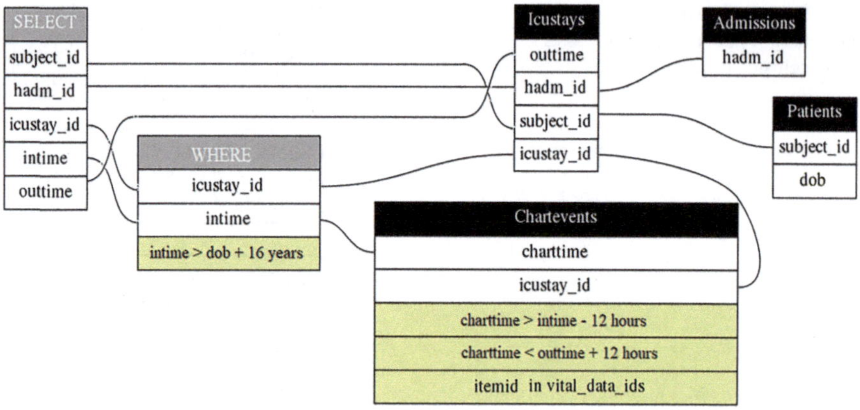

Fig. 23 MIMIC-III: query to extract all patients at ICU and all administration records

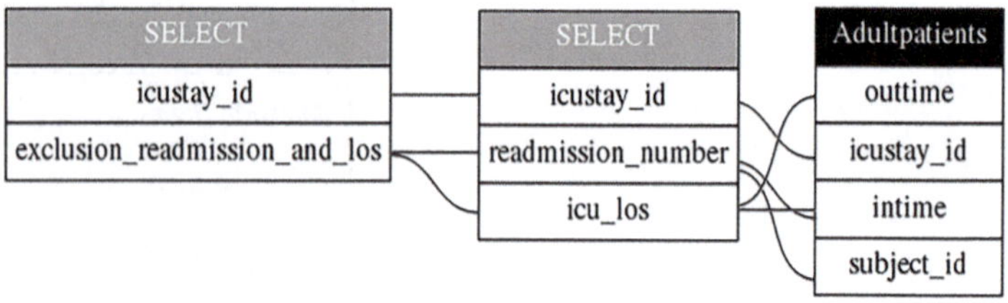

Fig. 24 MIMIC-III: query to exclude readmissions

3.9.1 Data Extraction from MIMIC-III for Catheterization and Mortality Study

This section provides a detailed guide to extracting and processing data from the MIMIC-III database for a study investigating the relationship between catheterization and mortality in intensive care units (ICUs). We will walk through each step of the data extraction pipeline, providing specific SQL queries and explanations to ensure reproducibility.

We begin by extracting vital data for patients over 16 years old. This involves joining multiple tables in the MIMIC-III database (Fig. 23). This query creates a table adult_patients containing the ICU stay identifier, subject identifier, hospital admission identifier, ICU admission and discharge times, date of birth, and chart time for heart rate measurements.

Next, we exclude readmissions and ICU stays shorter than 1 day (Fig. 24). This query creates a table eligible_stays that includes only ICU stays lasting at least 24 hours and marks the admission order for each patient.

We now identify patients who required mechanical ventilation within the first 24 hours of ICU admission (Fig. 25). This query identifies patients who received mechanical ventilation within 24 hours of ICU admission.

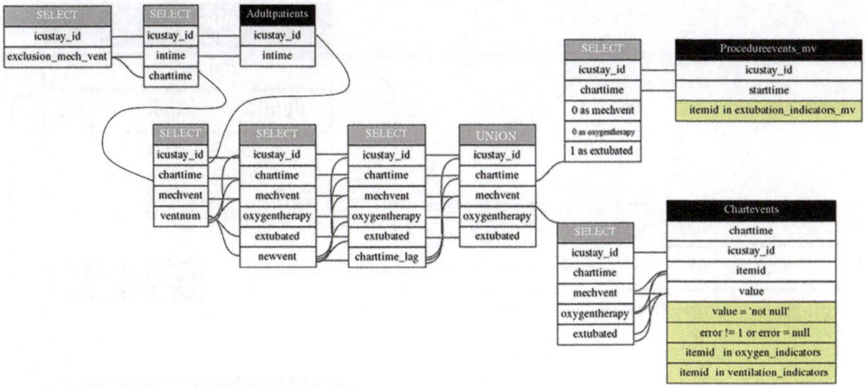

Fig. 25 MIMIC-III: query to include all patients with ventilation

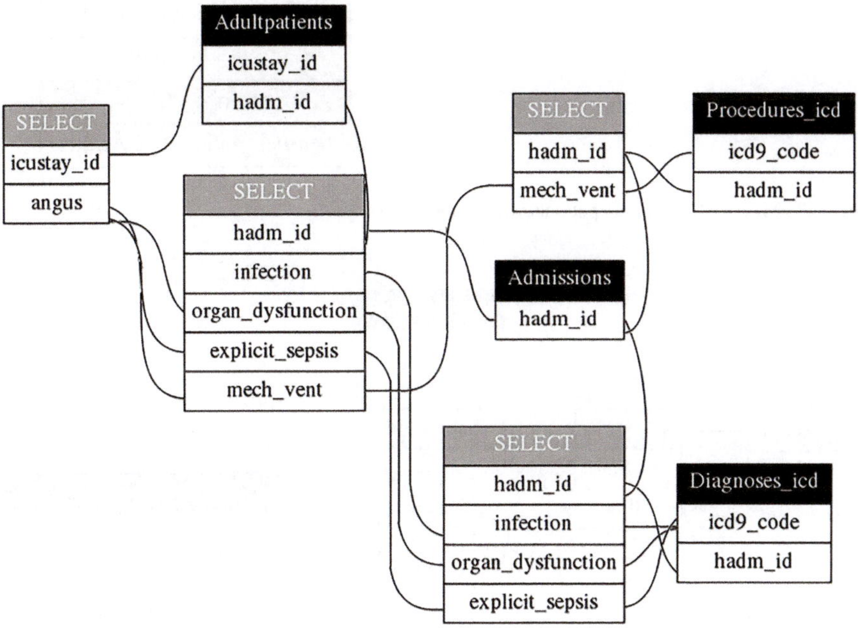

Fig. 26 MIMIC-III: exclude patients with sepsis

We apply the following exclusion criteria:

- We exclude patients diagnosed with sepsis using ICD-9 codes (Fig. 26).
- We exclude patients with specific medications (Fig. 27).
- We exclude patients with catheters placed before ICU admission (Fig. 28).
- Lastly, we exclude patients with prior surgery or Coronary Care Unit stays (Fig. 29).

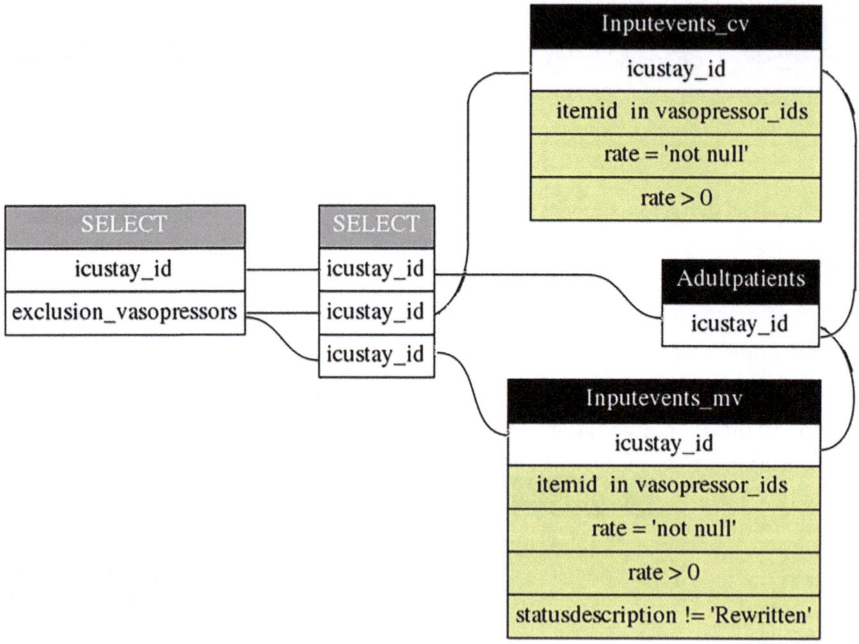

Fig. 27 MIMIC-III: exclude patients with vasopressors

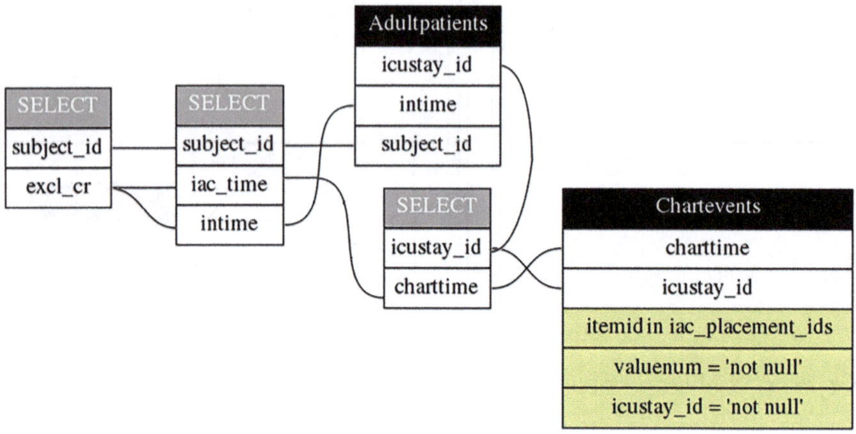

Fig. 28 MIMIC-III: extract patients with IAC placed

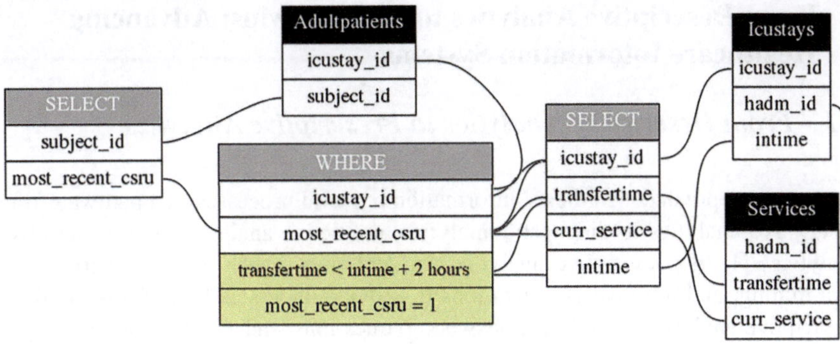

Fig. 29 MIMIC-III: exclude patients in CSRU or CCU

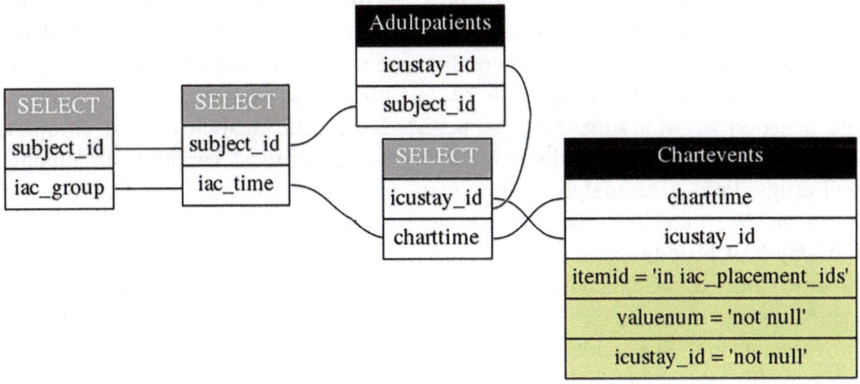

Fig. 30 MIMIC-III: divide cohort

We combine all criteria to create our final study cohort. Finally, we divide the cohort into catheterization and control groups (Fig. 30). This query assigns patients to either the catheterization group (1) or the control group (0) based on the presence of catheterization events in their records.

The data extraction process described in this section provides a comprehensive framework for selecting a study cohort from the MIMIC-III database. By following these steps, researchers can reproduce the data extraction process for studies investigating the relationship between catheterization and mortality in ICU patients.

4 From Descriptive Analytics to Digital Twins: Advancing Healthcare Information Systems

4.1 From Descriptive Analytics to Prescriptive Analytics

We can conceptualize healthcare information retrieval processes as a pathway from descriptive analytics to diagnostic analytics, predictive analytics, and prescriptive analytics [1, 6]. Descriptive analytics use techniques such as data aggregation, data mining, and intuitive visualizations to provide understanding of historic data. Descriptive analytics provide the answers to questions such as the following: How many patients were admitted to a hospital last year? How many patients died within 30 days? Or how many patients caught an infection? In other words, descriptive analytics offer intuitive ways to summarize the data via histograms and graphs and show the data distribution properties. Limitations of descriptive analytics are that it keeps limited ability to guide decision because it is based on a snapshot of the past.

Diagnostic analytics is a form of analytics that examines data to answer the question of why something happened. Diagnostic analytics could comprise correlation techniques that discover links between clinical variables, treatments, and drugs. Predictive analytics allow us to predict the outcome and likelihood of an event. We may like, for example, to predict the mortality risk of a patient, the length of hospitalization, or the risk for infection. Predictive analytics exploit historic values of the data with the aim to be able to provide useful information about critical events in the future. Predictive analytics are in demand because healthcare providers would like evidence-based ways to predict and avoid adverse events. Importantly, predictive analytics enable early intervention which can save patient lives and improve their quality of life. Prescriptive analytics aim to make decisions for optimal outcomes.

Prescriptive analytics attempt to quantify the effect of future decision to advise on possible outcomes before decisions are made. Therefore, they provide recommendations regarding actions that consider the results of predictive analytics. In other words, prescriptive analytics are important to transition a prediction model to a clinical decision support model.

4.2 Digital Twins

Patients increasingly expect seamless management of their information across their healthcare providers. Currently, traditional healthcare models rely on disconnected systems and multiple sources of information. The new digital healthcare model will transition toward an inherent capability to ensure seamless information exchange across both clinical system and real-time biomarkers of well-being (Fig. 31). In this way, they will update the records of each patient across all these systems. This

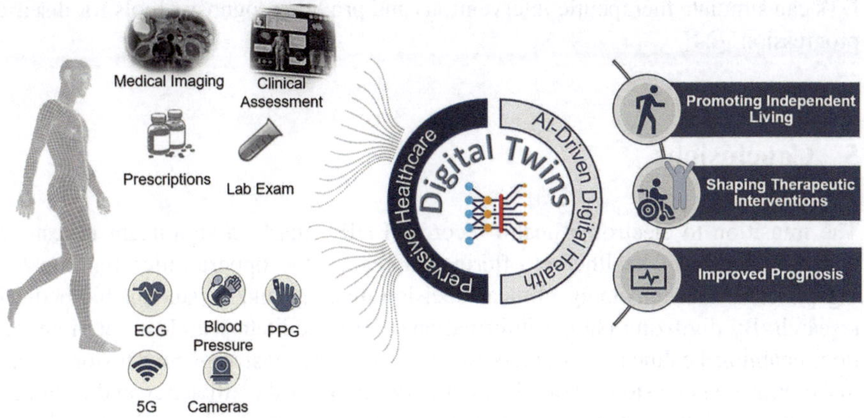

Fig. 31 Digital twins in healthcare

enables data mining and machine learning approaches to be successfully applied and advance our knowledge with relation to clinical decision-making systems.

"Digital twins" (DTs) is a concept very well known in industrial/manufacturing scale applications and it is used as a virtualization of physical assets and safety critical processes [17]. The aim is to simulate decision processes and optimize the design and the underlying operations in a safe and cost-efficient environment [21]. DTs definitions have been originally formalized by extensive work in large manufacturing companies and the NASA, which used the technology to develop digital copies of spacecrafts and monitor its status. In this approach, virtual models are made to duplicate every aspect from design to operation in a safe and cost-efficient way. Originally, DTs were modeled based on three entities, namely, physical entities, virtual models, and the connection between physical and virtual entities. This way though could not encompass recent advances in data science and it has been expanded to include DTs data and services, which ensure high prognostic fidelity [22].

It is only recently that it has been realized that DTs have enormous potentials in healthcare and in particular in precision medicine [24, 22, 5, 20, 17]. The reason behind this shift is that virtual copies of the patients can accelerate the analyses of large amounts of data derived from both clinical practice and sensing modalities in home environments. It has been emphasized that in chronic, multifaceted disorders and diseases such as multiple sclerosis and asthma, DTs can provide unprecedented capabilities of patient-centered disease preventive systems along with clinical decision support without increasing the burden on healthcare providers [24, 5]. For example, telemonitoring devices and sensing technologies in wearables and smart watches can provide digital patient support tools for several diseases such as asthma, multiple sclerosis, and diabetes [5]. In asthma and diabetes, DT technology encompasses machine learning models that will provide decision support to impel changes of lifestyle choices and other modifiable risk factors. In multiple sclerosis

DTs can simulate therapeutic interventions and provide prognostic tools for disease progression [24].

5 Conclusions

The transition to electronic health records (EHRs) marks a significant milestone in the evolution of healthcare, offering unprecedented opportunities for improving patient care, enhancing clinical decision-making, and advancing biomedical research. By digitizing patient information, EHRs facilitate seamless data integration, enabling healthcare providers to access comprehensive patient histories and make informed decisions. This shift not only improves the efficiency and accuracy of healthcare delivery but also supports the development of personalized medicine and digital twins in healthcare.

The adoption of EHRs in the UK and the USA highlights the complexities and challenges involved in transitioning from paper-based systems to digital records. Despite these challenges, the benefits of EHRs are evident in the improved quality of care, enhanced data management, and the potential for large-scale health data analysis. The integration of clinical registries and databases further enriches the healthcare landscape, providing valuable insights into patient outcomes and supporting evidence-based practice.

As we move toward a future where digital twins become an integral part of healthcare, the role of EHRs will continue to expand. These digital records will serve as the foundation for creating accurate and dynamic models of individual patients, enabling real-time monitoring and personalized treatment plans. The ongoing advancements in EHR technology and data analytics will undoubtedly drive innovation in healthcare, ultimately leading to better health outcomes and a more efficient healthcare system.

Acknowledgments Fani Deligianni is supported by funding from EPSRC (EP/W01212X/1) and Academy of Medical Sciences (NGR1/1678). She is also a member of the research team for NIHR (NIHR158303). This chapter builds upon content from the Coursera specialization "Informed Clinical Decision Making using Deep Learning."

References

1. Panesar Arjun. "Machine Learning and AI for Healthcare: Big Data for Improved Health Outcomes". In: *Apress: London, UK* (2019).
2. Amanda Bodell et al. "National quality register of congenital heart diseases-can we trust the data?" In: *Journal of Congenital Cardiology* 1 (2017), pp. 1–8.
3. Niall Boyce. "Bills of Mortality: tracking disease in early modern London". In: *The Lancet* 395.10231 (2020), pp. 1186–1187.

4. Centers for Disease Control and Prevention. *Classification of Diseases, Functioning, and Disability*. [Online; accessed 9-September-2024]. 2024. URL: https://www.cdc.gov/nchs/icd/index.html.
5. David Drummond, Jolt Roukema, and Mariëlle Pijnenburg. "Home monitoring in asthma: towards digital twins". In: *Current Opinion in Pulmonary Medicine* 29.4 (2023), pp. 270–276.
6. Aya Gamal, Sherif Barakat, and Amira Rezk. "Standardized electronic health record data modeling and persistence: A comparative review". In: *Journal of biomedical informatics* 114 (2021), p. 103670.
7. Nicholas R Hardiker et al. "Terminologies in the World Health Organisation Family of International Classifications (WHO-FIC)". In: ().
8. Douglas J Hsu et al. "The association between indwelling arterial catheters and mortality in hemodynamically stable patients with respiratory failure: a propensity score analysis". In: *Chest* 148.6 (2015), pp. 1470–1476.
9. Alistair EW Johnson et al. "MIMIC-III, a freely accessible critical care database". In: *Scientific data* 3.1 (2016), pp. 1–9.
10. The Lancet. *ICD-11*. 2019.
11. George B Moody. "Physionet: Research resource for complex physiologic signals". In: *Jpn. J. Electrocardiol* 29 (2009), pp. 1–3.
12. Iwao Milton Moriyama et al. "History of the statistical classification of diseases and causes of death". In: (2011).
13. Fouke Ombelet et al. "Creating the BELgian COngenital heart disease database combining administrative and clinical data (BELCODAC): rationale, design and methodology". In: *International journal of cardiology* 316 (2020), pp. 72–78.
14. World Health Organization. *International Statistical Classification of Diseases and related health problems: Alphabetical index*. Vol. 3. World Health Organization, 2004.
15. Fouzia F Ozair et al. "Ethical issues in electronic health records: A general overview". In: *Perspectives in clinical research* 6.2 (2015), pp. 73–76.
16. Sudha R Raminani. "International classification of diseases". In: *Encyclopaedia Britannica* (2015).
17. Adil Rasheed, Omer San, and Trond Kvamsdal. "Digital twin: Values, challenges and enablers". In: *arXiv preprint arXiv:1910.01719* (2019).
18. Mohammed Saeed et al. "MIMIC II: a massive temporal ICU patient database to support research in intelligent patient monitoring". In: *Computers in cardiology*. IEEE. 2002, pp. 641–644.
19. Mohammed Saeed et al. "Multiparameter Intelligent Monitoring in Intensive Care II: a public-access intensive care unit database". In: *Critical care medicine* 39.5 (2011), pp. 952–960.
20. Anastasios Loukas Sarris et al. "Towards a Digital Twin in Human Brain: Brain Tumor Detection Using K-Means". In: *Caring is Sharing-Exploiting the Value in Data for Health and Innovation*. IOS Press, 2023, pp. 1052–1056.
21. Fei Tao and Qinglin Qi. "Make more digital twins". In: *Nature* 573.7775 (2019), pp. 490–491.
22. Fei Tao et al. "Digital twin driven prognostics and health management for complex equipment'. In: *Cirp Annals* 67.1 (2018), pp. 169–172.
23. ET Vander Velde et al. "CONCOR, an initiative towards a national registry and DNA-bank of patients with congenital heart disease in the Netherlands: rationale, design, and first results". In: *European journal of epidemiology* 20 (2005), pp. 549–557.
24. Isabel Voigt et al. "Digital twins for multiple sclerosis". In: *Frontiers in immunology* 12 (2021), p. 669811.
25. Karen Wilson and Lara Khansa. "Migrating to electronic health record systems: a comparative study between the United States and the United Kingdom". In: *Health Policy* 122.11 (2018), pp. 1232–1239.

Artificial Intelligence-Driven Clinical Decision Support Systems

Muhammet Alkan, Idris Zakariyya, Samuel Leighton, Kaushik Bhargav Sivangi, Christos Anagnostopoulos, and Fani Deligianni

> *"The advance of technology is based on making it fit in so that you don't really even notice it, so it's part of everyday life."*
>
> *— Bill Gates*

Abstract As artificial intelligence (AI) becomes increasingly embedded in healthcare delivery, this chapter explores the critical aspects of developing reliable and ethical clinical decision support systems (CDSS). Beginning with the fundamental transition from traditional statistical models to sophisticated machine learning approaches, this work examines rigorous validation strategies and performance assessment methods, including the crucial role of model calibration and decision curve analysis. The chapter emphasizes that creating trustworthy AI systems in healthcare requires more than just technical accuracy; it demands careful consideration of fairness, explainability, and privacy. The challenge of ensuring equitable healthcare delivery through AI is stressed, discussing methods to identify and mitigate bias in clinical predictive models. The chapter then delves into explainability as a cornerstone of human-centered CDSS. This focus reflects the understanding that healthcare professionals must not only trust AI recommendations but also comprehend their underlying reasoning. The discussion advances in an analysis of privacy vulnerabilities in medical AI systems, from data leakage in deep learning models to sophisticated attacks against model explanations. The text explores privacy-preservation strategies such as differential privacy and federated learning while acknowledging the inherent trade-offs between privacy protection and model performance. This progression, from technical validation to ethical considerations, reflects the multifaceted challenges of developing AI systems that can be seamlessly

---

M. Alkan · I. Zakariyya · K. B. Sivangi · C. Anagnostopoulos · F. Deligianni (✉)
School of Computing Science at University of Glasgow, Glasgow, Scotland, UK
e-mail: fani.deligianni@glasgow.ac.uk

S. Leighton
School of Health and Well Being at University of Glasgow, Glasgow, Scotland, UK

B. Farahani et al. (eds.), *Smart and Connected Health: Applications and Real-World Use Cases*, https://doi.org/10.1007/978-3-032-09253-3_2

and reliably integrated into daily clinical practice while maintaining the highest standards of patient care and data protection.

1 Artificial Intelligence-Driven Clinical Decision Support Systems

1.1 From Machine Learning and Statistical Models to Clinical Decision Support Systems: An Overview

Clinical research demands a meticulous, multifaceted approach when evaluating predictive models, which extends beyond the traditional data science perspective. As we dive into this complex landscape, we must consider a series of key questions that shape the development and validation of machine learning models, ultimately determining their suitability as decision support systems in healthcare.

In prediction modeling, our main focus is on estimating the risk of adverse events based on a combination of factors. We seek to understand not only the predictive power of these factors but also their individual contributions to the model's decision-making process. This understanding is crucial, as it allows us to incorporate subject matter knowledge into the modeling pipeline, bridging the gap between data-driven insights and clinical expertise.

The foundations of the clinical prediction model validation process have been presented in Steyerberg et al. [47]. At the heart of this process lies the fundamental research question or hypothesis. For example, the choice of prediction outcome is paramount in clinical research. Outcomes such as mortality rates at 30 days are frequently relevant to various research questions. However, it's not just the nature of the outcome that matters but also its frequency within the dataset. This frequency effectively determines the sample size, which in turn influences the statistical power and reliability of the model.

The selection of patient data for the development of the model is a critical consideration. Often, these data are collected for purposes other than the study at hand, raising questions about their representativeness. We must carefully examine whether patient records truly reflect the population for which the study is intended. Additionally, the treatment of prognostic factors and their effects presents a unique challenge. Although traditional studies often consider treatment effects negligible compared to prognostic factors, there are instances where these effects warrant specific attention. Adjusting for baseline prognostic factors can offer significant advantages in estimating treatment effects applicable to individual patients.

The reliability and completeness of the predictor measurements pose another hurdle in model development. Incomplete datasets are common, with missing values for potential predictors. The approach to handling these missing data can significantly impact the model's performance and validity. Although complete case analysis—excluding patients with missing values—is a straightforward solution, it often results in the loss of significant information. More sophisticated methods, such as imputation techniques that leverage correlations between variables, offer a more nuanced approach to preserving data integrity. In addition, informative missingness, where the fact that data are missing is related to the outcome of interest, must be carefully considered to avoid biased results.

As we navigate these considerations, we recognize that the development of machine learning models for clinical decision support systems is a multifaceted process. It requires a delicate balance between statistical rigor, clinical relevance, and practical applicability. By addressing these key questions and challenges, we pave the way for more robust and reliable predictive models that can enhance clinical decision-making and, ultimately, patient care.

1.2 A Quick Overview of Model Development and Validation Strategies of Machine Learning Models

Model evaluation and selection are critical steps in the machine learning development process. In an ideal scenario, we would have access to data that perfectly represents the entire target population. In this case, we could train and test the machine learning model using the same data, and the error rate obtained would closely reflect the true error rate when the number of samples is very large. However, in reality, the error rate obtained when training and testing on the same dataset is positively biased. This is because the model has been exposed to the same data during both the training and testing phases, which can lead to an overly optimistic estimate of its performance. To address this issue, in real-world applications, it is important to split the available data into two separate sets: a training set and a testing set. Typically, around 70% of the data is used for training the model, while the remaining 30% is reserved for testing its performance on unseen examples. By separating the training and testing data, we can evaluate the model's ability to make accurate predictions on new, unseen data, which is crucial for deploying the model in real-world applications (Fig. 1).

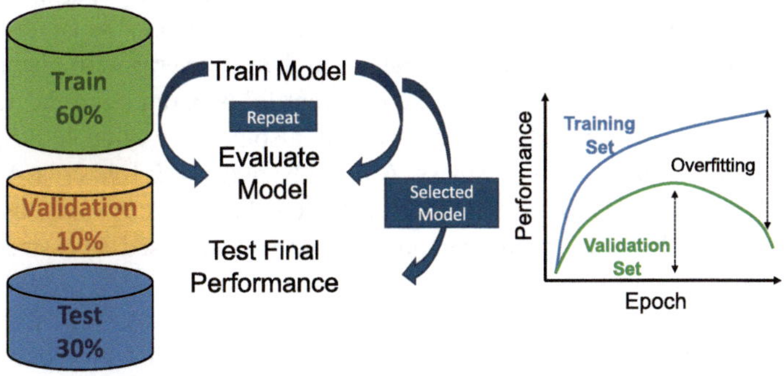

Fig. 1 Model development

In most cases, we estimate the empirical risk based on a limited number of samples. This involves measuring the loss function with respect to our trained classifier, as discussed in [20]. Empirical risk estimation is achieved by computing the average loss over the data points (m is the number of samples) according to a loss function L, which penalizes the differences between the predicted values $f(x)$ and the actual targets y:

$$R_S(f) = \frac{1}{m} \sum_{i=1}^{m} L(y_i, f(x_i)) \tag{1}$$

Variations in the empirical risk estimation can arise from several factors. These include random variations in the training and testing sets, the learning algorithm itself, and even the noise inherent in the data classes being considered. One key advantage of the holdout method (where the training and testing sets are independent) is that it provides some guarantees about the model's performance on data it has not been trained on before. However, we must also consider the confidence intervals around the empirical risk estimation. For example, when evaluating a machine learning algorithm, we should not assume a Gaussian distribution of the loss error, as the errors may be clustered near zero.

The error can be modeled using the Bernoulli distribution to calculate the error bound (Eq. 2), providing an indication of the potential deviation between the empirical risk estimation and the true risk with a probability of accuracy of $(1 - \delta)$. In Eq. 2, m' represents the sample size and δ represents the confidence parameter, which quantifies the probability that our estimate is accurate.

$$E = \sqrt{\frac{1}{2m'} \ln(\frac{2}{\delta})} \tag{2}$$

The k-fold cross-validation is one of the most popular error estimation approaches in machine learning model training and evaluation. In this method, the dataset is divided into k distinct parts or "folds" like in Fig. 2. During each iteration, one of these folds is reserved for testing, while the remaining $(k-1)$-folds are used for training the model. This process is repeated k times, with a different fold serving as the test set each time. By doing so, we obtain k separate estimates of the classifier's error rate. These estimates can then be averaged to give the mean performance of the algorithm across the different folds. Examining the variability of the error estimates across the k iterations can also provide valuable insights into the algorithm's stability and robustness. A key advantage of k-fold cross-validation is that the test samples are independent between the different folds, as there is no overlap. This helps to ensure a more reliable and unbiased assessment of the model's generalization capabilities.

A potential issue that can arise with standard k-fold cross-validation is that the data may not be evenly distributed across classes. This problem becomes worse when dealing with imbalanced class data, which is common in healthcare

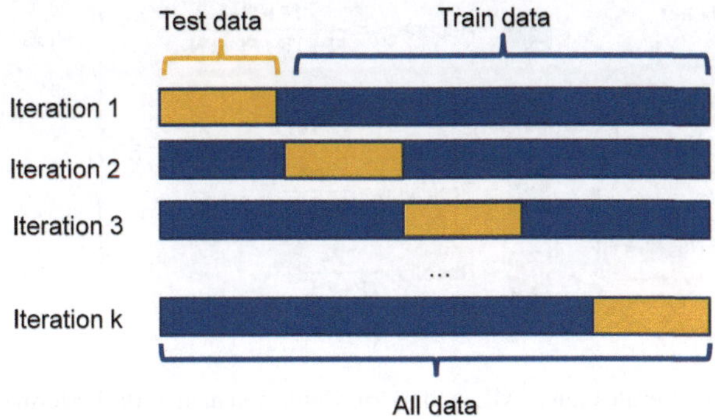

Fig. 2 k-fold cross validation

applications. To address this, we can employ a technique called stratified k-fold cross validation. In this approach, the folds are created in a way that ensures the class distribution within each fold closely matches the original class distribution in the overall dataset.

A special case of k-fold cross-validation is leave-one-out cross validation (LOOCV). In this case, the value k is set to the number of samples in the dataset, meaning that each sample is used as the test set once while the remaining $(k - 1)$ samples are used for training. LOOCV has the advantage of utilizing most of the available data, which can result in relatively unbiased estimate of the model's performance. However, this comes at the cost of significant computational expense, especially as the dataset size grows. Although, LOOCV may provide better performance estimates in datasets with extreme values, it is important to note that this is not a guarantee of an unbiased classifier, especially when dealing with small datasets. The underlying assumption of LOOCV is that the training set is representative of the true data distribution, which may not always hold.

One key aspect of validation techniques such as k-fold cross-validation and LOOCV is that the estimate is not based on a single, fixed classifier. Instead, the model is retrained each time, producing a new classifier with each iteration. This approach has both advantages and disadvantages. The primary advantage is that it allows us to assess the stability of machine learning models across different data partitions. However, the disadvantage is that when comparing the performance of different algorithms, we must remember that we are comparing the average performance estimates of various classifiers, rather than a single, fixed classifier as in the holdout method. To address this, nested cross-validation is often used, as it provides a more robust estimate of model performance by incorporating an additional layer of cross-validation to tune hyperparameters, thereby reducing the risk of overfitting.

One way to describe the performance of classification algorithms is through a confusion matrix as in Fig. 3b(a). This square matrix has rows and columns equal to the number of classes. The diagonal elements represent the true positives and true

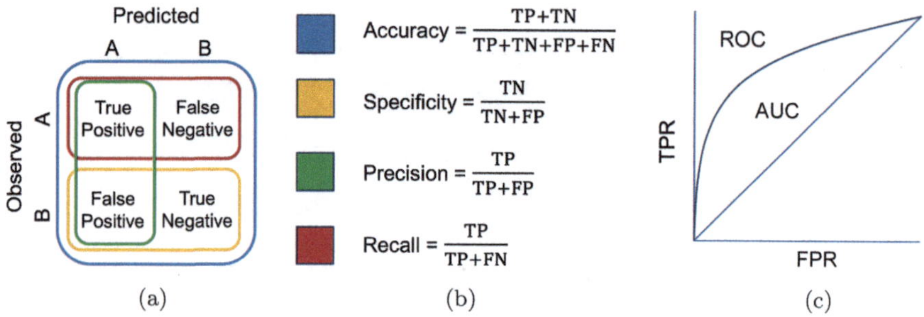

Fig. 3 Performance evaluation of ML models. (**a**) Confusion matrix. (**b**) Performance metrics. (**c**) The ROC curve

negatives, assuming "positive" refers to one class and "negative" to another. The off-diagonal elements indicate false positives and false negatives. From the confusion matrix, we can derive several performance metrics. For instance, accuracy is the ratio of correctly predicted observations to the total observations. Specificity, or the true negative rate, shows how well the classifier identifies negative cases, while recall also called sensitivity, or the true positive rate, indicates how well it identifies positive cases. Precision reflects the positive predictive value for a class. The F1 score combines recall and precision into a single metric, weighting them evenly.

An alternative method to assess the performance of a machine learning algorithm is by using a receiver operating characteristic (ROC) curve, as shown in Fig. 3c. The ROC curve plots the false positive rate on the horizontal axis and the true positive rate on the vertical axis. The true positive rate, or sensitivity, indicates how well the classifier identifies positive cases, while the false positive rate can be expressed as $(1 - \textit{specificity})$. Thus, the ROC curve illustrates the trade-off between sensitivity and specificity for the classifier. It is generated by varying the threshold for classifying positive and negative cases. This makes the ROC curve a comprehensive measure of performance, as it considers different threshold settings. It is used not only to analyze the behavior of machine learning algorithms but also for model selection by identifying the optimal threshold region. For a random classifier, the ROC curve would be a straight diagonal line. The area under the curve (AUC) summarizes the classifier's performance, with higher values indicating better performance. The AUC is often used to compare different classifiers. Additionally, there are various extensions of the ROC and AUC for multi-class scenarios.

When comparing the performance of one algorithm against another, or multiple algorithms across one or more datasets, it is common to use null hypothesis statistical testing. Several statistical tests can validate the performance of two algorithms, but it is crucial to consider the assumptions underlying these tests. For instance, the paired t-test assumes a normal distribution, independence of measurements, and an adequate sample size. If the normality assumption is violated, nonparametric tests are often used. One such test is the Wilcoxon signed-rank test, an alternative to the paired t-test. This test is based on the ranks of the absolute differences, making it more robust to outliers. It is important to note that both parametric and nonparametric tests can be manipulated by increasing the number of samples, potentially affecting the results of the null hypothesis statistical testing approach.

1.3 Performance Validation in Clinical Decision Support Systems

Validation of prediction models tailored for clinical decision support systems should occur through both internal and external methods [47, 41]. Internal validation, using techniques like split-sample validation, cross-validation, or bootstrapping, assesses reproducibility within the development population. External validation, involving patients from different populations, tests the model's generalizability across various settings and demographics.

While internal validation techniques provide valuable insights into a model's performance, external validation serves as a crucial complement in assessing predictive models for clinical use. This process involves testing the model on data that is entirely separate from the development dataset, often collected from different institutions or time periods. Despite the growing number of publications on prediction models, studies employing both internal and external validation remain relatively scarce. This highlights the challenges in establishing predictive models as reliable decision support systems.

Figure 4 illustrates a risk prediction tool designed to estimate the likelihood of symptom nonremission in first-episode psychosis [27]. The tool's internal validation performance was assessed using tenfold cross-validation yielding an AUC of 0.74. External validation was conducted with patient data from various sites, producing an AUC of 0.73 and confirming the model's generalizability. Successful external validation strengthens confidence in a model's clinical utility.

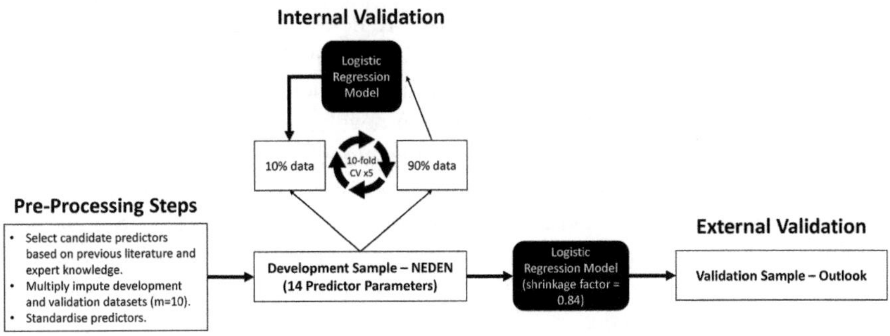

Fig. 4 A nonremission risk prediction model in first-episode psychosis (reproduced with permission [27])

It demonstrates that the model's predictions remain accurate across different patient populations and healthcare settings. This robustness is essential for establishing the model as a trustworthy component of clinical decision support systems. In other words, external validation offers a more rigorous test of a model's generalizability, revealing how well it performs in diverse real-world scenarios. It helps identify potential overfitting issues that may not be apparent through internal validation alone. In essence, external validation acts as a bridge between theoretical model development and practical clinical application. It provides the evidence needed to justify the integration of predictive models into healthcare decision-making processes, ultimately contributing to improved patient care and outcomes.

1.4 Calibration of Clinical Prediction Models

In clinical practice, the validation of machine learning models extends beyond traditional prediction performance metrics. While measures like AUC, precision, and F1 scores are crucial, they do not fully capture a model's clinical utility. Effective clinical decision support systems require assessment of underlying risk estimates and clinical usefulness, which can be subjective and application-dependent. Model calibration is a critical aspect of validation, referring to the agreement between observed outcomes and predictions [50]. For instance, if a model predicts a 15% risk of 30-day mortality, approximately 15 out of 100 patients with such a prediction should experience the outcome.

Calibration is typically assessed using flexible calibration curves, which plot estimated risks against observed proportions of events. Two key measures of calibration are the calibration in the large (alpha) and calibration slope (beta). A well-calibrated model should have an alpha close to zero and a beta close to one. However, these measures alone do not guarantee perfect calibration across all risk levels. Visualization through calibration plots helps identify areas of over- or underestimation. Figure 5 shows the calibration curve for the regression model presented in Fig. 4. This curve provides both qualitative and quantitative assessments of how well the developed risk prediction model aligns with the ideal calibration, represented by a blue line with a slope equal to 1.

Poor calibration can arise from various factors, including differences in patient characteristics or disease prevalence between development and validation populations, changes in healthcare practices over time, model overfitting, and measurement errors in medical data. Strategies to improve calibration include model refitting, continuous updating, and addressing population shifts dynamically. Sample size significantly impacts calibration assessment, with at least 200 events and nonevents recommended for precise evaluation. In smaller datasets, evaluating moderate calibration through intercept and slope calculations may suffice.

The importance of calibration in clinical settings cannot be overstated. A poorly calibrated model, even with high discrimination, can lead to misleading or potentially harmful clinical decisions. For example, in the risk prediction of

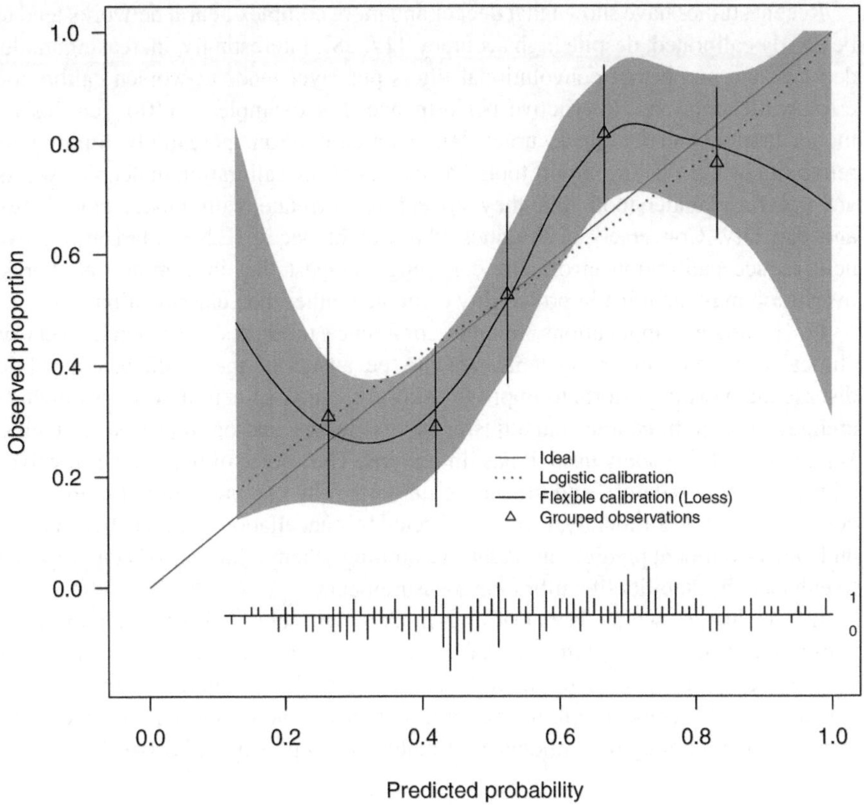

Fig. 5 Probability calibration plot (reproduced with permission [27])

nonremission in first-episode psychosis, overestimation could falsely suggest that patients do not need medication, thereby exposing them to unnecessary risks.

1.5 Calibration in Deep Learning Models for Clinical Decision Support

Calibration is also a critical aspect of deep learning models in clinical decision support systems, particularly for establishing trustworthiness with users. A well-calibrated model provides confidence estimates that accurately reflect the probability of correct predictions. For instance, a model with 90% confidence should be correct 90 out of 100 times. In practice, perfect calibration is unattainable, but we aim to approximate it. The expected calibration error (ECE) is a common metric used to assess calibration, measuring the difference between confidence and accuracy across prediction bins.

Recent studies have shown that deeper and more complex neural networks tend to be poorly calibrated, despite high accuracy [17, 38]. Interestingly, increasing model depth or the number of convolutional filters per layer tends to worsen calibration error while improving predictive performance. For example, a 110-layer ResNet model demonstrated high accuracy but poor calibration, potentially limiting its reliability as a decision support tool. The causes of miscalibration in deep networks are not fully understood, but they appear to correlate with model complexity and capacity. Conversely, techniques like weight decay (L2 regularization) can help reduce calibration error. These findings suggest that in complex networks, overfitting may manifest in probability estimates rather than classification errors.

For healthcare applications, reliable confidence measures are crucial. Users of clinical decision support systems need to be aware of the confidence level in disease diagnostics. Efforts to improve calibration in deep neural networks include architectural modifications and adjustments to training and optimization strategies. While the ECE is widely used, it has limitations. The choice of bin number involves a bias-variance trade-off, and the metric may not fully capture calibration in multi-class problems. Additionally, it can be affected by cancellation effects between over- and under-confident predictions. Adaptive binning schemes have also been proposed to enhance the stability of calibration measurements.

In conclusion, while deep learning models can achieve high accuracy, their calibration remains a significant challenge. Addressing this issue is essential for developing trustworthy and effective clinical decision support systems. Future research should focus on refining calibration metrics and developing techniques to improve the reliability of confidence estimates in complex neural networks.

1.6 Decision Curve Analysis

While accuracy metrics such as sensitivity, specificity, and area under the receiver operating characteristic curve are essential for evaluating prediction models, they fail to capture the clinical consequences of implementing these models in practice. Decision curve analysis (DCA) addresses this limitation by incorporating the concept of net benefit (NB), allowing for a more comprehensive assessment of a model's clinical utility [54, 51].

Net benefit is calculated across a range of threshold probabilities (ThresP), representing the point at which a clinician or patient would opt for intervention based on the model's prediction. This approach weighs the benefits of true positive

(TP) predictions against the harms of false positives (FP), taking into account the relative importance of these outcomes in a given clinical context.

$$NB = \frac{TP}{N} - \frac{FP}{N} \times \frac{\text{ThresP}}{1 - \text{ThresP}} \tag{3}$$

DCA plots the net benefit against threshold probabilities, comparing the prediction model's performance to alternative strategies such as treating all patients or treating none [53]. This visual representation allows stakeholders to assess the model's value across different risk thresholds, which may vary depending on the clinical scenario and individual preferences.

For instance, in cancer screening, a lower threshold probability might be preferred due to the severe consequences of missed diagnoses. Conversely, a higher threshold might be appropriate in situations where unnecessary interventions carry significant risks or costs. Figure 6 shows the DCA plot for the risk prediction model of nonremission presented in Fig. 4. The plot demonstrates that the net benefit of using the developed model is higher than the alternatives of treating all, treating no patients, or treating based on the duration of untreated psychosis (DUP).

The interpretation of DCA results requires careful consideration of the clinical context. A model demonstrating higher net benefit than alternative strategies within a clinically relevant range of threshold probabilities can be considered clinically useful. However, this assessment should be made in conjunction with expert opinion and patient preferences.

DCA can be applied to both continuous probability predictions and binary diagnostic tests. It is particularly valuable when evaluating models in external validation cohorts, as it provides insights into the model's generalizability and potential impact on clinical decision-making.

An illustrative example of DCA in practice is its application to a prediction model for outcomes in first-episode psychosis [28, 27]. By consulting specialist psychiatrists to determine clinically relevant threshold probabilities, researchers were able to demonstrate the model's superior net benefit compared to alternative strategies within the specified range.

It's important to note that while DCA offers valuable insights into clinical utility, it should not replace traditional measures of model performance. Instead, it should be viewed as a complementary tool in the comprehensive evaluation of prediction models, bridging the gap between statistical performance and clinical applicability.

In conclusion, decision curve analysis provides a robust framework for assessing the clinical value of prediction models. By incorporating the concept of net benefit and allowing for comparison across different decision thresholds, DCA enables more informed decisions about model implementation in clinical practice. As healthcare continues to move toward personalized medicine, tools like DCA will play an increasingly crucial role in translating prediction models into meaningful clinical support systems.

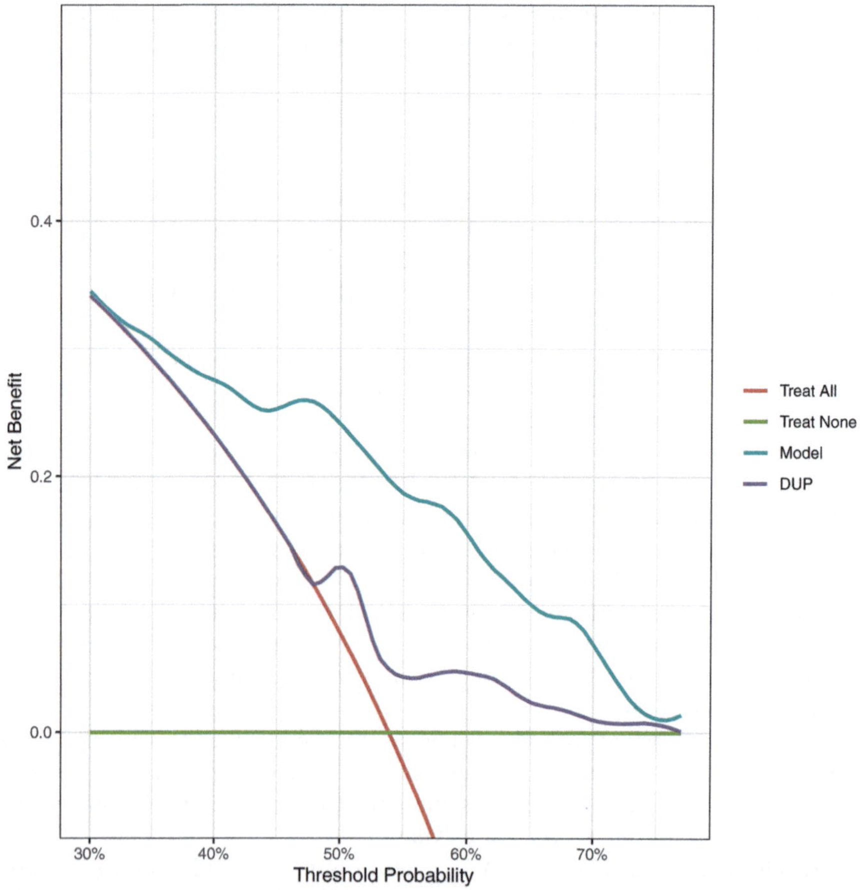

Fig. 6 Understanding clinical consequences of a risk prediction of non-remission in first-episode psychosis based on decision curve analysis (reproduced with permission [27])

1.7 Responsible Development of Artificial Intelligence-Driven Clinical Decision Support Systems

Previously, Steyerberg et al. [47] has established the importance of four steps to guide the development of machine learning models for healthcare applications: (A) calibration in the large, (B) calibration slope, (C) discrimination performance established both with internal and external validation, and (D) decision curve analysis. Recently, the integration of artificial intelligence (AI) and in particular deep learning in clinical decision support systems necessitates a more careful consideration of the principles behind responsible model development [19, 52]. These principles have extended guidelines to include additional safeguards that

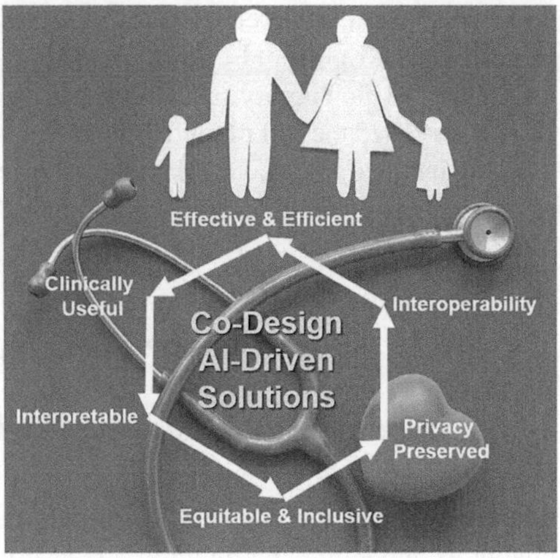

Fig. 7 Ethical considerations for the use of AI in clinical decision support systems

encompass fairness, explainability, privacy preservation, and interoperability, all of which are crucial for developing trustworthy AI applications in healthcare.

Figure 7 summarizes the concepts behind responsible AI models in healthcare. We already referred to the term "clinical usefulness," which encapsulates the necessity of an AI model to address a healthcare challenge and its capacity to evaluate if the advantages of an intervention justify the associated risks. "Effective and efficient" refers to the robustness of the underlying model and whether the outcome is reliably measured as highlighted by Steyerberg et al. [47]. "Interpretability" is important to understand what the underlying factors are that the model based its specific decision. Interpretability and transparency are cornerstones of responsible AI, enabling users to understand both the technical processes and human decisions involved in the system's operation. Accountability in AI systems requires mechanisms to minimize negative impacts and report adverse consequences. In this context, transparency is crucial for assessing accountability and fairness.

Healthcare practitioners have expressed concerns about overreliance on AI systems they cannot fully understand or explain. Addressing these concerns requires further training for practitioners and involving end-users in the design process. The next generation of human-centered clinical decision support systems should possess two key abilities: explaining the model's representation and decisions, and adapting based on feedback.

On the other hand, privacy and data governance are also paramount in clinical decision support systems. Developers must implement safeguards to protect user privacy, ensure data integrity, and control access to sensitive information. "Privacy-preserving" technologies are also essential to address concerns arising from the use of AI models in new applications that enable real-time tracking of human activity and health in home settings. Privacy-preserved AI model should leverage interoperable solutions that seamlessly integrate into the healthcare system.

In conclusion, adherence to responsible AI development guidelines is crucial in designing AI-powered clinical decision support systems. Emphasizing clinical usefulness, explainability, and human-in-the-loop designs is essential to mitigate risks and realize the full potential of AI in healthcare. The development of the right level of interactivity and explainability in medical applications remains an open research question, highlighting the ongoing challenges in this rapidly evolving field.

2 "Fairness" in Machine Learning Models

2.1 Assessing Bias in Clinical Predictive Models

Electronic health records (EHRs) present unique challenges and opportunities in predictive modeling, particularly in ensuring fairness and equity in healthcare outcomes. Biases can stem from flaws in study design, execution, or data analysis. To identify such biases, a comprehensive approach is necessary, considering the model's intended use, target population, predictors, and predicted outcomes.

Therefore, the evaluation of predictive models should extend beyond discrimination and calibration to encompass potential biases that may introduce systematic errors. A framework for assessing bias in clinical predictive models has been proposed, focusing on four key domains: participant selection, variable and predictor selection, outcome assessment, and analysis [55]. This framework emphasizes the importance of appropriate inclusion and exclusion criteria, consistent predictor definition and assessment across participants, and the use of standardized outcome definitions.

Additionally, the analysis should consider sample size adequacy and the handling of continuous and categorical predictors. Researchers are advised to avoid selecting predictors based solely on univariate analysis. Models can be categorized as having low, medium, or high risk of bias based on the assessment of these domains. Notably, prediction models developed without external validation should generally be considered high risk, except when based on very large datasets.

The concept of "informative presence" in EHRs refers to the potential information carried by the presence or absence of patient data at any given time point. For example, EHRs can be inherently biased because sicker individuals are monitored more frequently. This type of informative presence indicates that the frequency of health records can reflect a patient's health status. "Informative observation" extends this concept to the timing, frequency, and patterns of longitudinal observations in EHRs, which can provide insights into a patient's evolving health state [45]. While these phenomena can complicate causal or association studies, they also offer potential sources of implicit information that could be exploited in predictive models.

In conclusion, while identifying and addressing bias is crucial for developing robust clinical predictive models, the inherent characteristics of EHRs, such as informative presence and observation, present both challenges and opportunities. Researchers must carefully interpret results while also exploring innovative ways to leverage these implicit data patterns to improve prediction accuracy.

2.2 *Equity Challenges in Machine Learning for Healthcare Applications*

The pursuit of health equity is a global priority, as exemplified by the World Health Organization's vision of a society where all individuals enjoy long, healthy lives [3]. Machine learning algorithms have the power to identify unexpected patterns in data, but they can also inadvertently perpetuate and amplify existing biases. This is particularly concerning when these algorithms are used to support clinical decision-making, as they can systematically disadvantage certain population groups [6, 39].

Electronic health records, which serve as the foundation for many predictive models, often reflect historical biases in patient selection, policies, and societal circumstances. These biases can manifest in various ways, such as underrepresentation of minority groups in the data, systematic differences in feature availability across populations, or the compounding of initial biases over time.

The case of St. George Hospital in the UK serves as a cautionary tale [43]. In the 1980s, the hospital developed a computer program to streamline medical school admissions based on historical data. Unintentionally, this program formalized existing prejudices against racial minorities and women, demonstrating how algorithmic decision-making can perpetuate systemic biases.

Similar issues have been observed in other domains, such as facial recognition technology and online advertising. These examples highlight the potential for algorithms to discriminate based on race, gender, or age, often due to nonrepresentative training data or the inadvertent use of biased proxies.

In healthcare, a 2019 study revealed that a widely used risk prediction algorithm considered Black patients to be healthier than equally ill White patients [39]. This discrepancy arose because the algorithm used healthcare costs as a proxy for health needs, failing to account for disparities in healthcare access and utilization. Consequently, Black patients had to be significantly sicker than White patients to receive the same level of care.

This case underscores the importance of critically examining clinical decision support tools for potential biases. Even when sensitive attributes like race or gender are explicitly excluded from models, they can be implicitly correlated with other features, leading to discriminatory outcomes.

Discriminatory bias in healthcare algorithms can stem from various sources, including study design, data collection, clinician interactions, and patient behaviors. These biases may manifest as label bias, cohort bias, or various forms of data bias, such as minority underrepresentation or missing data for protected groups.

The interaction between clinicians and predictive models can also introduce biases. Automation bias may lead to overreliance on model predictions, while dismissal bias could result in ignoring alerts for certain groups. Patient interactions with healthcare systems can further complicate matters, with issues like privilege bias and informed mistrust affecting model effectiveness and fairness.

Addressing algorithmic bias in healthcare is challenging. Simply removing sensitive fields from the data is insufficient, as algorithms can identify and learn from proxy variables. Moreover, the complex nature of these biases makes them difficult to detect and eliminate, particularly when they reflect deeply ingrained societal prejudices.

In conclusion, while machine learning holds great promise for advancing healthcare, it is crucial to remain vigilant about the potential for algorithmic bias. Researchers and practitioners must actively work to quantify and mitigate these biases, ensuring that predictive models promote rather than hinder health equity. This requires ongoing investigation, diverse and representative datasets, and a commitment to fairness in algorithm design and implementation.

2.3 *Strategies to Ensure Fairness in Machine Learning Models for Healthcare*

The pursuit of fairness in machine learning for healthcare applications is crucial, as algorithmic bias can lead to discriminatory outcomes that impact patient care. This section explores systematic approaches to detect and mitigate such biases.

Recent legislation, such as the 2019 Algorithmic Accountability Act in the USA [30], has begun to address these concerns by requiring companies to assess and rectify algorithmic biases. However, the complexity and proprietary nature of many algorithms pose challenges for independent evaluation.

To mitigate discriminatory bias, several strategies have been proposed:

- Careful selection and representation of protected groups in the data
- Thorough investigation of potential healthcare disparities in historical data
- Incorporation of fairness goals into model training
- Continuous evaluation of fairness metrics and model performance across groups
- Vigilant monitoring of data and model reassessment during deployment

Furthermore, several fairness metrics have been proposed [7, 31]. Calibration techniques, applied within protected subgroups, can help identify algorithmic bias. Quantifying calibration bias within protected subgroups and assessing statistical parity are essential steps in evaluating model fairness. However, it is crucial to recognize that statistical parity may not be clinically meaningful when disease prevalence differs between subgroups. Other important metrics include independence, separation, and sufficiency [9]. Independence aims for classifier scores to be independent of group membership, while separation focuses on the independence of scores and sensitive variables conditional on the target variable. Sufficiency examines the independence of the target and sensitive variables given a particular score.

It is important to note that these fairness criteria cannot all be satisfied simultaneously when risk prevalence differs across groups. This highlights the need for careful consideration of trade-offs in AI model development. Therefore, further key objectives for fair decision-making in healthcare are defined that include achieving equal patient outcomes, equal model performance, and equal resource allocation across protected groups. However, these goals may sometimes conflict, necessitating thoughtful prioritization and stakeholder involvement in the design process.

In conclusion, ensuring fairness in machine learning models for healthcare is an ongoing challenge that requires continued research, vigilance, and collaboration among stakeholders. By addressing these issues systematically, we can work toward more equitable and effective healthcare systems that benefit all patients, regardless of their demographic characteristics.

3 Explainability as a Central Component of Human-Centered CDSS

3.1 *Interpretability vs. Explainability*

Interpretability and explainability, while often used interchangeably, have distinct meanings in machine learning. Interpretability is an inherent model property, whereas explainability involves methods to elucidate non-interpretable models. Figure 8 illustrates a network assessing patient risk based on factors like BMI, age, smoking habits, alcohol consumption, and blood pressure. Consider an 85-year-old female patient with a BMI of 32, high blood pressure, and no smoking or alcohol use. The system labels her "at risk" recommending medication. However,

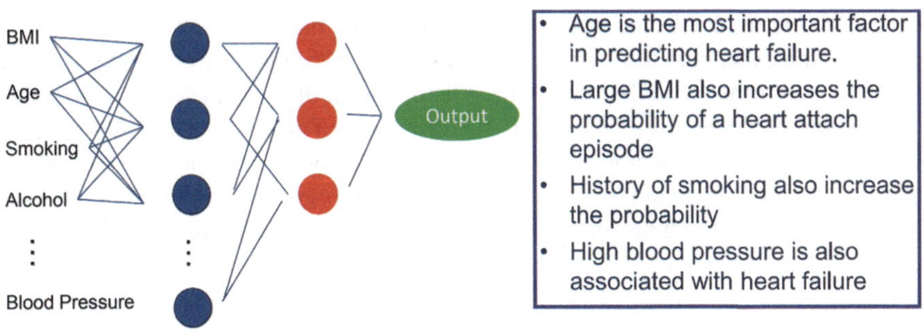

Fig. 8 An example network with some specific input factors

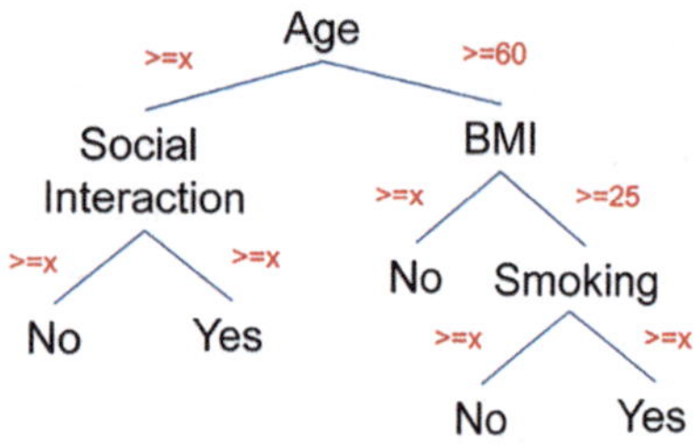

Fig. 9 An example of a decision tree

this output alone may not suffice for a doctor to trust and act upon the model's decision. Understanding the reasoning behind the model's output becomes crucial, highlighting the importance of explainability in complex models, especially in critical domains like healthcare where comprehending the decision-making process is essential for informed and ethical treatment decisions.

Explainability in machine learning models can address crucial questions about a model's performance, success conditions, and decision factors. For example, consider that in heart failure prediction age is a significant factor, with individuals over 60 years having an estimated likelihood of 60%. Furthermore, a BMI exceeding 25 increases risk by 20%, as does smoking for over a decade. High blood pressure is also correlated with heart failure. An explainable model should identify these key input factors and quantify their impact on the decision. This insight into the model's underlying function aids result interpretation, clarifies decision-making processes, and helps understand model failures in noisy conditions. Essentially, explainability provides transparency into the model's inner workings, enabling users to comprehend not just what the model predicts, but why it makes those predictions.

Decision trees are widely regarded as highly interpretable machine learning models due to their transparent and intuitive structure, as the example in Fig. 9 shows. The treelike representation clearly illustrates the decision-making process, with each node representing a specific decision point based on a particular feature. This allows users to easily follow the path from root to leaf, understanding how the model arrives at its predictions. The clear criteria used at each node for splitting the data provide insight into precisely how decisions are made at each step. This hierarchical nature lends itself well to visual representation, making it easier for both experts and nonspecialists to grasp the model's logic. For any given prediction, one can trace the exact path through the tree, observing which features were considered and how they influenced the final output. This traceability enhances accountability and facilitates debugging. Moreover, the structure of the tree inherently reveals which features are most important for classification or regression, as they appear closer to the root and in more splits. Unlike more complex models, decision trees can be explained to individuals without a background in machine learning or statistics, facilitating communication between data scientists and stakeholders. However, it is important to note that interpretability can decrease as tree depth increases. Very deep trees may become more challenging to interpret, potentially approaching the complexity of "black-box" models.

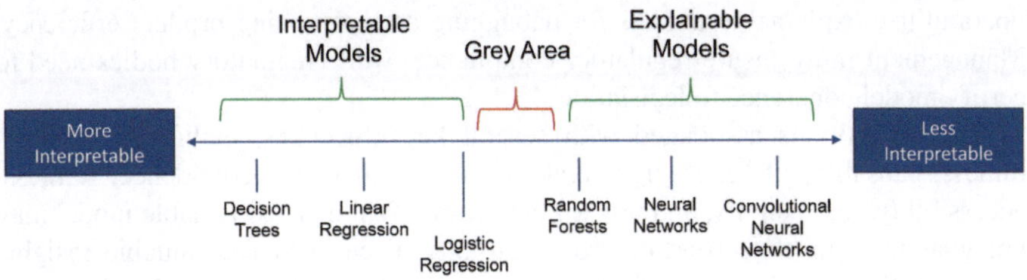

Fig. 10 Interpretable vs. explainable models

Figure 10 provides a simplified overview of interpretable and explainable models. On the left, we see inherently interpretable models such as decision trees, linear regression, and logistic regression. These models have been widely used in clinical practice and decision-making due to their simple construction and easily understandable results. The straightforward nature of these models allows practitioners to readily interpret their outputs and understand the reasoning behind decisions. Consequently, these models do not require additional methods to explain their results, as their decision-making process is transparent by design. This intrinsic interpretability distinguishes them from more complex models that may require additional explanation techniques to elucidate their outputs.

While linear regression and decision trees offer high interpretability, they often fall short in performance, especially given the complexity of modern datasets and available computational resources. Sacrificing predictive accuracy for inherent interpretability is increasingly seen as suboptimal in many applications. Instead, the focus has shifted toward developing methods to explain high-performing, complex models. This approach aims to harness the superior predictive power of sophisticated algorithms while still providing insights into their decision-making processes.

3.2 "Explainability" in Healthcare Applications

In healthcare applications, explainability is particularly critical for assessing model stability, visualizing relationships affecting outcomes, and enabling ethical analysis, especially concerning minority groups. It also facilitates patient involvement in decision-making processes and allows for the evaluation of privacy risks associated with complex model representations. This transparency is essential for maintaining the confidence of healthcare professionals, patients, and end-users. Moreover, explainability aids in monitoring model performance over time, as data distributions may shift and affect outcomes.

The significance of explainability extends to various stakeholders. Clinicians require trustworthy models that contribute to scientific knowledge. Patients need assurance of fair treatment and absence of hidden biases. Data scientists and developers utilize explainable models for debugging and improving product efficiency. Management must ensure regulatory compliance, while regulatory bodies need to certify model adherence to legislation.

Explainability is associated with several key objectives, including trustworthiness, causality inference, transferability, informativeness, confidence, fairness, accessibility, interactivity, and privacy awareness. While an explainable model may not guarantee absolute trust or prove causality, it can provide valuable insights into potential causal relationships and help validate results from other inference techniques.

3.3 Taxonomy of Explainability Methods

Some of the most common categorizations of explainability methods are local versus global, model agnostic versus model specific, data modality specific versus data modality agnostic, and ad hoc versus post hoc explanations [4, 36, 48]. Local versus global explanation is a very common distinction that translates on whether we get an explanation that relates to the overall function of the model or an explanation related to specific decisions. Model-agnostic versus model-specific categorization is also very common [42]. An important difference is that model-agnostic explanations are not bound to a specific machine learning algorithm, whereas model-specific explanations are derived directly based on the machine learning model employed.

Explainability methods can also be categorized based on the modality in the form of data modality-specific versus data modality-agnostic methods. In this case, the complexity of data representation is highlighted. For example, explaining decisions with relation to imaging data require a completely different approach than explaining differences in tabular data as they do not have any spatiotemporal dependency between variables. Another categorization is for ad hoc versus post hoc explanations. While ad hoc explanations aim to achieve interpretability by restricting the complexity of the machine learning model (intrinsic explainability), post hoc methods aim to achieve interpretability by applying methods that analyze the model after training.

Figure 11a presents a two-dimensional taxonomy of explainability methods, organized along two axes: local versus global explanations, and model-specific versus model-agnostic approaches. For example, Shapley Additive Explanation (SHAP) provides both local and global interpretability by analyzing variable impacts through perturbation—locally for individual predictions and globally for overall model behavior. In contrast, class activation maps and integrated gradients provide model-specific interpretability by analyzing neural network activations. These methods trace how specific input features influence the network's decision-making process for individual samples, revealing the internal reasoning patterns of the model.

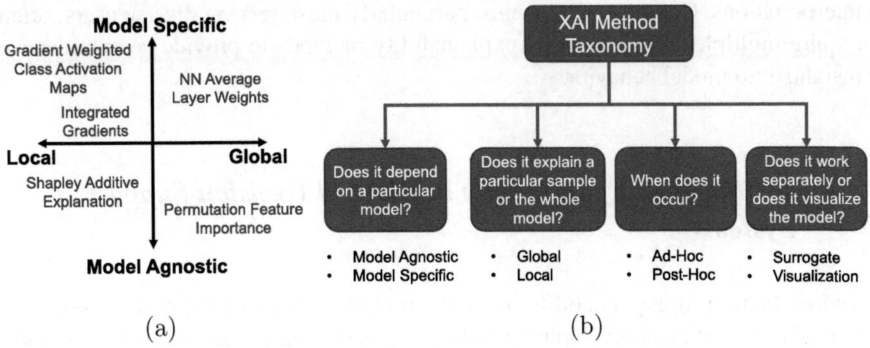

Fig. 11 Explainability methods (**a**) Model-specific versus model-agnostic explanations. (**b**) Overview of the different explainability methods

Permutation feature importance is one of the simplest, yet powerful approaches that links the input variable with the outcome across all the samples and for this reason it is considered a global explainability method. Permutation-based approaches such as permutation feature importance permute the input values and features in order to understand the importance of each input variable [35]. In other cases, they can apply a surrogate model, an inherently interpretable model that is simple enough to understand its behavior in a local space. The local importance highlights the important features for any individual prediction. It is very helpful to illustrate the local behavior of the underlying model. Sometimes we can aggregate an average result from local methods in order to get a better idea of how the model functions globally [22, 33].

The advantage that a model-agnostic explanation provides is that we can compare different models with the same data and get an idea of the quality of the prediction as well as the explanation. Model-agnostic tools are usually applied after the model has been trained, so they are considered as post hoc explanations. In ad hoc explanations, the model has been designed to be intrinsically explainable. Such an example is representation learning where the model identifies latent factors that intuitively have an explanation and they relate to the function of the model.

On the other hand, we can also categorize methods based on whether they are surrogate models or attribution-based methods. Surrogate models basically try to use a simpler model to approximate the behavior of a black box. Attribution methods can also be categorized based on whether they are perturbation-based or backpropagation or gradient-based methods.

Figure 11b illustrates the key dimensions for categorizing explainability methods in AI systems. These dimensions include model dependency (agnostic vs. specific), scope of explanation (global vs. local), timing (intrinsic vs. post hoc), and approach (surrogate models vs. visualization). Each dimension guides the selection of appropriate explainability techniques based on specific analysis requirements. The diverse taxonomy of explainability approaches yields varied types of model interpretations. Complex AI systems, particularly those serving diverse users, often require multiple complementary explainability methods to provide comprehensive insights into model behavior.

3.4 Evaluating Explainability in Clinical Decision Support Systems

The evaluation of explainability methods in clinical decision support systems is a crucial aspect of their development and deployment. This process considers not only the technical aspects of the explanations but also their effectiveness for end-users, whether they are healthcare providers, clinicians, patients, or carers.

Explanations typically provide three types of information: the importance of features or attributes to the model, including their interactions; the reasoning behind specific predictions; and an approximation of the complex model using a simpler, interpretable surrogate model such as rule-based systems, decision trees, or linear models.

The evaluation of explanations involves assessing both the model's intrinsic interpretability and the quality of its approximation by interpretable explanations. Key aspects of this assessment include clarity (consistency of rationale for similar instances), parsimony (complexity and compactness of the explanation), fidelity (accuracy in describing the task model), and soundness (truthfulness to the task model).

Attribution-based explanations, which identify the input features most relevant to the model's decision, are common in post hoc explanation methods. While these explanations may not fully satisfy the sufficiency property, they often meet the parsimony criterion if the features are understandable to humans. For clinicians, such explanations are valuable in comparing the model's decision-making process with their own clinical knowledge.

User-based evaluation, both quantitative and qualitative, is crucial in understanding how trust in AI models affects overall system performance in human-in-the-loop scenarios. This approach bridges the gap between technical performance and practical utility in clinical settings, ensuring that explainable AI systems not only perform well mathematically but also integrate effectively into clinical workflows and decision-making processes. Toward this end, quantitative evaluations often utilize metrics based on questionnaires assessing the usefulness, satisfaction, and interest provided by the system's explanations. They may also measure human-machine task performance in terms of accuracy and response time.

4 Privacy Concerns in Clinical Decision Support Systems: A Brief Overview

4.1 Leakage and Privacy Concerns of Deep Learning Models

The integration of machine learning in healthcare applications, particularly in clinical decision support systems (CDSS), necessitates understanding and addressing significant privacy concerns. Advanced machine learning models can memorize and inadvertently expose sensitive information, even when identity data has been removed or pseudo-anonymized. Furthermore, with state-of-the-art sensing technologies, it is now possible to monitor people 24/7 at home for healthcare applications [56]. This capability introduces new privacy challenges, as information about people's activities and physiology can be leaked in real time [57]. For example, Wi-Fi and radar-based human motion sensing has gained significant attention for offering unobtrusive observation in sensitive environments like assisted living facilities, hospitals, and residential settings. However, emerging research has uncovered a critical privacy concern: advanced radar systems can now accurately identify individuals through their unique walking patterns. These findings challenge the initial assumption of radar sensing as a privacy-preserving technology, highlighting the urgent need for robust privacy protection mechanisms in sensing systems, even when the raw data may not be visually comprehensible to humans. This section explores the inherent risks in deep learning models and discusses strategies to mitigate privacy threats in the context of healthcare applications.

Deep learning models excel at processing complex, correlated sensing data, which has led to substantial improvements in fields such as computer vision and information retrieval. However, these models inadvertently memorize training data within their weights, making it possible to reconstruct parts of the original dataset from the algorithm itself [18]. Traditional anonymization techniques, such as removing personally identifiable information or using pseudo-anonymization, are insufficient protection against sophisticated privacy attacks against deep neural networks. These powerful algorithms can determine an individual's identity by exploiting similarities to other datasets or inferring information from remaining data points. This capability has led to large-scale reidentification attacks.

Privacy threats in machine learning can manifest in various sophisticated ways, even with limited model access. For example, memorization of the training data might be reflected in assigning high likelihood to specific input samples. An attacker might gain insights into a model through its architecture, weights, training data, or even just its output logits. In the context of membership inference attacks (MIA), adversaries can potentially determine whether a specific data record was part of the original training dataset. The attack typically involves an attacker model that learns to predict the outcomes of the target model, with the goal of reconstructing sensitive input data.

While seemingly straightforward, MIA attacks serve as a foundational technique for more advanced data extraction methods via model inversion and reconstruction.

The attack process often employs a shadow model, which is a surrogate model with similar architecture as the target model but different parameters [44]. By generating synthetic data and evaluating its performance against the shadow model, the attacker can identify data likely similar to the original training set. This method is particularly powerful when the attacker lacks direct access to the target model's training data. A high prediction score from the shadow model suggests that the synthetic data closely resembles the original target data [11].

To address these concerns, innovative approaches are being developed. One promising method involves disentangling latent representations in data, separating key features from identity information [32, 29]. This approach has shown success in improving classification performance while protecting individual privacy, as demonstrated in recent work on human pose data. Other potential mitigation mechanisms include incorporating robust privacy techniques during the development of the target model to enhance its resistance to data leakage attacks.

In summary, the powerful memorization capabilities of deep neural networks create inherent vulnerabilities that can compromise patient privacy. It is essential to safeguard user privacy by integrating robust privacy protection mechanisms early in the development of decision support systems. Tools like Privacy Meter available at [37] can facilitate this process by measuring how an AI system may potentially leak sensitive information. By separating biometric data from features of interest and filtering out identity information early in the processing pipeline, it is possible to enhance both the effectiveness and the privacy-preserving qualities of these systems.

4.2 Defenses Against Privacy Attacks

Privacy preservation in healthcare machine learning applications is crucial, both at the data and model levels. This section explores various approaches to protect patient privacy, contrasting centralized and federated learning methods.

4.2.1 Differential Privacy Against Privacy Attacks

Differential privacy (DP) is a mathematical framework designed to protect sensitive data used in training AI models. DP algorithms mitigate the risk of data leakage by introducing calibrated noise into computations. This is particularly valuable when handling sensitive information, such as healthcare data, in the development of decision support systems. The procedure involves adding noise based on two privacy budget parameters, ϵ and δ [14]. The parameter ϵ controls the amount of noise added, while δ defines the probability of the mechanism failing to maintain privacy. Stronger privacy guarantees are achieved with smaller values of ϵ and δ.

Definition 1 *A randomized function F provides pure ϵ-DP if for all neighboring input datasets D_1 and D_2 differing on at most one element and $\forall S \subseteq Range(F)$*

[14], satisfying Eq. 4, where Pr in Eqs. 4 and 5 represents a probability measure.

$$\Pr[F(D_1) \in S] \leq e^{\epsilon} \Pr[F(D_2) \in S] \tag{4}$$

For (ϵ, δ)-DP, the guarantee is relaxed as follows:

$$\Pr[F(D_1) \in S] \leq e^{\epsilon} \Pr[F(D_2) \in S] + \delta \tag{5}$$

Thus, when $\delta = 0$, the relaxed guarantee simplifies to the pure ϵ-DP condition. The noise introduced can follow either a Gaussian or Laplace distribution, depending on the desired balance between privacy guarantees and utility. DP techniques are extensively utilized to develop robust machine learning models that protect training data from leakage, especially in the presence of MIA. In this context, noise can be introduced to either the data or the model gradients during training.

DP methods that focus on input data features often rely on the addition of Laplace noise [40, 15, 57]. Proper implementation of additive noise injection on specific features effectively safeguards sensitive data against MIA [57].

Another widely adopted mechanism for creating DP-compliant models is applying (ϵ, δ)-DP to model gradients during training [1, 13, 8]. This approach involves clipping gradients and adding Gaussian noise proportional to ϵ at each training iteration. The clipping norm, a fixed threshold (C), bounds the gradient magnitudes for individual data points, managing the effect of large gradients [24]. The model is iteratively trained using mini-batches of the input data samples and a stochastic gradient descent (SGD) algorithm, commonly referred to as the DP-SGD procedure. Algorithm 1 outlines the DP-SGD training procedure, incorporating gradient clipping and noise addition to ensure DP guarantee.

Algorithm 1 Differentially Private Stochastic Gradient Descent (DP-SGD)

Input: Dataset $\mathcal{D} = \{x_1, \ldots, x_n\}$, loss function $\mathcal{L}$, learning rate η, clipping norm C, noise scale σ, batch size B, number of iterations T.
Output: Model parameters θ_T.

Initialize model parameters θ_0.
for $t = 1$ to T **do**
  Sample a random mini-batch $\mathcal{B}_t \subset \mathcal{D}$ of size B.
  for each $x_i \in \mathcal{B}_t$ **do**
    Compute the gradient $\mathbf{g}_i = \nabla_\theta \mathcal{L}(\theta_{t-1}, x_i)$.
    Clip the gradient: $\tilde{\mathbf{g}}_i = \mathbf{g}_i / \max\left(1, \frac{\|\mathbf{g}_i\|_2}{C}\right)$.
  end for
  Aggregate the clipped gradients: $\bar{\mathbf{g}} = \frac{1}{B} \sum_{i \in \mathcal{B}_t} \tilde{\mathbf{g}}_i$.
  Add noise: $\hat{\mathbf{g}} = \bar{\mathbf{g}} + \mathcal{N}(0, \sigma^2 C^2 \mathbf{I})$.
  Update the model: $\theta_t = \theta_{t-1} - \eta \hat{\mathbf{g}}$.
end for
return θ_T

4.2.2 Federated Learning and Defenses Against Privacy Attacks

Traditionally, machine learning models in healthcare have been developed in centralized settings, where both data and models reside within the same environment [2]. For instance, the MIMIC database, a clinical database integrating information from thousands of patients, exemplifies such centralized approaches. Similarly, technology corporations like Google collect usage data directly from mobile devices for training models. While these methods are effective in leveraging large datasets, they also raise significant privacy concerns.

Federated learning has emerged as an alternative decentralized approach for building machine learning models. This paradigm involves a single server and multiple clients, each retaining their own data. Instead of centralizing data, the algorithm is distributed to where the data resides. Training iterations occur locally on the clients, with model parameters shared with a centralized server that aggregates these updates to create a global model [34]. A commonly used federated learning algorithm is federated averaging (FedAvg) [34]. FedAvg enables the creation of a global model by using a central server to average the weights of local models from various client devices. This process is iterative: during each communication round, the server interacts with clients to update the global model until convergence is achieved.

Federated learning allows data to remain with its owners while still enabling collaborative algorithm training. However, it can be communication and memory intensive. Additionally, federated learning alone does not inherently guarantee security and privacy, necessitating additional protective measures. Both centralized and edge models are vulnerable to privacy and adversarial attacks [23, 46, 25].

Model-level defenses include machine unlearning (or "forgetting"), which aims to remove specific data without retraining the entire model. This approach, while conceptually appealing, faces challenges in implementation and verification. Adversarial defenses, such as adding targeted noise to confidence score vectors, have been proposed to protect against MIA, though they lack theoretical privacy guarantees [21].

DP has gained significant attention in federated learning due to its ability to provide a guaranteed maximum privacy loss. In federated learning, this method involves adding calibrated noise to model updates during local training on each device, enhancing both generalization and patient privacy. However, determining appropriate privacy thresholds for local model updates remains a challenge.

Homomorphic encryption represents another promising avenue for secure AI in healthcare, allowing data processing without decryption [23]. However, its computational complexity is prohibitive for practical applications and it also suffers from a lack of explainability and transparency. Despite these challenges, it offers strong privacy guarantees and it is considered a key element in next-generation secure and private AI systems.

Ultimately, the goal is to strike an optimal balance between accuracy, explainability, fairness, and privacy in healthcare AI systems. While federated learning and other privacy-preserving methods offer promising solutions, they must be carefully combined and implemented to ensure both the utility and security of sensitive healthcare data. Several open questions remain in this field. Researchers are exploring whether decentralized data storage and federated learning can enable cross-institutional research while preserving privacy.

4.3 Adversarial Attacks Against Explanations in Deep Learning

Explainability techniques in deep neural networks, aimed at improving interpretability and trust, are increasingly recognized as being susceptible to adversarial attacks [5]. This section examines the vulnerabilities of these techniques and their potential implications for healthcare applications, where trust and reliability are paramount.

Recent research has revealed that explanations for deep learning models can be manipulated in ways that are difficult to detect. Two main types of attacks have been identified: those that manipulate the model's loss function to produce misleading explanations while preserving performance, and those that introduce small, nearly imperceptible perturbations to input data to alter explanations significantly. Popular adversarial attacks targeting explainability models include both white-box and black-box approaches [5, 49]. White-box attacks operate under the assumption that the adversary has complete knowledge of the model's architecture, parameters, and gradients. Examples of white-box attacks include the fast gradient sign method (FGSM) [16]; its iterative variant, the projected gradient descent (PGD) [26]; and the Carlini & Wagner (C&W) attack [10]. These methods leverage detailed internal information to craft adversarial examples that maximize model vulnerability.

In contrast, black-box attacks are designed for scenarios where the attacker lacks access to the model's internal parameters or architecture. These attacks are often query-based, relying on input-output observations to infer model behavior. A notable example is the zeroth order optimization (ZOO) attack [12], which uses iterative querying to approximate gradients and generate adversarial examples.

Both attack paradigms aim to manipulate model gradients, creating adversarial inputs that lead to misclassifications. Such attacks pose significant risks to the security and robustness of AI systems, particularly in sensitive applications where model integrity is important.

Gradient-based explanation methods, such as Grad-CAM, are particularly susceptible to manipulation. Researchers have demonstrated various techniques to fool these methods, including:

- Location fooling: Making the explanation highlight a specific region of the input, regardless of its relevance
- Top-k fooling: Reducing the importance of pixels that originally had the highest interpretation scores
- Center-mass fooling: Optimizing the heatmap to diverge as much as possible from the original without affecting classification performance

- Active fooling: Swapping explanations between different target classes entirely

These manipulations are achieved by introducing additional terms to the loss function during training, balancing between classification accuracy and explanation manipulation.

Research has identified potential defenses. One approach involves smoothing explanations at the network level by replacing rectified linear unit (ReLU) activation functions with softplus functions. The smoothness of these functions, controlled by a beta parameter, can be adjusted to make networks more resistant to data perturbation-based manipulations. DP can also help protect against privacy attacks on explanations.

Understanding and modeling the susceptibility of neural networks to explanation manipulation is crucial. It helps not only in preventing malicious attacks but also in detecting when explanations might be misleading due to other limitations. This knowledge can guide the development of more robust explainability techniques, which is particularly important in healthcare applications where the reliability of AI-assisted decisions is paramount.

4.4 Trade-Offs Between Privacy Protection and Model Performance

The implementation of privacy-preserving techniques in machine learning introduces a fundamental tension between data protection and algorithmic performance. This delicate balance is particularly critical in healthcare applications, where both privacy and predictive accuracy are paramount. Privacy-preserving methods fundamentally challenge the traditional approach to machine learning by introducing mechanisms that deliberately obscure and modify data or alter the convergence of the model training process. At the core of this challenge lies the complex interplay between protecting individual privacy and maintaining the integrity of predictive models. When privacy techniques are applied, they invariably reduce the richness and depth of available information, creating a nuanced landscape of compromises.

Feature representation suffers particularly from privacy sanitization techniques. By removing or masking sensitive identifying information, models lose critical contextual information that might be essential for accurate predictions. In personalized medicine, this can mean overlooking subtle but important correlations that could significantly impact diagnostic accuracy.

Empirical research has demonstrated concrete performance implications. Differential privacy can reduce model accuracy by 5–20%, depending on the privacy budget. They are also notoriously difficult to train from scratch, and thus, they depend on pretrained models. Federated learning models often show a 3–15% performance reduction compared to centralized training approaches. Synthetic data generation presents even more significant challenges, potentially leading to up to 30% loss in predictive power if not meticulously implemented.

Despite these technical challenges, the ethical imperative remains clear. The potential psychological and social harm from data breaches far outweighs marginal improvements in predictive accuracy. Privacy protection must remain a fundamental consideration, not an afterthought. Ultimately, the goal is not to choose between privacy and performance, but to develop intelligent systems that can maintain both.

5 Conclusions

The integration of artificial intelligence into clinical decision support systems represents both a transformative opportunity and a complex challenge in healthcare. While these systems offer unprecedented capabilities for improving patient care, their successful implementation demands a careful balance of multiple critical factors. The development pathway must address not only technical excellence in model performance but also robust validation, proper calibration, and thorough decision curve analysis. As we look to the future, the success of AI in healthcare will depend on our ability to navigate challenges in explainability, causality, and bias while maintaining the trust of both healthcare providers and patients.

The complexity of healthcare AI systems raises three paramount concerns that must be carefully balanced. First, there is the critical challenge of ensuring fairness and addressing bias, particularly given the inherent "informative presence" in electronic health records and their potential to reflect historical societal disparities. Second is the need for transparency and explainability, requiring sophisticated approaches that go beyond simple interpretability to provide meaningful insights for healthcare professionals and patients alike. Third is the fundamental requirement to protect patient privacy, particularly given the potential vulnerabilities created by deep learning models' ability to memorize training data. While techniques such as differential privacy, federated learning, and homomorphic encryption offer promising solutions, they often require careful trade-offs between privacy protection and model performance.

Success in this evolving landscape requires a holistic approach that considers these interconnected challenges. The path forward demands ongoing collaboration between technical experts, healthcare providers, and policymakers to develop systems that are not only technically sophisticated but also ethically sound, clinically useful, and privacy-preserving. As AI continues to transform healthcare, our ability to balance these competing demands while maintaining focus on improved patient outcomes will determine the ultimate impact of these innovations on the future of medicine.

Acknowledgments Fani Deligianni is supported by funding from EPSRC (EP/W01212X/1) and Academy of Medical Sciences (NGR1/1678). She is also a member of the research team for NIHR (NIHR158303). This chapter builds upon content from the Coursera specialization "Informed Clinical Decision Making using Deep Learning."

References

1. Martin Abadi et al. "Deep learning with differential privacy". In: *Proceedings of the 2016 ACM SIGSAC conference on computer and communications security*. 2016, pp. 308–318.
2. Abdullah Alanazi. "Using machine learning for healthcare challenges and opportunities". In: *Informatics in Medicine Unlocked* 30 (2022), p. 100924.
3. Michelle M Amri et al. "Scoping review of the World Health Organization's underlying equity discourses: apparent ambiguities, inadequacy, and contradictions". In: *International Journal for Equity in Health* 20.1 (2021), p. 70.
4. Alejandro Barredo Arrieta et al. "Explainable Artificial Intelligence (XAI): Concepts, taxonomies, opportunities and challenges toward responsible AI". In: *Information fusion* 58 (2020), pp. 82–115.
5. Hubert Baniecki and Przemyslaw Biecek. "Adversarial attacks and defenses in explainable artificial intelligence: A survey". In: *Information Fusion* (2024), p. 102303.
6. Solon Barocas and Andrew D Selbst. "Big data's disparate impact". In: *Calif. L. Rev.* 104 (2016), p. 671.
7. Alex Beutel et al. "Putting fairness principles into practice: Challenges, metrics, and improvements". In: *Proceedings of the 2019 AAAI/ACM Conference on AI, Ethics, and Society*. 2019, pp. 453–459.
8. Franziska Boenisch et al. "Have it your way: Individualized Privacy Assignment for DP-SGD". In: *Advances in Neural Information Processing Systems* 36 (2024).
9. Alycia N Carey and Xintao Wu. "The statistical fairness field guide: perspectives from social and formal sciences". In: *AI and Ethics* 3.1 (2023), pp. 1–23.
10. Nicholas Carlini and David Wagner. "Towards evaluating the robustness of neural networks". In: *2017 ieee symposium on security and privacy (sp)*. Ieee. 2017, pp. 39–57.
11. Nicholas Carlini et al. "Membership inference attacks from first principles". In: *2022 IEEE Symposium on Security and Privacy (SP)*. IEEE. 2022, pp. 1897–1914.
12. Pin-Yu Chen et al. "Zoo: Zeroth order optimization based black-box attacks to deep neural networks without training substitute models". In: *Proceedings of the 10th ACM workshop on artificial intelligence and security*. 2017, pp. 15–26.
13. Christophe Dupuy et al. "An efficient dp-sgd mechanism for large scale nlu models". In: *ICASSP 2022-2022 IEEE International Conference on Acoustics, Speech and Signal Processing (ICASSP)*. IEEE. 2022, pp. 4118–4122.
14. Cynthia Dwork, Aaron Roth, et al. "The algorithmic foundations of differential privacy". In: *Foundations and Trends®in Theoretical Computer Science* 9.3–4 (2014), pp. 211–407.
15. Ryusei Fujimoto, Yugo Nakamura, and Yutaka Arakawa. "Differential Privacy with Weighted ϵ for Privacy-Preservation in Human Activity Recognition". In: *2023 IEEE International Conference on Pervasive Computing and Communications Workshops and other Affiliated Events (PerCom Workshops)*. IEEE. 2023, pp. 634–639.
16. Ian J Goodfellow, Jonathon Shlens, and Christian Szegedy. "Explaining and harnessing adversarial examples". In: *arXiv preprint arXiv:1412.6572* (2014).
17. Chuan Guo et al. "On calibration of modern neural networks". In: *International conference on machine learning*. PMLR. 2017, pp. 1321–1330.
18. John Hartley et al. "Neural networks memorise personal information from one sample". In: *Scientific Reports* 13.1 (2023), p. 21366.
19. Anne AH de Hond et al. "Guidelines and quality criteria for artificial intelligence-based prediction models in healthcare: a scoping review". In: *NPJ digital medicine* 5.1 (2022), p. 2.
20. Nathalie Japkowicz and Mohak Shah. *Evaluating learning algorithms: a classification perspective*. Cambridge University Press, 2011.
21. Jinyuan Jia et al. "Memguard: Defending against black-box membership inference attacks via adversarial examples". In: *Proceedings of the 2019 ACM SIGSAC conference on computer and communications security*. 2019, pp. 259–274.

22. Yola Jones, Fani Deligianni, and Jeff Dalton. "Improving ECG classification interpretability using saliency maps". In: *2020 IEEE 20th International Conference on Bioinformatics and Bioengineering (BIBE)* IEEE. 2020, pp. 675–682.
23. Georgios A. Kaissis et al. "Secure, privacy-preserving and federated machine learning in medical imaging". In: *Nature Machine Intelligence* (2024).
24. Weiwei Kong and Andres Munoz Medina. "A unified fast gradient clipping framework for DP-SGD". In: *Advances in Neural Information Processing Systems* 36 (2024).
25. Kummari Naveen Kumar, Chalavadi Krishna Mohan, and Linga Reddy Cenkeramaddi. "The impact of adversarial attacks on federated learning: A survey". In: *IEEE Transactions on Pattern Analysis and Machine Intelligence* 46.5 (2023), pp. 2672–2691.
26. Alexey Kurakin, Ian Goodfellow, and Samy Bengio. "Adversarial machine learning at scale". In: *arXiv preprint arXiv:1611.01236* (2016).
27. Samuel P Leighton et al. "Development and validation of a nonremission risk prediction model in first-episode psychosis: an analysis of 2 longitudinal studies". In: *Schizophrenia bulletin open* 2.1 (2021), sgab041.
28. Samuel P Leighton et al. "Development and validation of multivariable prediction models of remission, recovery, and quality of life outcomes in people with first episode psychosis: a machine learning approach". In: *The Lancet Digital Health* 1.6 (2019), e261–e270.
29. Zhi Li et al. "Mutual information-based preference disentangling and transferring for non-overlapped multi-target cross-domain recommendations". In: *Proceedings of the 47th International ACM SIGIR Conference on Research and Development in Information Retrieval* 2024, pp. 2124–2133.
30. Mark MacCarthy. *An examination of the Algorithmic Accountability Act of 2019*. SSRN, 2020.
31. Suvodeep Majumder et al. "Fair enough: Searching for sufficient measures of fairness". In: *ACM Transactions on Software Engineering and Methodology* 32.6 (2023), pp. 1–22.
32. Matthew Malek–Podjaski and Fani Deligianni. "Towards Explainable, Privacy-Preserved Human-Motion Affect Recognition". In: *2021 IEEE Symposium Series on Computational Intelligence (SSCI)* 2021, pp. 01–09. https://doi.org/10.1109/SSCI50451.2021.9660129.
33. Joshua Mayourian et al. "Pediatric ECG-Based Deep Learning to Predict Left Ventricular Dysfunction and Remodeling". In: *Circulation* 149.12 (2024), pp. 917–931. https://doi.org/10.1161/CIRCULATIONAHA.123.067750. eprint: https://www.ahajournals.org/doi/pdf/10.1161/CIRCULATIONAHA.123.067750. URL: https://www.ahajournals.org/doi/abs/10.1161/CIRCULATIONAHA.123.067750
34. Brendan McMahan et al. "Communication-efficient learning of deep networks from decentralized data". In: *Artificial intelligence and statistics*. PMLR. 2017, pp. 1273–1282.
35. Xinlei Mi et al. "Permutation-based identification of important biomarkers for complex diseases via machine learning models". In: *Nature communications* 12.1 (2021), p. 3008.
36. Christoph Molnar. *Interpretable machine learning* Lulu. com, 2020.
37. Sasi Kumar Murakonda and Reza Shokri. "Ml privacy meter: Aiding regulatory compliance by quantifying the privacy risks of machine learning". In: *arXiv preprint arXiv:2007.09339* (2020).
38. Jeremy Nixon et al. "Measuring Calibration in Deep Learning." In: *CVPR workshops*. Vol. 2. 7. 2019.
39. Ziad Obermeyer et al. "Dissecting racial bias in an algorithm used to manage the health of populations". In: *Science* 366.6464 (2019), pp. 447–453.
40. NhatHai Phan et al. "Adaptive laplace mechanism: Differential privacy preservation in deep learning". In: *2017 IEEE international conference on data mining (ICDM)*. IEEE. 2017, pp. 385–394.
41. Chava L Ramspek et al. "External validation of prognostic models: what, why, how, when and where?" In: *Clinical Kidney Journal* 14.1 (2021), pp. 49–58.
42. Marco Tulio Ribeiro, Sameer Singh, and Carlos Guestrin. "Model-agnostic interpretability of machine learning". In: *arXiv preprint arXiv:1606.05386* (2016).
43. Oscar Schwartz. "Untold history of AI: Algorithmic bias was born in the 1980s". In: *IEEE Spectrum* 15 (2019).

44. Reza Shokri et al. "Membership inference attacks against machine learning models". In: *2017 IEEE symposium on security and privacy (SP)* IEEE. 2017, pp. 3–18.
45. Rose Sisk et al. "Informative presence and observation in routine health data: a review of methodology for clinical risk prediction". In: *Journal of the American Medical Informatics Association* 28.1 (2021), pp. 155–166.
46. Mengkai Song et al. "Analyzing user-level privacy attack against federated learning". In: *IEEE Journal on Selected Areas in Communications* 38.10 (2020), pp. 2430–2444.
47. Ewout W Steyerberg and Yvonne Vergouwe. "Towards better clinical prediction models: seven steps for development and an ABCD for validation". In: *European heart journal* 35.29 (2014), pp. 1925–1931.
48. Gregor Stiglic et al. "Interpretability of machine learning-based prediction models in healthcare". In: *Wiley Interdisciplinary Reviews: Data Mining and Knowledge Discovery* 10.5 (2020), e1379.
49. Jon Vadillo, Roberto Santana, and Jose A Lozano. "Adversarial Attacks in Explainable Machine Learning: A Survey of Threats Against Models and Humans". In: *Wiley Interdisciplinary Reviews: Data Mining and Knowledge Discovery* (2024), e1567.
50. Ben Van Calster et al. "Calibration: the Achilles heel of predictive analytics". In: *BMC medicine* 17.1 (2019), p. 230.
51. Ben Van Calster et al. "Reporting and interpreting decision curve analysis: a guide for investigators". In: *European urology* 74.6 (2018), pp. 796–804.
52. Maarten Van Smeden et al. "Critical appraisal of artificial intelligence-based prediction models for cardiovascular disease". In: *European heart journal* 43.31 (2022), pp. 2921–2930.
53. Andrew J Vickers, Ben van Calster, and Ewout W Steyerberg. "A simple, step-by-step guide to interpreting decision curve analysis". In: *Diagnostic and prognostic research* 3 (2019), pp. 1–8.
54. Andrew J Vickers and Elena B Elkin. "Decision curve analysis: a novel method for evaluating prediction models". In: *Medical Decision Making* 26.6 (2006), pp. 565–574.
55. Robert F Wolff et al. "PROBAST: a tool to assess the risk of bias and applicability of prediction model studies". In: *Annals of internal medicine* 170.1 (2019), pp. 51–58.
56. Shufan Yang et al. "The human activity radar challenge: benchmarking based on the 'radar signatures of human activities' dataset from Glasgow University". In: *IEEE Journal of Biomedical and Health Informatics* 27.4 (2023), pp. 1813–1824.
57. Idris Zakariyya et al. "Differentially Private Integrated Decision Gradients (IDG-DP) for Radar-based Human Activity Recognition". In: *arXiv preprint arXiv:2411.02099* (2024).

Pandemic Resilience: Smart and Connected Solutions in Healthcare

Bahar Farahani, Farshad Firouzi, Mahmoud Daneshmand, and JaeSeung Song

"In the middle of difficulty lies opportunity."

—Albert Einstein

Abstract The global challenges posed by pandemics, exemplified by the COVID-19 crisis, underscore the critical importance of innovative and integrated healthcare solutions. This chapter explores the role of Smart and Connected Health (SCH) technologies in addressing pandemic scenarios, leveraging lessons learned from COVID-19 to illustrate their transformative potential. By harnessing the capabilities of the Internet of Things (IoT), Artificial Intelligence (AI) (including machine learning and big data analytics), robotics, and blockchain, SCH systems can enhance disease management, support frontline efforts, and improve preparedness for future health crises. This chapter provides an in-depth examination of how these technological domains contribute to pandemic response within the broader SCH framework. Drawing from the COVID-19 experience, this chapter highlights the synergistic potential of SCH technologies to advance healthcare systems and prepare for emerging health challenges. By presenting real-world use cases and multidisciplinary approaches, it aims to inspire further exploration of smart and connected solutions for resilient and adaptive healthcare systems.

This chapter is adapted from Firouzi et al. [1].

B. Farahani (✉)
Cyberspace Research Institute, Shahid Beheshti University, Tehran, Iran
e-mail: b_farahani@sbu.ac.ir

F. Firouzi
Arizona State University, Tempe, AZ, USA

M. Daneshmand
Stevens Institute of Technology, Hoboken, NJ, USA

J. Song
Department of Computer Engineering, Sejong University, Seoul, South Korea

B. Farahani et al. (eds.), *Smart and Connected Health: Applications and Real-World Use Cases*, https://doi.org/10.1007/978-3-032-09253-3_3

1 Introduction

Pandemics have historically challenged humanity, highlighting vulnerabilities in healthcare systems and emphasizing the need for innovative solutions. The recent COVID-19 pandemic, caused by the SARS-CoV-2 virus, was a stark reminder of the critical role that multidisciplinary and technology-driven approaches play in managing public health crises. Although COVID-19 is now largely under control, the lessons learned have underscored the transformative potential of *Smart and Connected Health (SCH)* technologies in enhancing preparedness and resilience for future pandemics and broader healthcare challenges [1].

Smart and Connected Health (SCH) technologies—including the Internet of Things (IoT), Generative IoT (GIoT), Artificial Intelligence (AI), robotics, and Distributed Ledger Technologies (DLTs)—have played a pivotal role in addressing complex healthcare challenges [2]. These innovations enable real-time data collection, integration, and analysis, fostering personalized, patient-centric healthcare ecosystems. The increasing adoption of Wearable IoT (WIoT) devices and biosensors has laid the foundation for highly connected health systems, where real-time data streams are leveraged by AI and deep learning to predict risks, optimize outcomes, and accelerate scientific discovery [3–5]. Furthermore, DLTs enhance data security, privacy, and traceability, addressing critical limitations of traditional client/server IoT models [1].

Throughout history, pandemics have evolved with human society, often fueled by urbanization and increased mobility. As illustrated in Fig. 1, the spread of diseases such as smallpox, leprosy, and malaria followed trade and migration pathways. In the modern era, the rapid growth of global populations and interconnected societies has amplified the frequency and severity of epidemics and pandemics [6]. Unlike earlier centuries, where the slow pace of travel hindered the spread of diseases, today's high-speed transportation networks facilitate the swift proliferation of infectious agents, presenting new challenges for containment and mitigation [1].

In the last century, numerous epidemics, including those caused by influenza viruses (H1N1, H2N2, and H3N2) and coronaviruses (SARS-CoV, MERS-CoV), have highlighted the persistent threat of emerging pathogens [7]. Each outbreak has brought unique challenges, and the COVID-19 pandemic was no exception. Emerging in December 2019 in Wuhan, China, COVID-19 rapidly escalated into a global health crisis, declared a pandemic by the World Health Organization (WHO) on March 11, 2020. Its highly contagious nature, extended incubation period, and asymptomatic transmission contributed to its global impact [7].

Despite the unprecedented challenges posed by COVID-19, it also catalyzed innovation in SCH technologies. IoT-enabled devices, AI-driven insights, and blockchain-based solutions demonstrated their potential to support contact tracing, remote monitoring, early diagnosis, and vaccine development. These technologies empowered healthcare systems to respond more effectively, offering real-time surveillance, predictive analytics, and robust decision-making tools. While COVID-

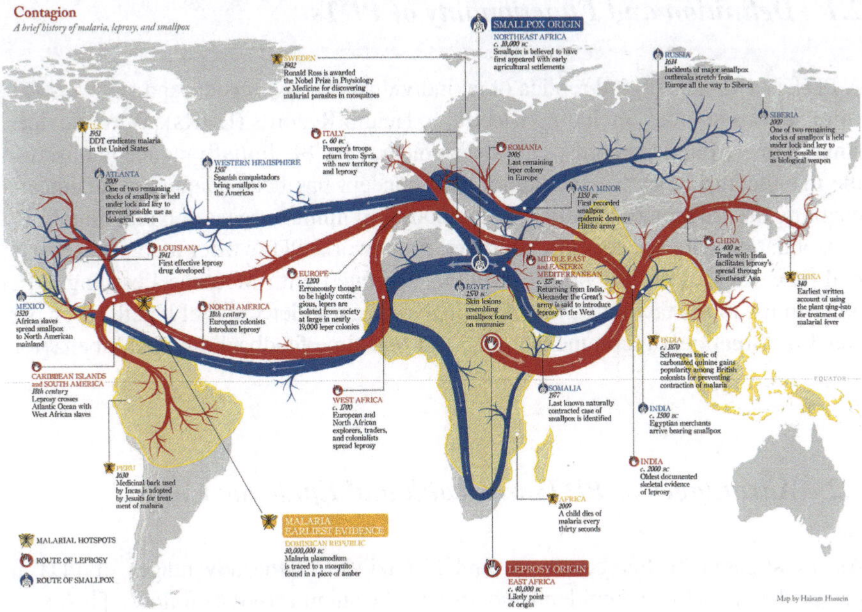

Fig. 1 Origin and spread of smallpox, leprosy, and malaria around the globe. Macrorepresentation (Image credit: Doug Belshaw blog) [6]

19 has receded, the importance of these technologies persists as the foundation for resilient and adaptive healthcare systems [1].

2 Personal Digital Twins (PDT)

The interconnected and densely populated nature of modern society creates ideal conditions for the rapid spread of infectious diseases. While recent pandemics, such as COVID-19, have subsided, they have revealed critical vulnerabilities in global health systems and underscored the urgent need for advanced technological solutions to manage future outbreaks. Early detection and swift response remain crucial in mitigating the impact of emerging infectious agents, particularly those that appear suddenly and spread unpredictably [1].

Historical contagion data, as shown in Fig. 1, provide valuable insights into the patterns and pathways of disease transmission. However, real-time forecasting is essential for proactively managing future epidemics. Advances in digital healthcare now enable the integration of data from diverse sources, utilizing sophisticated epidemic models to predict the dynamics and velocity of disease spread across communities. In this context, Personal Digital Twins (PDTs) emerge as a transformative innovation, offering a personalized, dynamic, and scalable approach to epidemic monitoring and control [1].

2.1 Definition and Functionality of PDTs

A PDT is a virtual, digital replica of an individual that aggregates and analyzes data from various sources, including Electronic Health Records (EHRs), WIoT devices, smartphones, public records, and social media [6, 8]. Initially conceptualized in the manufacturing sector, the digital twin concept has been adapted to healthcare, enabling the creation of detailed virtual models of human organs and individuals. By consolidating data streams into a unified framework, PDTs offer a comprehensive, real-time view of an individual's health. Integrated with Machine Learning (ML) algorithms, PDTs facilitate predictive analytics, early detection of health risks, personalized treatment recommendations, and real-time feedback on health behaviors [1].

2.2 Advantages of PDTs in Health and Epidemic Control

As investigated by Saracco et al. and Firouzi et al., the advantages of PDT in personalized health and epidemic control can be summarized as follows [1, 6]:

- **Self-generation of alerts**: Automated alerts help individuals remain vigilant about potential health risks, enabling timely preventive actions.
- **Widespread analytics**: Aggregating data across communities or nations supports predictive models to anticipate large-scale health trends and potential outbreaks.
- **Focused interventions**: Targeted analytics minimize broad restrictions, allowing stricter measures in high-risk areas while reducing disruptions elsewhere.
- **Cost-effectiveness**: Precision interventions lower the economic burden associated with indiscriminate restrictions.
- **Dynamic adaptability**: PDTs enable individuals and groups to adjust dynamically to evolving health scenarios with personalized recommendations.
- **Enhanced personal awareness**: Real-time insights empower individuals to make informed decisions about their health and behavior.
- **Real-time feedback**: Continuous data monitoring allows rapid evaluation of actions, ensuring effective decision-making.
- **Efficient monitoring**: Automation reduces the effort and cost required for large-scale health surveillance and management.
- **Service innovation**: PDTs facilitate the creation of customized health services, including virtual support communities for those affected.
- **Optimized resource use**: Data-driven insights enable the efficient allocation of resources to address competing needs effectively.
- **Accelerated triage**: PDTs streamline access to appropriate medical support and services, enhancing response times (Fig. 2).

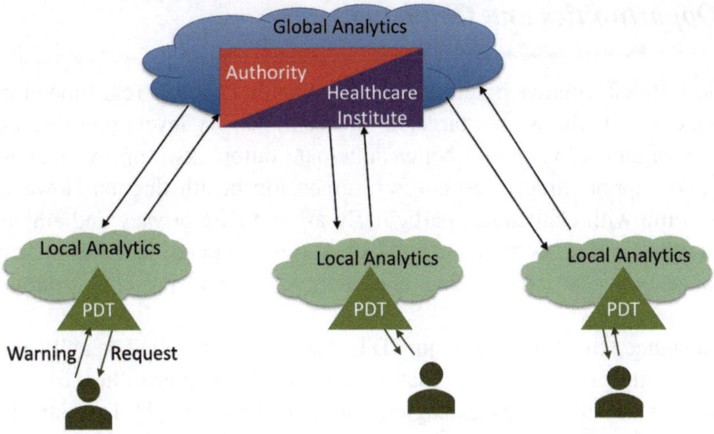

Fig. 2 Schematic representation of PDTs in epidemic control. PDTs integrate multiple data streams, provide actionable insights, and interact with healthcare systems to facilitate timely interventions while ensuring data privacy [1, 6]

2.3 The Role of PDTs in Public Health Monitoring

PDTs represent an innovative way to track various aspects of an individual's health, movement, and interactions in geophysical locations. For instance, a person's movements can be monitored using smartphone location data, while health metrics such as resting heart rate or temperature can be gathered via smartwatch sensors. PDTs can also be programmed to detect and alert users to patterns indicative of potential health risks. For example, an alert could be generated if data show elevated body temperature and rapid breathing—signs that might indicate the onset of an illness. These alerts could then be transmitted to healthcare organizations, which analyze trends, assess local patterns, and correlate individual data with broader public health concerns [1, 6].

Although the concept of comprehensive epidemic monitoring and control using PDTs remains in its early stages, related strategies have already been applied successfully. During past health crises, such as the COVID-19 pandemic, countries like South Korea demonstrated how digital tools and data analytics can enable efficient containment. By leveraging smartphone location data, telecommunications records, and public security cameras, South Korea created a micro-level contagion map. This, combined with rapid testing, allowed targeted interventions and reduced the need for broad lockdowns, which were common in other countries. Comparing South Korea's approach to the potential use of PDTs provides insight into how emerging technologies can enhance public health strategies [1, 6].

2.4 Opportunities and Challenges

PDTs offer transformative potential in public health, enabling real-time monitoring, early detection of illness, and targeted interventions. By leveraging data analytics, healthcare organizations could better anticipate outbreaks, improve resource allocation, and support timely responses to emerging health threats. However, these benefits come with challenges, particularly around data privacy and ethical use of personal information. As digital health technologies expand, societies must navigate the balance between safeguarding individual rights and protecting public health [1, 6].

For instance, the South Korean PDT model, which relied heavily on micro-level interventions, has shown that precision-driven approaches can minimize disruptions while effectively managing public health risks [1]. In contrast, macro-level strategies such as nationwide lockdowns often impose significant social and economic costs. Both approaches highlight the need for careful consideration of trade-offs between individual freedoms and societal benefits [1, 6].

2.5 Sustaining Digital Health Innovations

While PDTs represent an exciting frontier, their long-term success depends on addressing sustainability challenges. Health-related systems developed during crises often face a decline in use once the immediate threat passes. For instance, during the COVID-19 pandemic, Korea's mask inventory management app was widely used during initial shortages but quickly became obsolete once the supply stabilized. In contrast, platforms such as the Epidemic Investigation Support System (EISS) employed in South Korea demonstrated their adaptability by integrating telecommunications and financial data into broader smart city systems, extending their utility beyond a single crisis [1].

To ensure lasting relevance, public health technologies must evolve from being crisis-specific tools to general-purpose platforms capable of addressing a range of challenges. PDTs, for example, could be integrated into broader health monitoring systems, allowing them to play a continuous role in detecting and managing illnesses, promoting preventive care, and supporting individual well-being [6].

2.6 Balancing Privacy and Public Health

As PDTs gain prominence, it is critical to address privacy concerns. Technology can bridge the gap between personal privacy and societal needs by transmitting only essential metadata to healthcare organizations while safeguarding individual details. Blockchain technology, for instance, holds promise for monitoring data flows and enhancing privacy protections. By defining clear regulatory frameworks, societies can ensure that digital health tools like PDTs meet both public health goals and ethical standards. In this way, PDTs can help create a balance between individual privacy and societal safety, enabling healthier, more resilient communities [1, 6].

3 The Role of IoT in Managing COVID-19

The IoT revolution has reshaped the healthcare landscape, providing significant opportunities and advantages during the COVID-19 pandemic. Although the acute phase of the pandemic has subsided, the lessons learned have highlighted the immense potential of IoT technologies in managing large-scale health crises. IoT played a vital role in increasing the availability and accessibility of diagnostics and treatments, reducing hospital visits, alleviating healthcare workers' fatigue, minimizing the risk of infection for medical staff, and lowering costs and interactions. These applications were instrumental in managing COVID-19 across three main phases: early diagnosis, quarantine period, and post-recovery [7, 9]. The major applications of IoT during these phases are summarized below [1]:

- **Contact Tracing:** IoT enabled the identification of individuals who had contact with an infected person, helping to control the virus' spread efficiently.
- **Rapid Screening and Early Diagnosis:** Early detection was crucial in preventing the spread of COVID-19. WIoT devices facilitated remote monitoring and detection of symptoms such as fever, dry cough, tiredness, aches and pains, sore throat, diarrhea, headache, loss of taste or smell, skin rash, discoloration of fingers or toes, and difficulty breathing. This approach proved faster and more efficient than traditional methods while minimizing infection risk for medical personnel.
- **Remote Monitoring (RM):** IoT-based remote monitoring collected medical data from biosensors or WIoT devices to track patient conditions, such as heart rate variability, blood pressure, and body temperature, outside clinical settings. RM was essential for providing ongoing care to patients, reducing healthcare workers' fatigue, and lowering their exposure to infection.
- **Emergency Alerts:** IoT devices issued real-time alerts to authorities, healthcare providers, and families during emergencies, ensuring swift response and care.
- **Social Distancing Control:** IoT technology was utilized to monitor and enforce social distancing policies through sensors and automated systems, reducing the need for manual interventions.

The experience of using IoT during the pandemic demonstrated its transformative potential in healthcare. These applications not only helped mitigate the immediate impacts of COVID-19 but also provided a blueprint for leveraging IoT in future public health emergencies. More recently, the emergence of GIoT—which integrates Generative Artificial Intelligence (GenAI), including Large Language Models (LLMs), with IoT systems—further extends these capabilities by enabling the generation of real-time insights, predictions, and adaptive responses from continuously collected data [2]. In this context, LLMs play a key role in enhancing human—machine interaction, translating complex health data into understandable and actionable information, and supporting clinical decision-making. Together, these advances mark a shift from connected devices toward intelligent, interactive, and proactive healthcare ecosystems [2, 10].

3.1 Contact Tracing

Contact tracing refers to the process of identifying individuals who have been exposed to or infected by a virus. The extended incubation period of COVID-19 and the limited availability of widespread testing posed significant challenges for authorities in accurately quantifying the number of infections. According to the World Health Organization (WHO), contact tracing involves three essential steps [7]:

- Identifying individuals who have had contact with an infected person
- Documenting the details of these contacts
- Testing these individuals as quickly as possible

During the spring of 2020, several smartphone-based contact tracing applications emerged, spearheaded by projects like TraceTogether, developed by the Singaporean government. TraceTogether leverages Bluetooth technology to facilitate anonymous, close-proximity smartphone-to-smartphone communication. Many other contact tracing applications have also adopted Bluetooth as their core technology. A key technical challenge in these applications is achieving precise distance measurement and reliable discernment of close interactions within a meter-scale accuracy. Most implementations use a 2-meter distance standard for identifying close contacts. The implementation of contact-tracing applications relies on one or more of the following technologies [1]:

- **Bluetooth:** Bluetooth-based applications estimate the distance between two individuals by calculating the received signal strength indicator (RSSI) between devices. These applications store the historical Bluetooth connection data, including the duration of each interaction. If an individual tests positive for COVID-19, the application can use their Bluetooth connection history to identify and notify individuals who were exposed to the infected person [7].
- **GPS:** Some government agencies utilized GPS data to monitor the real-time locations of COVID-19 patients and analyze their historical movement patterns. This information proved invaluable for tracing potential exposure and understanding transmission pathways.
- **Ultrasonic:** While Bluetooth has limitations in accurately measuring distances and GPS can inadvertently log interactions through walls or ceilings, ultrasonic technology offers a promising alternative. Ultrasound calculates device distance

by measuring the time it takes for sound to travel between devices, achieving greater precision. For instance, the NOVID application, launched in April 2020, integrates ultrasonic technology with Bluetooth. Experiments with NOVID in various real-world environments demonstrated its effectiveness. The corresponding findings highlight the significant benefits of combining ultrasonic technology with traditional methods for enhancing contact tracing accuracy.

3.2 *Remote Patient Monitoring (RPM)*

WIoT technologies played a pivotal role during the COVID-19 pandemic, offering unique, patient-centric insights into health and well-being in everyday settings. Unlike traditional monitoring solutions, which are conducted intermittently, WIoT provided continuous, real-time physiological data, reshaping the healthcare landscape. Furthermore, wearables introduced innovative methods to incentivize or "gamify" self-monitoring, fostering greater patient engagement [1].

Stay-at-home orders and social distancing measures during COVID-19 accelerated the adoption of WIoT technologies. These devices were widely used in clinical trials to remotely monitor individuals who required hospital admission, those recently discharged, or individuals in quarantine. They also enabled the early detection of symptoms by tracking systemic markers of infection [11, 12]. Wearable devices demonstrated their value in alerting patients and healthcare providers to potential symptoms before severe illness developed. Their noninvasive design, coupled with the ability to monitor infections both at home and in healthcare settings, made them invaluable during the pandemic [1].

Looking forward, the shift from hospital-based monitoring to home-based care signals optimism for addressing future public health challenges. Wearable technologies with clinical-grade accuracy continue to illustrate their benefits through ongoing studies. Emphasis on data sharing and interoperability is driving the development of predictive algorithms that can be generalized across diverse populations. Beyond COVID-19, skin-integrated devices and sensors placed at optimal body locations address the critical need for continuous, objective, and sensitive systems to detect and monitor various health conditions [1].

Generally, after exposure to a virus, the human body undergoes various physiological changes that can be monitored for treatment purposes. These changes manifest as biochemical, electrical, and biosignals derived from different body parts [13, 14]. Monitoring these physiological signs allows for the prediction of a patient's health status, enabling timely intervention. Based on these measurements, appropriate medications can be administered, or individuals can be transferred to hospitals when necessary. However, current commercial wearables and WIoT technologies can only monitor and collect a limited subset of physiological data and biomarkers (see Fig. 3) [5, 15–24].

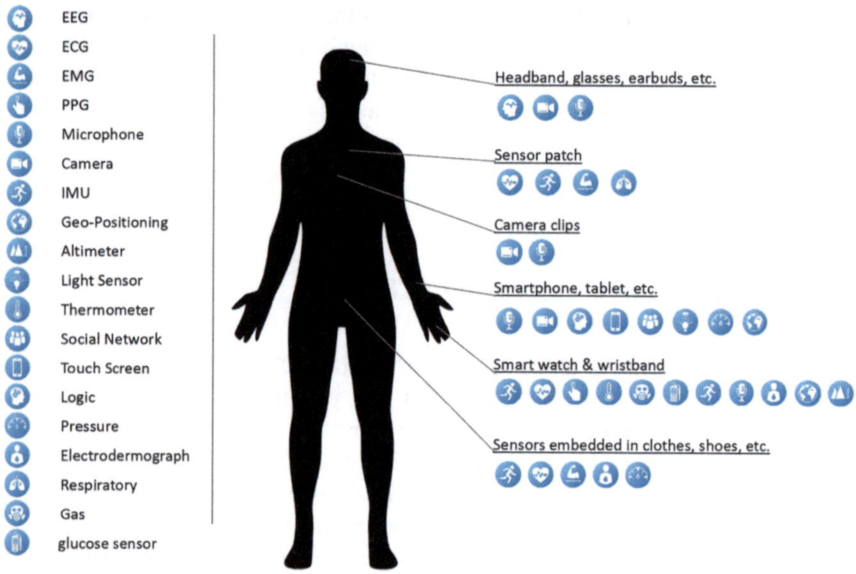

Fig. 3 Currently available wearables or portable platforms [3]

The sensor modalities employed to measure physiological symptoms at the onset of viral infections include, but are not limited to, the following:

- **Temperature**: Measured using contact or infrared sensors
- **Heart Rate (HR)**: Measured via electrocardiogram (ECG) or pulse plethysmography (PPG)
- **Heart Rate Variability (HRV)**: Assessed using ECG or PPG
- **Blood Pressure**: Derived from PPG readings
- **Respiratory Rate**: Monitored using ECG, PPG, or accelerometers
- **Oxygen Saturation (SpO2)**: Determined via PPG
- **Sleep Patterns**: Measured through accelerometry
- **Cough Detection**: Monitored using mechanical or piezoelectric sensing

While wearables, implantable devices, and smart textiles have garnered significant attention in the literature and numerous consumer wearables are now commercially available, the potential of WIoT remains largely untapped. This gap can be attributed to several technological limitations, as outlined below [25–30]:

- **Limited Measurement Precision and Modalities**: Current wearables often lack the capability to measure critical parameters such as body temperature, high-fidelity respiratory rates, or precise pulse oximetry.
- **Nonmedical-Grade Devices**: Many popular wearables, such as the Oura Ring and Fitbit sensors, lack FDA approval for remote monitoring. Even FDA-approved devices, such as the Apple Watch Series 4, are limited in functionality. For example, they can perform episodic ECGs and notify users of uneven heart

rhythms in individuals aged over 22 but are not substitutes for clinical diagnostic systems.
- **Lack of Interoperability**: Existing solutions often do not meet interoperability standards, limiting their ability to connect, exchange data, and work seamlessly with other systems.
- **Inconsistent Sensor Performance**: Data collected by wearables are frequently impacted by motion artifacts and measurement inaccuracies, resulting in unreliable readings over time. These limitations can undermine the potential of wearables to monitor, predict, and track viral infections effectively.
- **Challenges in Data Management and Analytics**: Continuous monitoring with WIoT generates vast quantities of diverse health data, creating significant challenges in storage, processing, and analysis. Wearable sensor systems must integrate secure and accessible information backends to manage this data effectively. Furthermore, these systems need to be interoperable with additional data sources, such as EHRs, to enhance data utility. By incorporating machine learning techniques, physiological data combined with clinical outcomes can improve the detection of asymptomatic infections and enable predictive analytics. This integration supports the development of digital biomes tailored to individual health profiles, facilitating personalized recovery plans and informed decision-making for healthcare staff.

In summary, the experience of using WIoT during COVID-19 has laid the foundation for its broader adoption in healthcare. These devices enable rapid diagnosis and ongoing tracking of vital biological signs, empowering healthcare providers to take timely and appropriate actions. By reducing the risk of severe health deterioration and supporting quicker recovery, WIoT devices represent a significant step forward in modern healthcare.

4 The Role of AI/ML in Managing COVID-19

During the global public health crisis caused by COVID-19, the healthcare industry relied on AI/ML technologies to monitor, control, and mitigate the spread of the virus. These tools were instrumental in tracing the virus, identifying at-risk individuals, and predicting infection rates in real time. AI/ML further supported population screening and notification efforts, contributing to improved treatment outcomes for COVID-19 patients through evidence-based decision-making. This section reviews and discusses the most significant AI/ML use cases developed to address the challenges of the pandemic [1].

4.1 Multimodal Diagnosis of COVID-19

Early diagnosis and prognosis of COVID-19 cases were critical for managing limited healthcare resources. AI/ML models were deployed to analyze data from multiple modalities, including triage records, CT scans, ECG signals, and cough recordings. For example, a dataset from Sina Hospital in Tehran enabled ML techniques to predict infection risk, classify patient severity, and guide clinical decisions [1]. These models demonstrated the power of combining diverse data sources to improve diagnostic accuracy and reduce the burden on healthcare systems.

4.2 Image-Based Diagnosis of COVID-19

AI/ML played a transformative role in diagnosing COVID-19 through medical imaging. Chest CT scans, combined with clinical data and symptomatology, were analyzed by AI algorithms to identify COVID-19-specific patterns, such as ground-glass opacities. For instance, a dataset of 4173 CT scans from Brazilian hospitals was used to train interpretable models like xDNN [31], which generated transparent diagnostic rules. These tools offered rapid, accurate diagnostics, aiding clinicians in early detection and treatment planning [1].

4.3 Forecasting Spread of COVID-19

AI/ML forecasting models were essential for monitoring and predicting the spread of COVID-19. These models used a variety of real-time data sources, including search engine trends, social media activity, mobility data, and historical infection patterns, to provide timely and actionable insights. The main forecasting approaches included the following [1]:

- **Nowcasting:** Nowcasting focuses on estimating the current state of an outbreak by analyzing real-time data. This method integrates dynamic data streams, such as Internet search trends, social media activity, and hospital reports, to produce near-instantaneous snapshots of COVID-19 prevalence. Nowcasting provides a near-instantaneous snapshot, enabling healthcare systems to respond promptly to emerging hotspots. However, during the pandemic, panic-driven behavior altered online activity patterns, as many individuals searched for coronavirus-related information regardless of whether they were symptomatic. This presented a challenge for algorithms, which had to distinguish between genuine signals of illness and unrelated noise. Nowcasting was further complicated by the lack of historical data for COVID-19, necessitating adaptive models capable of learning correlations between real-world conditions and data signals.
- **Machine Learning-Based Forecasting:** ML models predicted future trends in COVID-19 cases by analyzing historical infection data, mobility patterns, and epidemiological variables. These models forecasted key milestones, such as peak infection periods and hospital capacity requirements, enabling proactive

resource allocation. However, pandemics like COVID-19 are far less predictable than cyclical diseases like the flu. For instance, the H1N1 pandemic in 2009 affected younger populations, whereas COVID-19 disproportionately impacted older individuals. Additionally, the data tracking systems available during H1N1 were less developed, limiting their predictive accuracy. In response to these challenges, researchers incorporated data from countries with earlier outbreaks and continuously updated ML models as new data became available. Weekly revisions to case trajectories, such as those provided by the CDC, allowed laboratories to refine their forecasts and address gaps in rolling statistics.

- **Crowd-Sourced Opinion Prediction:** This approach leveraged collective intelligence by aggregating forecasts from experts and nonexperts. Participants provided predictions about case trajectories, which were combined to create consensus forecasts. Despite the simplicity of this approach, crowd-sourced predictions often matched the accuracy of sophisticated ML models, offering valuable insights into public perceptions and behaviors. These forecasts were particularly effective in integrating qualitative assessments with quantitative data to address uncertainties in COVID-19 transmission patterns.

4.4 Risk Prediction

AI-based risk prediction models were critical in stratifying patients during the COVID-19 pandemic based on their likelihood of infection, the severity of symptoms, and specific treatment needs. These models leveraged data from various sources, such as EHRs and clinical observations, to enhance patient care and optimize resource allocation during critical phases of the pandemic [1].

- **Risk of Infection:** AI models utilized demographic, environmental, and clinical data to predict an individual's susceptibility to COVID-19. These predictions were instrumental in guiding testing priorities and isolation measures, particularly in vulnerable populations and high-risk areas.
- **Risk of Severe Symptom Development:** ML algorithms were applied to identify individuals at higher risk of developing severe complications, such as respiratory failure or organ dysfunction. Early identification enabled healthcare providers to allocate resources, such as ICU beds and ventilators, more effectively.
- **Risk of Specific Treatment Use:** AI models predicted the likelihood of a patient requiring specific interventions, such as mechanical ventilation or antiviral therapies. These insights helped hospitals manage inventory and ensure that critical supplies were available when and where they were needed.

The integration of AI-based risk prediction into healthcare demonstrated significant benefits:

- **Personalized Care:** Patients received targeted interventions based on their individual risk profiles, improving treatment outcomes.

- **Resource Optimization:** Early identification of high-risk cases allowed for efficient allocation of ICU beds, ventilators, and medical personnel.
- **Proactive Management:** AI models facilitated early detection of potential complications, enabling timely interventions to prevent deterioration.
- **Clinical Decision Support:** Risk prediction models supported clinicians by providing evidence-based recommendations tailored to patient needs.

By leveraging advanced methodologies and addressing key challenges, AI-driven risk prediction models played a transformative role in enhancing healthcare delivery during the COVID-19 pandemic and beyond.

4.5 Voice-Based Diagnosis of COVID-19

AI-powered voice analysis applications, such as the Corona voice detect app, utilized forensic voice technology to identify patterns in speech and tone indicative of respiratory conditions. These tools offered a noninvasive and accessible screening method for COVID-19, especially useful for preliminary diagnosis and triaging [1].

4.6 Respondent-Driven Sampling (RDS)

Respondent-driven sampling (RDS) leverages social networks for targeted population surveys through a chain-referral process, where participants recruit others within their circles. This method is particularly effective in reaching underrepresented populations with limited healthcare access or higher disease risk. During public health crises like COVID-19, RDS enables efficient community-level sampling to identify undiagnosed cases, assess infection spread, and guide early interventions. An example of this methodology is the SNOWBALL Study, a CDC-funded initiative. It utilized RDS to identify COVID-19 cases through the active engagement of social networks. The study aimed to improve understanding of SARS-CoV-2 transmission and targeted underrepresented populations by using referral-based recruitment, surveys, and testing to identify and interrupt transmission chains. Outcomes included improved data on community spread, better resource allocation, and a framework for rapid-deployment sampling in future outbreaks.

The main benefits of RDS can be summarized as follows [1]:

- **Targeted Sampling Efficiency:** By focusing on social networks, RDS ensures that testing resources are directed to areas with the highest likelihood of undiagnosed infections.
- **Community Representation:** RDS reaches marginalized and diverse populations, improving equity in public health interventions.

- **Scalability and Adaptability:** The methodology can be scaled to different regions and adapted to varying epidemiological contexts, making it a robust approach for managing infectious diseases.

4.7 Early Warnings and Alerts

AI technologies played a critical role in early warnings and alerts during the COVID-19 pandemic. AI-enhanced infrared cameras combined temperature scanning with facial recognition to detect individuals with elevated body temperatures in crowded settings. These systems were deployed in public spaces such as airports, hospitals, and healthcare facilities, helping to minimize the exposure of frontline staff while automating risk assessments [1]. During the pandemic, these systems were instrumental in monitoring quarantine compliance, assessing risks in public areas, and facilitating real-time alerts for violations. For example, technologies like the FindFace system, integrated with CCTV cameras, provided real-time identification and movement tracking, supporting effective quarantine enforcement and infection control. Their capabilities included [1]:

- **Social Interaction Analysis:** AI systems analyzed social networks and historical data to evaluate potential transmission risks by identifying patterns of interaction.
- **Real-Time Violation Monitoring:** These systems flagged quarantined individuals who violated restrictions and alerted authorities, even if masks or coverings were worn.
- **Tracking Quarantine Compliance:** Advanced algorithms monitored individuals across multiple camera feeds, recognizing silhouettes when facial imaging was unavailable.
- **Age Detection:** AI estimated age, allowing for targeted monitoring of vulnerable populations, such as those over 60, who were at higher risk of severe outcomes during the pandemic.

While the COVID-19 pandemic has ended, the deployment of AI-driven technologies has left a legacy of enhanced preparedness for future public health challenges. These tools have demonstrated their potential to automate risk assessments, ensure compliance with health policies, and improve the effectiveness of large-scale surveillance systems in preventing the spread of infectious diseases.

4.8 Social Control and Fake News Detection

The rapid dissemination of misinformation during the COVID-19 pandemic posed a significant challenge to public health efforts. Fake news, defined as deliberately inaccurate or misleading information, proliferated through social media and other online platforms. This misinformation often promoted unverified treatments, supplements, or products, misleading individuals and undermining trust in public health guidelines. Stressful situations, such as pandemics, tend to amplify the influx of both information and misinformation, further complicating the public's ability to discern credible sources. The ease of spreading content online and the lack of centralized control over information exacerbated this issue, creating confusion and potentially harmful behaviors [1].

To address this challenge, AI tools were deployed to detect, classify, and counteract fake news. These systems used machine learning and natural language processing (NLP) to evaluate the language, structure, and content of news articles and social media posts. For example, the Joint Research Centre (JRC) developed an AI-based classifier capable of identifying misinformation with high accuracy. By analyzing patterns in text and assessing the credibility of sources, such tools helped flag misleading content and promote actionable, accurate information [1]. Key components of the solution included [1]:

- **Detection and Classification:** AI algorithms identified misleading content by analyzing linguistic and contextual cues in online articles and posts.
- **Source Identification:** These systems traced the origins of misinformation, helping to understand its propagation pathways and enabling targeted countermeasures.
- **Scalability and Speed:** AI systems processed vast volumes of data in real time, allowing for rapid detection and intervention to limit the spread of harmful narratives.

While AI tools like the JRC ML classifier could not achieve 100% accuracy, their integration into public health communication strategies significantly mitigated the impact of misinformation.

4.9 Vaccine and Drug Development

AI/ML technologies revolutionized vaccine and drug discovery by accelerating the identification of therapeutic targets and optimizing development pipelines. By analyzing genomic and molecular data, these tools provided critical insights into SARS-CoV-2, significantly reducing the time required for vaccine and drug development [1]. For instance, AI-enabled vaccine development utilized tools like Vaxign-ML, which predicted potential vaccine candidates by analyzing the entire proteome of SARS-CoV-2, including nonstructural proteins and the spike (S) protein. In drug discovery, deep learning models such as the molecule transformer-drug target interaction (MT-DTI) demonstrated the ability to identify existing drugs with potential efficacy against COVID-19. For example, MT-DTI highlighted atazanavir, an antiretroviral drug used to treat HIV, as a candidate for repurposing [1].

4.10 Communication and Chatbots

AI-driven communication tools, such as medical chatbots, played a vital role in managing healthcare demands during the COVID-19 pandemic. These systems reduced the burden on physicians by automating patient interactions, enabling early symptom identification, and providing essential health education. Moreover, the integration of chatbot systems with broader AI-driven monitoring frameworks enabled real-time data analysis, contributing to public health efforts in tracking and mitigating the spread of the virus. Medical chatbots used AI algorithms to assist patients in several ways, such as [1]:

- **Symptom Assessment:** Chatbots guided users through symptom-checking processes, helping them identify potential COVID-19 symptoms and advising whether further medical attention was needed.
- **Health Education:** These tools disseminated information about preventive measures, such as hand hygiene, mask usage, and social distancing, ensuring the public was well-informed about best practices to limit virus spread.
- **Referral for Treatment:** When symptoms were identified as severe or worsening, chatbots referred patients to appropriate medical services, reducing delays in accessing necessary care.
- **Monitoring and Data Collection:** Integrated phone applications allowed patients to record and track vital signs, such as temperature and respiratory symptoms. This data could be used by AI algorithms to monitor health trends and provide alerts for clinical follow-up if necessary.

4.11 Generative AI and Large Language Models

Recent advances in GenAI and LLMs have significantly expanded the capabilities of AI-driven healthcare systems [2, 10]. Unlike earlier rule-based chatbots and conventional Natural Language Processing (NLP) tools used during the COVID-19 pandemic, LLMs can synthesize large volumes of biomedical literature, support clinical decision-making, assist in triage interactions, and generate timely public health guidance. These models also enable more advanced conversational systems, including intelligent medical chatbots and personal digital health assistants that can provide personalized health information, symptom assessment, and patient guidance. In addition, GenAI technologies show promise in accelerating drug discovery, summarizing rapidly evolving scientific knowledge, and improving communication between healthcare providers and the public. As these technologies continue to mature, they are expected to play an increasingly important role in pandemic preparedness, health information dissemination, and next-generation digital health services.

5 The Role of Robotics in Managing COVID-19

The COVID-19 pandemic underscored the transformative potential of robotics in public health. During the crisis, robots played a critical role in reducing human exposure to the virus, optimizing healthcare processes, and enhancing operational efficiency. Although the pandemic has ended, the lessons learned highlight how robotics can continue to shape future public health strategies and improve preparedness for global health challenges. Robotics demonstrated versatility in addressing a range of pandemic-related needs [1]:

- **Disease Prevention:** Robotic ultraviolet (UV) disinfection systems were widely deployed to sanitize high-touch surfaces, as SARS-CoV-2 could survive on materials such as glass, plastic, and metal for days. Autonomous robots equipped with UV light provided a fast, effective, and low-risk solution, eliminating the need for manual disinfection. Future innovations could integrate intelligent navigation and area sensing, enabling robots to autonomously sanitize high-traffic zones continuously.
- **Crowd Surveillance:** Ensuring social distancing during the pandemic required robust monitoring mechanisms. Drones were utilized by countries like India and China to monitor gatherings and enforce compliance with distancing measures, showcasing the potential of robotics in public safety management.
- **Public Announcements:** Drones equipped with broadcasting systems delivered vital public health messages, particularly in remote or underserved areas. This approach ensured that critical information about prevention and safety reached all segments of the population.
- **Mass Screening:** Robotics enabled efficient and contactless crowd screening through the use of thermal cameras, night vision, computer vision systems, and specialized sensors. These technologies allowed the monitoring of body temperature and heart rate data, aiding in early detection and reducing the burden on healthcare personnel.
- **Essential Supply Delivery:** Robotic systems, including drones, were instrumental in delivering medical supplies and essential goods to remote or high-risk areas. Automated solutions for oral and nasal swabbing further streamlined testing processes, minimized infection risks for healthcare workers, and freed up staff for more complex tasks.
- **Screening and Diagnosis:** Mobile robots equipped with thermal sensors and vision algorithms were employed to measure body temperature in public spaces and healthcare facilities. These systems enhanced screening coverage while reducing direct contact. When integrated with healthcare IT systems, robots provided real-time data for monitoring hospitalized patients or outpatients. Advanced applications linked facial recognition and security systems for faster contact tracing, though strict privacy regulations remain essential.
- **Remote Access and Outreach:** Drones and unmanned aerial vehicles (UAVs) were vital in reaching remote areas with limited infrastructure. They facilitated the delivery of essential supplies, supported surveillance efforts, and minimized

human contact. China's extensive use of drone technology during the pandemic demonstrated its effectiveness, inspiring global collaborations to explore further applications for public health emergencies.

The deployment of robotics during COVID-19 showcased their capability to enhance safety, efficiency, and scalability in public health responses. As the pandemic recedes into history, these innovations remain relevant, paving the way for the integration of robotics into broader healthcare systems. Lessons learned during this time will inform future strategies, ensuring that robotics play a central role in building resilience against potential global health crises [1].

6 The Role of Blockchain in Managing COVID-19

The COVID-19 pandemic demonstrated the transformative potential of blockchain technology in addressing critical challenges in healthcare. As the pandemic unfolded, blockchain emerged as a robust solution for improving data security, transparency, and efficiency across various healthcare systems. While the immediate crisis has ended, the applications of blockchain technology during this period provide valuable insights for future preparedness and innovation in healthcare [1].

Blockchain, particularly when integrated with the IoT, offered solutions to many vulnerabilities in healthcare data systems. Traditional cloud-based IoT systems often face security risks, but blockchain's decentralized architecture mitigates these issues by providing tamper-resistant data storage and secure communication networks. Key features such as consensus mechanisms ensured that compromised nodes could not contribute erroneous data, preserving the integrity of information. Additionally, blockchain's ability to manage device identities and create immutable audit trails significantly enhanced the security and functionality of IoT-based healthcare systems [1]. During the pandemic, blockchain technology was deployed in several impactful ways, such as [1]:

- **Healthcare Data Security and Privacy:** Blockchain allowed for secure and decentralized data sharing among healthcare providers, reducing reliance on centralized third parties. By enabling pseudonymous and secure transactions, patients retained greater ownership and control of their health data. Smart contracts facilitated permissions for data sharing, ensuring compliance with GDPR and HIPAA regulations while maintaining transparency.
- **Verifiable Health Certificates:** Blockchain provided a means to issue tamper-proof COVID-19 testing and vaccination certificates. These verifiable blockchain records ensured authenticity, enabling individuals to confirm their health status while protecting privacy. This application was pivotal for lifting restrictions and targeting quarantine measures more effectively.
- **Patient Monitoring and Containment:** Blockchain, in conjunction with IoT devices, enabled the monitoring of infected or exposed individuals. Wearable

IoT devices tracked movements to support effective containment strategies while safeguarding user privacy through secure data sharing protocols.

- **Remote Patient Monitoring:** Blockchain enhanced remote patient monitoring by securely storing and sharing physiological data from wearable devices. This ensured compliance with privacy regulations while providing healthcare providers with reliable, tamper-proof access to patient data.
- **Supply Chain Management:** Blockchain-based platforms like VeChain played a crucial role in tracking vaccine production during the pandemic. Distributed ledgers documented every stage of vaccine production, from materials and packaging to final distribution. This transparency mitigated the risk of data tampering, ensured high-quality vaccine development, and provided a dependable framework for supply chain management.

Blockchain technology not only addressed immediate challenges during the COVID-19 pandemic but also demonstrated its potential to revolutionize healthcare systems. By enabling secure, decentralized, and transparent operations, blockchain has paved the way for more resilient healthcare infrastructures capable of responding to future global health crises with greater efficiency and trust.

7 Conclusions

The COVID-19 pandemic served as a pivotal turning point, revealing the critical role of smart and connected health technologies in enhancing healthcare system resilience during global crises. As discussed throughout this chapter, Personal Digital Twins, IoT, artificial intelligence and machine learning, robotics, and blockchain each contributed in distinct yet complementary ways to disease monitoring, early diagnosis, remote care, risk prediction, resource optimization, secure data management, and public health response. Beyond their individual applications, the true value of these technologies lies in their integration into a connected, intelligent, and adaptive healthcare ecosystem capable of supporting timely decision-making and personalized interventions. In the post-COVID era, these innovations should no longer be viewed only as emergency tools, but as core components of next-generation healthcare systems. Looking to the future, continued advances in interoperability, trustworthy AI, digital health infrastructure, and human-centered design will be essential for building more resilient, equitable, and pandemic-ready health ecosystems.

Acknowledgments The authors acknowledge the use of Generative AI (GenAI) tools to assist in the editing and preparation of this book chapter.

References

1. F. Firouzi, B. Farahani, M. Daneshmand, K. Grise, J. Song, R. Saracco, L. L. Wang, K. Lo, P. Angelov, E. Soares *et al.*, "Harnessing the power of smart and connected health to tackle covid-19: Iot, ai, robotics, and blockchain for a better world," *IEEE Internet of Things Journal*, vol. 8, no. 16, pp. 12 826–12 846, 2021.
2. F. Firouzi, A. Ray, B. Farahani, M. Daneshmand, J. Song, S. Wu, K. Chakrabarty, "Generative IoT (GIoT): Advancing IoT with Generative AI and Large Language Models," *Digital Communications and Networks*, Elsevier, 2026.
3. F. Firouzi, B. Farahani, M. Barzegari, and M. Daneshmand, "Ai-driven data monetization: The other face of data in iot-based smart and connected health," *IEEE Internet of Things Journal*, 2020.
4. F. Firouzi, K. Chakrabarty, and S. Nassif, *Intelligent Internet of Things: From Device to Fog and Cloud*. Springer, 2020.
5. B. Farahani, F. Firouzi, V. Chang, M. Badaroglu, N. Constant, and K. Mankodiya, "Towards fog-driven iot ehealth: Promises and challenges of iot in medicine and healthcare," *Future Generation Computer Systems*, vol. 78, pp. 659–676, 2018.
6. R. Saracco, "Personal digital twins and their role in epidemics control," 2020.
7. V. Chamola, V. Hassija, V. Gupta, and M. Guizani, "A comprehensive review of the covid-19 pandemic and the role of iot, drones, ai, blockchain, and 5g in managing its impact," *IEEE Access*, vol. 8, pp. 90 225–90 265, 2020.
8. H. Van Houten, "How a virtual heart could save your real one," 2018.
9. M. Nasajpour, S. Pouriyeh, R. M. Parizi, M. Dorodchi, M. Valero, and H. R. Arabnia, "Internet of things for current covid-19 and future pandemics: An exploratory study," *Journal of healthcare informatics research*, pp. 1–40, 2020.
10. A. Mahmoudi, G. Farahani, P. Domanski, B. Farahani, F. Firouzi, K. Chakrabarty, "DiabLLM: An LLM-based framework for blood glucose prediction in Type 1 diabetes," *IEEE Journal of Biomedical and Health Informatics*, pp. 1–14, IEEE, 2026. https://doi.org/10.1109/JBHI.2026.3658588
11. H. J. Davies, I. Williams, N. S. Peters, and D. P. Mandic, "In-ear spo2: A tool for wearable, unobtrusive monitoring of core blood oxygen saturation," *Sensors*, vol. 20, no. 17, p. 4879, 2020.
12. V. Singh, H. Chandna, A. Kumar, S. Kumar, N. Upadhyay, and K. Utkarsh, "Iot-q-band: A low cost internet of things based wearable band to detect and track absconding covid-19 quarantine subjects," *EAI Endorsed Transactions on Internet of Things*, vol. 6, no. 21, 2020.
13. M. Boyett, Y. Wang, and A. D'Souza, "Crosstalk opposing view: heart rate variability as a measure of cardiac autonomic responsiveness is fundamentally flawed," *The Journal of physiology*, vol. 597, no. 10, p. 2599, 2019.
14. A. M. Ghosh, D. Halder, and S. A. Hossain, "Remote health monitoring system through iot," in *2016 5th International Conference on Informatics, Electronics and Vision (ICIEV)*. IEEE, 2016, pp. 921–926.
15. M. Ghijsen, T. B. Rice, B. Yang, S. M. White, and B. J. Tromberg, "Wearable speckle plethysmography (spg) for characterizing microvascular flow and resistance," *Biomedical optics express*, vol. 9, no. 8, pp. 3937–3952, 2018.
16. P. Farzam, Z. Starkweather, and M. A. Franceschini, "Validation of a novel wearable, wireless technology to estimate oxygen levels and lactate threshold power in the exercising muscle," *Physiological reports*, vol. 6, no. 7, p. e13664, 2018.
17. G. Hu, Q. Zhang, V. Ivkovic, and G. E. Strangman, "Ambulatory diffuse optical tomography and multimodality physiological monitoring system for muscle and exercise applications," *Journal of biomedical optics*, vol. 21, no. 9, p. 091314, 2016.
18. P. Zakharov, M. Talary, and A. Caduff, "A wearable diffuse reflectance sensor for continuous monitoring of cutaneous blood content," *Physics in Medicine & Biology*, vol. 54, no. 17, p. 5301, 2009.

19. R. Istfan, S. LaRochelle, R. Chaudury, and D. Roblyer, "A miniature frequency domain diffuse optical optode for quantitative wearable oximetry," in *Optical Tomography and Spectroscopy of Tissue XIII*, vol. 10874. International Society for Optics and Photonics, 2019, p. 108742B.
20. J. W. Kang, Y. S. Park, H. Chang, W. Lee, S. P. Singh, W. Choi, L. H. Galindo, R. R. Dasari, S. H. Nam, J. Park *et al.*, "Direct observation of glucose fingerprint using in vivo raman spectroscopy," *Science Advances*, vol. 6, no. 4, p. eaay5206, 2020.
21. M. Shokrekhodaei and S. Quinones, "Review of non-invasive glucose sensing techniques: optical, electrical and breath acetone," *Sensors*, vol. 20, no. 5, p. 1251, 2020.
22. H. Wu, Z. Ji, and M. Li, "Non-invasive continuous blood-pressure monitoring models based on photoplethysmography and electrocardiography," *Sensors*, vol. 19, no. 24, p. 5543, 2019.
23. K. Plarre, A. Raij, S. M. Hossain, A. A. Ali, M. Nakajima, M. Al'Absi, E. Ertin, T. Kamarck, S. Kumar, M. Scott *et al.*, "Continuous inference of psychological stress from sensory measurements collected in the natural environment," in *Proceedings of the 10th ACM/IEEE international conference on information processing in sensor networks*. IEEE, 2011, pp. 97–108.
24. M. Gjoreski, H. Gjoreski, M. Luštrek, and M. Gams, "Continuous stress detection using a wrist device: in laboratory and real life," in *proceedings of the 2016 ACM international joint conference on pervasive and ubiquitous computing: Adjunct*, 2016, pp. 1185–1193.
25. R. Fan and T. L. Andrew, "Perspective–challenges in developing wearable electrochemical sensors for longitudinal health monitoring," *Journal of The Electrochemical Society*, vol. 167, no. 3, p. 037542, 2020.
26. A. J. Bandodkar, W. J. Jeang, R. Ghaffari, and J. A. Rogers, "Wearable sensors for biochemical sweat analysis," *Annual Review of Analytical Chemistry*, vol. 12, pp. 1–22, 2019.
27. T. R. Ray, J. Choi, A. J. Bandodkar, S. Krishnan, P. Gutruf, L. Tian, R. Ghaffari, and J. A. Rogers, "Bio-integrated wearable systems: A comprehensive review," *Chemical reviews*, vol. 119, no. 8, pp. 5461–5533, 2019.
28. R. Săndulescu and C. Cristea, "Editorial overview: Exploring new ideas and challenges in electrochemical sensing: Towards implantable or wearable devices and nanorobots," 2018.
29. J. Heikenfeld, A. Jajack, J. Rogers, P. Gutruf, L. Tian, T. Pan, R. Li, M. Khine, J. Kim, and J. Wang, "Wearable sensors: modalities, challenges, and prospects," *Lab on a Chip*, vol. 18, no. 2, pp. 217–248, 2018.
30. M. A. Yokus, T. Songkakul, V. A. Pozdin, A. Bozkurt, and M. A. Daniele, "Wearable multiplexed biosensor system toward continuous monitoring of metabolites," *Biosensors and Bioelectronics*, vol. 153, p. 112038, 2020.
31. P. Angelov and E. Soares, "Towards explainable deep neural networks (xdnn)," *Neural Networks*, vol. 130, pp. 185–194, 2020.

Stress Monitoring with Machine Learning: Detecting and Managing Mental Health

Shiyi Jiang, Yuan Li, Farshad Firouzi, Krishnendu Chakrabarty, and Eric Elbogen

"It's not stress that kills us; it is our reaction to it."

– Hans Selye

Abstract Recent advancements in artificial intelligence/machine learning (AI/ML), wearable devices, and the Internet of Things (IoT) have significantly influenced electronic healthcare (eHealth), particularly in remote health monitoring. This approach, which involves data sensing, analysis, and feedback, is essential for early disease identification and can reduce travel costs and diagnosis times compared to traditional healthcare. Stress monitoring is one of its important applications. Conventionally, data is sent to remote cloud servers for analysis, but this incurs high time and energy costs. While edge devices enable faster analysis, their limited resources cannot guarantee satisfactory prediction precision. Additionally, direct data transmission raises privacy concerns and heterogeneity among user data can hinder the effectiveness of universal models, highlighting the need for personalized approaches. In this chapter, we present an edge-cloud hierarchical solution that balances prediction accuracy, latency, and energy during data transmission for stress monitoring. Additionally, we suggest personalized

This chapter is adapted from the following work: Jiang et al. [1–3].

S. Jiang (✉) · E. Elbogen
ECE, Duke University, Durham, NC, USA
e-mail: shiyi.jiang@duke.edu

Y. Li
Duke Kunshan University, Kunshan, Jiangsu, China

F. Firouzi
Duke University, Durham, NC, USA

Arizona State University, Tempe, AZ, USA

K. Chakrabarty
Arizona State University, Tempe, AZ, USA

B. Farahani et al. (eds.), *Smart and Connected Health: Applications and Real-World Use Cases*, https://doi.org/10.1007/978-3-032-09253-3_4

stress monitoring solutions framed under federated learning by leveraging latent space clustering, multi-domain learning, and graph neural networks to mitigate inter-client and intra-client data heterogeneity.

Keywords Stress monitoring · Wearable IoT · Internet of medical things (IoMT) · Internet of things (IoT) · Edge computing · Cloud computing · Machine learning · Signal processing · Biosignal analysis · Privacy-preserving machine learning · Federated learning

1 Introduction

Stress is a major societal health concern today. Long-term stress can lead to a wide variety of ailments, e.g., depression, eating disorders, personality disorders, substance abuse, and schizophrenia [4, 5]. It can also alter the physiology of the brain, leading to an increased risk of obesity, diabetes, and heart disease [6, 7]. Early treatment of stress is therefore critical; however, today's physician-centered healthcare model is inadequate for addressing stress-related concerns because it tends to be reactive. In particular, the physician-centered model requires a patient to travel to a clinic multiple times a week, resulting in high cost for the patient and high dropout rates. Another major challenge associated with this model is that of limited observations, which leads to inefficient diagnosis and treatment. Recent advances in eHealth based on the convergence of mobile technology, wearable Internet of Things (WIoT), and artificial intelligence (AI) enable new ways to potentially track and address barriers to stress monitoring and treatment [8–12]. These technologies improve patient outcomes by offering a holistic, collaborative, and measurement-based solution that utilizes patient-generated insights, real-time remote stress tracking and monitoring, just-in-time intervention, and personalized treatment.

Stress is induced when an individual encounters harmful physical or mental environments, resulting in a fight-or-flight response that releases hormones, which change the behavior of the central nervous system in human bodies [5]. Stress is usually accompanied by increased heart rate, quicker breathing, and tighter muscles [5]. Physiological signals such as electrocardiogram (ECG), electromyogram (EMG), electrical dermal activities (EDA) (or GSR—galvanic skin response), respiratory signal, and photoplethysmogram (PPG) can be measured as stress indicators [13–17]. In addition to physiological signals, daily activities, body movements or gestures, and environmental factors are other potential elements for stress evaluation [18]. Recent studies in the area of stress monitoring have led to machine learning (ML) models for stress evaluation and classification. Aqajari et al. [19] and Oskooei et al. [20] proposed effective ways of physiological signal feature extraction through neural networks such as convolutional neural networks (CNN) and convolutional autoencoders (CAE). Others such as Tripathi and Beigi [21], Busso et al. [22], and Jaiswal et al. [23] proposed to use images and audio information captured from recorded videos with predesigned scenes for stress classification. Additionally, utilizing text messages posted on social media for stress identification was another area explored by Guntuku et al. [24] and others [25, 26]. In addition to using ML methods such as random forest (RF) [19], recent work has also focused on multilayer perceptron (MLP) and CNNs [27–32].

However, most of the existing state-of-the-art methods for stress recognition are centralized cloud-based approaches that involve the collection of data from each participant and performing model training in the cloud. Such a centralized training approach raises significant privacy concerns. To tackle this problem, researchers have integrated techniques for privacy-preserving machine learning such as multiparty computation [33], differential privacy [34], and homomorphic encryption [35]. In addition, an alternative approach is to utilize federated learning (FL), a decentralized learning framework that preserves privacy and is communication-efficient [36, 37].

Instead of transferring private data to the remote central server, FL conducts model training locally and only uploads local model parameters to the server, with all raw data remaining local. A general FL algorithm employs an iterative approach. Each iteration includes selecting participating clients, conducting model training using local data at each client, uploading client model parameters to the central server, aggregating parameters as a global model, and sending them back to the clients. Nowadays, the FL algorithm is essential in the biomedical field, especially in facilitating data sharing between hospitals [38]. FL has also been applied to various IoT applications [39–41]. In addition, studies show that FL achieves better performance compared to conventional centralized training [42]. Chhikara et al. [42] conducted human emotion classification via FL and achieved an accuracy of 71.64% and an accuracy of 85.04% using facial images and speech signals. Their results outperform centralized state-of-the-art methods. Chen et al. [43] proposed FedHealth that enables client model personalization via transfer learning. FedHealth outperforms existing work on human activity recognition by a 5.3% increase in classification accuracy.

Conventional FL algorithms assume that local data from each client is independent and identically distributed (i.i.d.) [44]. However, in practice, a dataset is typically composed of different domains (i.e., data distributions), which are nonindependent and not identically distributed (non-i.i.d.). For instance, different styles and backgrounds of the images and various languages used in texts are considered distinct domains. Such domain difference, or interchangeably, data heterogeneity, adversely impacts the generalization ability of the global FL model across all data, resulting in degradation in task performance [45].

In this chapter, we provide background information on stress/emotion recognition, the Internet of Medical Things, and privacy-preserving federated learning in Sect. 2. Section 3 describes the hierarchical edge-cloud solution for stress monitoring. We explain personalized stress monitoring solutions that mitigate the inter-client and intra-client data heterogeneity in Sects. 4 and 5, respectively. Finally, we conclude the chapter in Sect. 6.

2 Background

2.1 Stress/Emotion Recognition

Stress/emotion recognition has been a popular research area for decades. Researchers initially used images and audio information for emotion recognition and later shifted their focus to stressed emotion identification using machine learning methods. Research on basic emotion (happiness, sadness, and neutral) recognition using speech and facial expressions began two decades ago with statistical models such as the hidden Markov model [46]. Motivated by the lack of natural interactions between participants and insufficient contextualization of previously developed emotional databases, Busso et al. [22] created an interactive emotional dyadic motion-capture database (IEMOCAP), which recorded improvised scenes by two actors with specific emotions: happiness, anger, sadness, frustration, and neutral state. They further developed an MSP-IMPROV database similar to IEMOCAP with four designed emotions: anger, sadness, happiness, and neutral. They applied support vector machine (SVM) with sequential minimal optimization on speech and facial expression separately, and they reached an optimal accuracy of 70% using facial expression. Accuracy, which ranges from 0% to 100%, and the F1 score, which ranges from 0 to 1, are common metrics used in machine learning for model performance evaluation. The higher these values are, the better performance the model has. Formal definitions are presented in Sect. 3.3.2.

Tripathi and Beigi [21] employed IEMOCAP and applied multiple CNN and long short-term memory (LSTM) on speech, text, and image separately and merged them using a fully connected layer for final four-class emotion detection. The final accuracy was reported to be over 70%. Jaiswal et al. [23] created a multimodal dataset of stressed emotion (MuSE) with audio, visual, lexical, thermal, and physiological information, and performed multiple binary stress classifications using unimodal deep neural networks. They reported an accuracy of 67% using audio information. Hossain et al. [47] integrated emotion recognition with IoT by feeding audio-visual data to models trained in the cloud.

Researchers have also used plaintext scraped from social media for stress detection. Lin et al. [25] applied CNN in combination with a factor-graph model to text streamed from Weibo, resulting in a 6–9% increase in the F1 score compared to the use of logistic regression, SVM, RF, and decision tree. Turcan and McKeown [26] approached this problem by applying classifiers such as gated recurrent neural network (RNN), CNN, and logistic regression using a dataset from Reddit. They achieved an F1 score of 0.79 using logistic regression. Guntuku et al. [24] manipulated text from Facebook and Twitter with regularized linear regression and SVM, demonstrating that the addition of social media language outperformed other sociodemographic features.

Additionally, researchers have utilized physiological signals for stress classification. Schmidt et al. [17] developed a wearable stress and affect detection (WESAD) dataset with multiple physiological signals, such as EDA, ECG, and blood volume pressure (BVP), and built five models for 3-class (stressed, neutral, amusement) and binary classification from extracted features. AdaBoost provided the highest performance with around 80% accuracy for 3-class classification, and linear discriminant analysis (LDA) generated the highest accuracy (over 92%) for binary classification [17]. Lin et al. [29] processed the WESAD dataset for 3-class stress classification by building multiple CNN models using single or small groups of physiological signals and the result was then fed to RF for a final decision—an accuracy of 85%. Mozafari et al. [30] utilized EDA, PPG, abdominal and thoracic respiration signals from the mental arousal level recognition competition dataset and performed five-level stress classification using the nondominated sorting genetic algorithm II (NSGA-II) with a set of machine learning methods: fine tree, bagged tree, cubic SVM, and ensemble discriminant subspace. The best performance using ensemble discriminant subspace was over 68% [30].

In recent years, attempts have also been made to increase classification accuracy based on a single type of signal and data fusion method. Aqajari et al. [19] proposed a CNN-based feature extraction on EDA signals and showed significant binary classification accuracy enhancement using K-nearest neighbors (KNN), SVM, RF, and naïve Bayes (NB) compared to the use of statistically extracted features. KNN was found to be the most effective model with over 90% accuracy [19]. Oskooei et al. [20] suggested an unsupervised method for binary stress classification using heart-rate variability (HRV) extracted through CAE. Sarkar and Etemad [31] proposed a self-supervised ECG learning for classifying emotion arousal, valence, and stress by first training a multitasking CNN using transformed signals and then transferring the network for the emotion prediction task. They used four public datasets and achieved 95% accuracy on the WESAD dataset [31]. Stewart et al. [32] explored the effectiveness of meta-learning methods. They chose a neural process, composed of a classification network and an encoder network, for binary stress classification using statistical features extracted from the ECG and EDA signals from the WESAD and Drivedb datasets [32]. They evaluated the performance based on the area under the curve (AUC) and showed improvement compared to other baseline classifiers: Lasso, SVM, and KNN [32].

Kumar et al. [48] used 1D CNN sub-subnetworks for feature extraction and performed model-level fusion from the outputs of two 1D CNN subnetworks using the WESAD dataset. A CNN was utilized after model-level fusion for 3-class classification. An accuracy of 87.7% and an F1 score of 0.83 were obtained by the model. Khowaja et al. [49] developed novel local maxima and minima features and designed a voting and similarity-based fusion method for binary stress classification using the Drivedb dataset. The proposed method resulted in 92.05% accuracy and an F1 score of 0.9041. Nath and Thapliyal [50] proposed an IoT-based framework for stress evaluation using cortisol level as the ground truth. Physiological data was collected from human subjects by wearable edge devices and was sent to the cloud for analysis using RF. By applying data fusion, an accuracy of 94% and an F1 score of 0.92 were achieved.

Researchers also utilize data with various modalities that can provide auxiliary information for stress prediction. Paredes et al. [18] collected multimodal data such as demographics, weather, daily activity, questionnaires, geolocations, and body movements, and used RF to predict the stress level and provide recommendations. Alharthi et al. [27] used ECG in combination with location, weather, and daily activity data for 3-class stress detection (stressed, normal, relaxed) with an NB solution implemented in the cloud and reached over 78% accuracy.

Aside from the aim of improving the performance/accuracy of stress prediction, there has been some work investigating other aspects of the task, such as data preprocessing, feature importance, and model generalizability. For instance, Jiang et al. [51] utilized nuclear norm minimization to impute missing data that occurred during data acquisition from wearable devices, resulting in improved stress-level predictions. Can et al. [52, 53], Dalmeida et al. [54], and Coutts et al. [55] showed that features derived from EDA and HRV are crucial indicators for stress prediction. Smets et al. [56] explored associations between physiology, context, and behavior information to better understand stress in daily life. Vos et al. [57] proved that using the ensemble method for stress prediction provides better model generalizability compared to using a single prediction model.

At the end of this section, we summarize publicly available datasets used in stress/emotion recognition literature in Table 1.

2.2 *Internet of Medical Things (IoMT)*

The Internet of Things (IoT) is a network that involves a variety of elements such as sensors, computing devices, software, humans, and physical objects, wherein the elements in the network can communicate and exchange data with each other [9, 10]. The Internet of Medical Things (IoMT), or Healthcare Internet of Things (HIoT), is a specific type of IoT focusing on medical applications such as seamless health monitoring and proactive disease prevention [58–61]. By leveraging advances in wearable IoT, machine learning, and edge/fog computing, IoMT creates a new paradigm with an emphasis on research problems such as task performance optimization, resource allocation, cost minimization, etc. [62–64].

Rahmani et al. provide a general overview of fog computing and demonstrate the efficiency of an intermediate level of architecture between the edge and the cloud via a medical case study [65]. A variety of healthcare applications have been developed under the IoMT framework, e.g., human activity recognition (HAR) [62, 66, 67], seizure detection [68], ECG arrhythmia classification [69], and COVID-19-related symptom detection [70].

Due to the limitations associated with battery life, energy consumption, and computational capability of endpoint IoT devices, performing all the computations locally is not viable. A subset of tasks, therefore, needs to be offloaded, either to a nearby data center or to the cloud that has abundant computational resources. Furthermore, when multiple endpoint devices are involved, algorithms that allocate resources in an efficient manner are essential. Aazam et al. [71] presents a review of recently proposed offloading schemes. Offloading frameworks using reinforcement

Table 1 Public stress/emotion datasets used in prior work

Dataset	Average time recorded per per subject (min)	No. of subjects	Modality	Attribute	No. of classes
MSR-IMPROV [22]	–	12	Audio, visual	Emotion	4
WESAD [17, 19, 28, 29, 31, 32]	100	15	Physiological	Stress	3
IEMOCAP [21, 22]	72	10	Audio, visual	Emotion	10
SWELL [28, 31]	75	25	Lexical, physiological, visual	Stress	3
Mental arousal level recognition competition dataset [30]	10	30	Physiological	Stress	5
DREAMER [31]	18	23	Physiological	Arousal/valence	5
Dreaddit [26]	–	–	Lexical	Stress	2
AMIGOS [31]	–	40	Audio, physiological, visual	Arousal/valence	9
MuSE [23]	45	28	Audio, lexical, physiological, visual	Stressed emotion	3
eNTERFACE'05 database [47]	1.5	42	Audio, visual	Emotion	6
Berlin emotional speech database [47]	2.2	10	Audio	Emotion	7
Kanade-Cohn emotional face database [47]	–	182	Visual	Emotion	6
Drivedb [32]	50	24	Physiological	Stress	3

learning [72, 73], game theory [64], and other algorithms [63, 74–76] have been proposed and shown to be effective in reducing latency and energy cost.

Additional research topics in the context of IoMT include streaming data preprocessing [77], location prediction of endpoint IoT devices [78], and data privacy [79]. In addition, recent advances in neuromorphic computing highlight the potential of integrating edge devices within the IoMT framework [80].

2.3 Privacy-Preserving Federated Learning

The concept of federated learning was first proposed by McMahan et al. [36]. To avoid information leakage associated with the sending of private data to a remote server, they suggested decentralized training so that only model parameters were uploaded to the remote server, with data kept locally, as shown in Fig. 1, where the parameters ω_1, ω_2, and ω_3 represent local client model parameters, and the parameter ω represents the aggregated global model parameters. A general FL algorithm starts with the selection of participating clients. In each communication round, a subset of clients is selected for federated learning. The selected clients train their models locally and send model parameters to the central server. The server performs parameter aggregation and sends the updated global model parameters back to selected clients. This process is repeated for a predefined number of communication rounds. McMahan et al. proposed the first FL algorithm (FedAvg), in which the central server aggregates client model parameters by averaging. Since these early advances, researchers have proposed many additional FL algorithms [37].

One of the challenges in federated learning is data heterogeneity. In actual practice, the data distribution may not be consistent across all clients, and therefore aggregating all client model parameters might result in worse model performance than training the model locally. To tackle this problem, researchers have proposed clustering of clients based on their similarities. Briggs et al. [82] suggested clustering client local updates via hierarchical clustering. They argued that their proposed method achieved comparable classification performance as FedAvg [36],

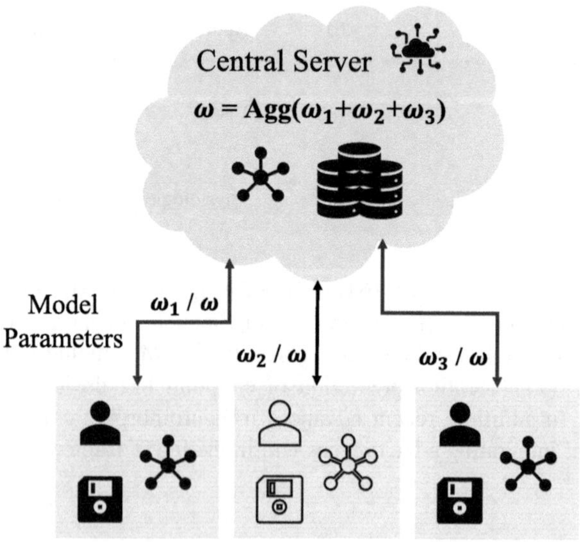

Fig. 1 General federated learning framework [81]

but with fewer communication rounds. Ghosh et al. [83] proposed an iterative federated clustering algorithm that clusters clients based on minimum loss to the cluster model. Results showed that their algorithm achieved superior performance compared to both the global and the local models. Kim et al. [84] proposed clustering clients based on local data using a generated adversarial network (GANs) named ClusterGAN and updating clusters based on client model performance. Their results show that they outperform existing methods on time-series forecasting tasks by 43%.

Additionally, researchers have proposed many other algorithms to address data heterogeneity and other outstanding concerns in FL, such as lack of privacy, high communication cost, and device heterogeneity. Li et al. [85] suggested LotteryFL based on the lottery ticket hypothesis to alleviate data heterogeneity and high communication costs. Furthermore, Li et al. [86] proposed Hermes, which achieves high accuracy, low communication costs, and high efficiency during inference via structured pruning and selective model parameter sharing. McMahan et al. [87] developed an algorithm that provides a privacy guarantee for each user by combining differential privacy with federated learning for language models. Their results indicate the algorithm preserves user-level privacy with minimum utility degradation. Wei et al. [88] proposed the noising before aggregation FL (NbAFL) algorithm that utilizes differential privacy and provided theoretical analysis on the trade-off between participating client number selection, model convergence, and privacy levels. Zhang et al. [89] developed BatchCrypt for FL across organizations using homomorphic encryption with reduced encryption and communication overhead via quantization and encoding techniques. Some studies mitigate communication overhead in FL from the perspective of wireless communication. For instance, Xu et al. developed a two-level bandwidth resource allocation algorithm [90]. Yang et al. proposed an algorithm to minimize energy consumption during client model computation and data transmission [91]. Chen et al. established an algorithm via integer linear programming and convex optimization methods to optimize network resource allocation [92]. Moreover, Smith et al. [93] suggested a federated multitask learning algorithm, MOCHA, that tackles data/device heterogeneity and communication efficiency. Chen et al. [94] developed an asynchronous FL algorithm, ASO-Fed, for streaming data to tackle device heterogeneity among clients.

3 A Resilient and Hierarchical IoT-Based Solution for Stress Monitoring in Everyday Settings

Current stress monitoring solutions are episodic; the status of a subject may not be precise due to the lack of information before and after the examination. Real-time IoT-based monitoring is therefore essential. In addition, the current reactive, symptom-based approach can be converted into a proactive solution, with

IoT-assisted measurement and AI-powered assessments that provide just-in-time interventions. Additionally, an IoT-based solution provides personalization that captures subject-to-subject variations. However, existing works are not suitable for implementation on endpoint IoT devices. The complexity and the number of features associated with these methods require high computational effort, which makes them unsuitable for endpoint IoT devices where the computation resources, memory, and power source are severely constrained. Also, high latency and power-hungry data transmission to remote servers require a careful trade-off between computing at the edge and in the cloud; this trade-off impacts energy consumption, battery lifetime, end-to-end response time, as well as the accuracy/reliability of the ML model [71]. Another challenge that has not been adequately tackled is related to missing data, which is a common problem for data gathered from sensors that measure physiological signals. Missing data can potentially result in ineffective stress-level classification and long-term stress monitoring. Finally, the limited availability of labeled training data leads to reduced accuracy with today's solutions.

In this section, we describe a robust and hierarchical IoT-based solution that enables real-time monitoring and early detection of stress situations outside of conventional clinical settings. We provide some key points of our work as follows:

- We present a holistic edge-assisted IoT-based hierarchical solution for processing multimodal data that achieves a balance between accuracy, latency, and energy and significantly outperforms six state-of-the-art classification methods for stress analysis. The included offloading technique also ensures low-overhead data offloading and is especially effective for real-time stress monitoring.
- We exploit design-space exploration to optimize model performance and energy efficiency at the edge during data inference. We utilize recently proposed algorithms that have been shown to be effective in the literature. We employ a spiking neural network (SNN), a neural network with conditionally parameterized convolutions (CondConv), and a support vector machine (SVM) for comparison. The SVM model achieves the highest accuracy with the lowest computational complexity.
- We present a novel approach that uses a few-shot learning model for stress classification to alleviate the shortcoming of a limited sample size. Our model achieves a significant increase in the accuracy and the F1 score compared to state-of-the-art models for stress evaluation.
- We perform real-time stress monitoring simulation to compare our approach with solutions that utilize either only edge-based or cloud-based models. Our approach achieves an accuracy of 91.25% with a significant amount of reduction in response time and energy consumption compared to the IoT-cloud solution. These results further highlight the merits of our method compared to other nonhierarchical models.
- We present a resilient technique to impute the missing data that is likely to occur during data acquisition in connected IoT systems for seamless stress-level monitoring.

The organization of this section is as follows. In Sect. 3.1, we present signal preprocessing, missing value imputation, and a general overview of our methods for each level of the hierarchy. We demonstrate details associated with feature extraction, feature selection, design-space exploration, and model training for both levels of the hierarchy and their interplay in Sect. 3.2. We finally present experimental results and evaluation methods in Sect. 3.3.

3.1 Description of Our Solution

A number of IoT systems have been proposed for healthcare and medicine [62, 65, 96]. Today's IoMT frameworks are heterogeneous in nature, and currently, there is no consensus with respect to any reference model/architecture [9–11]. We adapt the architecture introduced in our earlier work [9, 61, 97, 98] shown in Fig. 2, which follows an edge-fog-cloud pattern in which the fog layer extends from the edge to the cloud. The edge layer is where wearable devices, portable sensors, and smartphones are leveraged to monitor a subject's health status by collecting real-time multimodal data (e.g., physical activities, EDA, and ECG). Endpoint IoT devices can be either directly connected to the fog/cloud layer or connected via gateways (e.g., smartphones) as bridges in between. The fog layer is designed based on a multilayer hierarchical cluster-based pattern, allowing each fog node to communicate vertically and horizontally. Fog nodes locate between the endpoint IoT devices and the cloud, which provide scalable distributed computing, storage, and network connectivity [99]. Finally, data storage, big data processing, and analytics

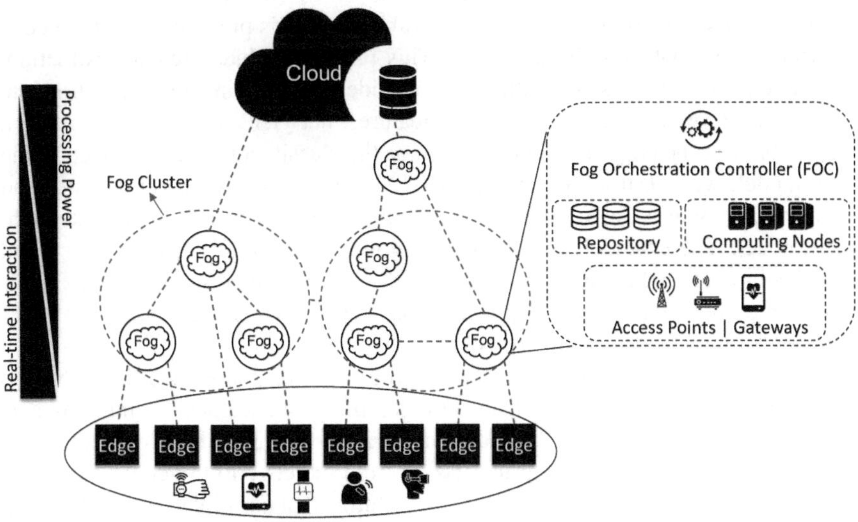

Fig. 2 Illustration of our general framework for healthcare applications [95]

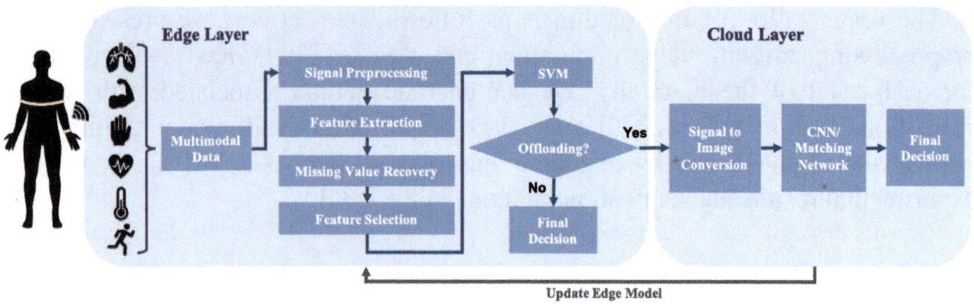

Fig. 3 Illustration of our framework for stress evaluation [95]

take place in the cloud. The design of this general-purpose architecture ensures that any healthcare application can be mapped to it while satisfying the application requirements.

In this study, we present a two-tier scheme for stress monitoring by treating the fog layer as part of the cloud. Given the small-sized dataset in our study, the fog alone has sufficient computational resources to perform advanced data analysis. To highlight the importance of the edge-fog-cloud interplay, an offloading mechanism is also presented. The presented solution is general and can be expanded by incorporating advanced offloading techniques (e.g., Markov decision process and deep reinforcement learning-based offloading schemes) [72, 73]. Although our solution is appropriate for our small dataset, a full edge-fog-cloud architecture that considers task offloading between the fog and the cloud layer will be integrated into future work with larger datasets.

Figure 3 presents our stress-level evaluation framework using a hierarchical IoT architecture with an edge layer and a cloud layer. Multimodal data such as ECG and respiratory signals collected from wearable sensors is preprocessed at the edge to ensure clean data for the next step. This process includes feature extraction. Imputation recovers missing values that are identified during feature extraction. Feature selection determines a subset of features that yields the best classification performance at the edge. In order to optimize the classification performance of the edge model, we construct a recurrent SNN and a CondConv neural network and compare them with SVM for design-space exploration to select the best lightweight edge model. These models are compared in terms of classification accuracy and model computational complexity. Details of the models are presented in Sect. 3.2.1. Due to the constrained computational power at the edge, we utilize an SVM for lightweight classification using the selected features.

We can obtain predictions at the edge with rapid response time, but the edge prediction may suffer from unsatisfactory classification accuracy. Training models and performing inference in the cloud ensure high performance, but they lead to high time and energy costs for data transmission from the edge to the cloud. To tackle this problem, we describe a hierarchical IoT solution that partitions the data between the edge and the cloud to obtain enhanced overall performance, while avoiding the time and energy costs associated with unnecessary communication and cloud computation. The criteria used for data partitioning are elaborated in Sect. 3.2.3. Samples extracted from the data segments that satisfy the criteria remain at the edge for lightweight classification and prediction. Samples that do not satisfy the criteria are sent to the cloud and are then converted to images for advanced analysis. In the cloud, we utilize a CNN and a matching network model and assess their performance. The model that has the better performance is selected for prediction.

3.1.1 Biosignals

Here we explain the biosignals that we utilized in the study of stress-level assessment. ECG measures the electrical activity of the heart in millivolts (mV) using multiple electrodes placed on the body. EDA reflects the activity of the human sympathetic network via sweat glands, which can be measured as the conductance in microsiemens (μS). Respiratory-signal measures displacement variations based on volume changes of the thorax or abdomen during inhalation and exhalation. EMG measures electrical signals associated with the muscle fibers in millivolts (mV) sent from motor sensory neurons in the brain. Accelerometer data tracks body movements in X, Y, and Z coordinates. BVP is obtained from PPG, which measures blood volume change optically by recording changes in light absorption in the skin. The skin temperature is measured in Celsius.

3.1.2 Signal Preprocessing

We apply downsampling, filtering, min-max normalization, and segmentation for signal preprocessing.

Downsampling We perform downsampling to reduce processing time and memory usage at the edge. ECG, EDA, EMG, respiratory signal, temperature, and accelerometer data were downsampled from 700 Hz to 100 Hz. BVP was maintained at the original 64 Hz sampling rate.

Filtering, Normalization, and Segmentation We calculated the moving average with a 1-second window and min-max normalization to all signals to remove noise caused by environmental differences in participants as suggested by [19]. All signals were sliced with a 90-second window for feature extraction. For cloud analysis, EDA and respiratory signals were converted to images of size 128 × 128 based on 90-second window segments. ECG was first split into 90-second windows, and for each 90-second window, it was further divided into images of size 128 × 128 that contain one R-peak per image using the R-peak detection algorithm [100].

3.1.3 Missing Value Recovery

Many methods exist for resolving missing values in a dataset [101]. The simplest one is deletion by directly removing sample points. Imputation-based methods such as KNN imputation estimate missing values by weighing the previous k nearest samples. Model-based methods construct either statistical models such as maximum likelihood estimation or machine learning models such as SVM and neural networks based on previous data. According to [101], when the missing window is small (below 1 hour), KNN imputation achieves the lowest root-mean-square error compared to other types of methods. The above methods are categorized as single imputation, where the missing value is replaced by one value. However, single imputation may lead to high bias since it fails to reveal sampling variability [102].

We therefore utilize multiple imputation, which is useful for reducing bias and increasing analysis precision and is resilient to outliers [103]. Multiple imputation generates n distinct datasets with missing values in the original dataset filled by n potential values. To impute the dataset, we utilize sequential regression imputation by estimating one feature with missing values at a time based on regression models established using the other features. Creating separate regression models is beneficial to adjusting for covariates in targeted analysis and for bias reduction [104]. We repeat the imputation n times to obtain n distinct datasets. We train a lightweight classification model at the edge for each dataset and obtain the predictions. We average the predictions from the n models as the final result of multiple imputation. The overall workflow is shown in Fig. 4. We used a baseline obtained by filling missing data with values chosen uniformly from the dataset for comparison.

To obtain customized imputations, auxiliary data (signals collected from other sensors, information acquired from surveys, etc.) can be utilized in combination with the outputs of multiple imputation. However, due to the limited sample size of the selected dataset, we were not able to perform customization for missing value imputation.

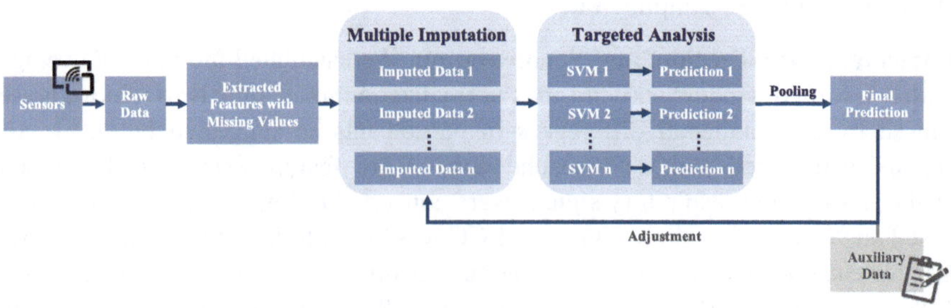

Fig. 4 Steps associated with the recovery of missing values [95]

3.2 *Stress Recognition Using Hierarchical IoT*

3.2.1 Stress Recognition at the Edge

Feature Extraction We computed maximum, minimum, sum, standard deviation, mean, skewness, and kurtosis as statistical features for each segment of all seven signals. Furthermore, additional features based on the specific properties of signals were computed. For accelerometer data, AUC and the mean peak amplitude along each axis were calculated. The slope and dynamic range were calculated for temperature. The respiratory signal was filtered to distinguish between cycles of inhalation and exhalation, and we obtained the mean, standard deviation of inhalation and exhalation, ratio of inhalation to exhalation, range of respiration, and mean respiratory rate as features. The skin conductance response (SCR) was extracted from EDA based on [105] and [106]. The mean, standard deviation, sum amplitude, number of SCR peaks, AUC, slope, and dynamic range were used as features. The dynamic range, AUC, median, 10th percentile, 90th percentile, mean, standard deviation, number of peaks, and sum of the peak amplitude of EMG were also evaluated. The time intervals between R-peaks (N-N intervals) were identified using the peak detection algorithm [107]. They were then utilized to compute a series of features such as relative and normalized powers and low-frequency to high-frequency ratio (LF/HF) using four frequency bands: ultralow frequency (ULF) (0–0.003 Hz), very low frequency (VLF) (0.003–0.04 Hz), LF (0.04–0.15 Hz), and HF (0.15–0.4 Hz) based on [108] and [109]. Figure 5 shows some examples of

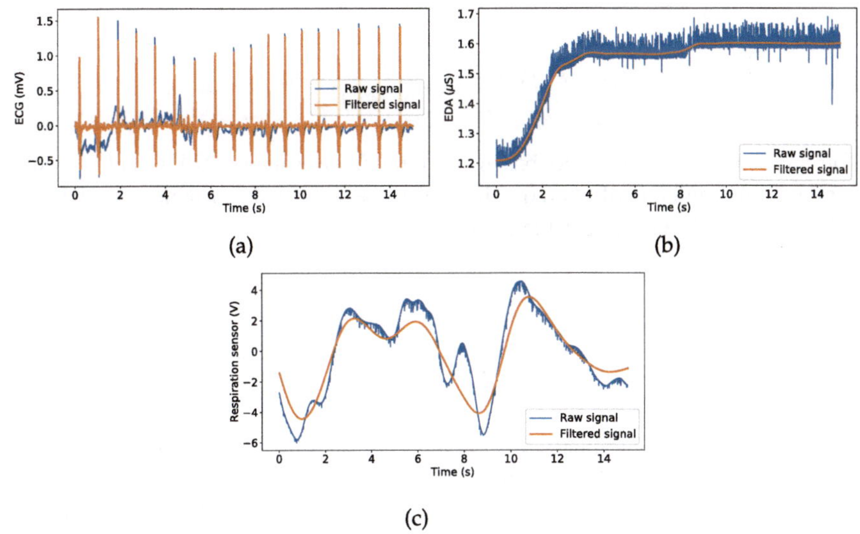

Fig. 5 Comparison between raw and filtered ECG, EDA, and respiratory signal [95]. (**a**) Raw and filtered ECG signal. (**b**) Raw and filtered EDA signal. (**c**) Raw and filtered respiratory signal

Table 2 Additional features extracted from signals

Signals	Additional features
Accelerometer data	AUC and mean peak amplitude for each axis
Temperature	Slope, dynamic range
Respiratory signal	Mean, std of inhalation and exhalation, inhalation/exhalation ratio, stretch, mean respiratory rate
EDA	Slope, dynamic range, mean, std of SCR peaks, # of SCR peaks, sum of SCR amplitude, AUC
EMG	Dynamic range, AUC, median, 10th percentile, 90th percentile, mean, std of peaks, # of peaks, sum of peak amplitude
ECG	Sum of peaks, LF/HF ratio, total power from ULF, VLF, LF and HF, relative and normalized powers, RMSSD

signals before and after filtering. All the computed features are listed in Table 2 for each signal.

Feature Selection We applied the Fisher score [110], joint mutual information (JMI) [111], and double input symmetrical relevance (DISR) [112] to the extracted features described above, and compared their effectiveness in selecting crucial features for stress assessment. We selected these methods based on their computational efficiency and ability to minimize overfitting. The Fisher score computes a score for every feature independently based on Eq. (1), where $\mathcal{F}(j)$ is the Fisher score for feature j, N_k is the number of samples in class k, μ_j^k and σ_j^k are the mean and standard deviation of feature j in samples from class k, and μ_j is the mean of feature j among all samples.

$$\mathcal{F}(j) = \frac{\sum N_k(\mu_j^k - \mu_j)^2}{\sum N_k(\sigma_j^k)^2}. \tag{1}$$

The JMI measure is derived based on mutual information (MI) [113], which computes the amount of shared information between two variables $\mathcal{X}$ and $\mathcal{Y}$ as defined in Eq. (2) where $\mathcal{H}$ is the entropy [113].

$$\begin{aligned}\mathcal{I}(\mathcal{X};\mathcal{Y}) &= \mathcal{H}(\mathcal{X}) - \mathcal{H}(\mathcal{X}|\mathcal{Y}) \\ &= \sum_{x\in\mathcal{X}}\sum_{y\in\mathcal{Y}} p(xy)\log\frac{p(xy)}{p(x)p(y)}.\end{aligned} \tag{2}$$

Note that JMI measures shared information between the target variable $\mathcal{Y}$ and the set of variables $\mathcal{X}_i$ and $\mathcal{X}_j$ as indicated in Eq. (3), where S is the set of currently selected features. It evaluates whether the addition of feature $\mathcal{X}_i \notin S$ expands the amount of shared information.

$$\mathcal{I}_{\text{JMI}}(\mathcal{X}_i) = \sum_{\mathcal{X}_j \in S} \mathcal{I}(\mathcal{X}_i, \mathcal{X}_j; \mathcal{Y}) \tag{3}$$

DISR is a method built upon JMI that measures the symmetrical relevance of all combinations of two variables in a set. It first selects a feature that has the maximum MI with target variable $\mathcal{Y}$. It then performs a greedy step of adding features one by one to the selected features that yield a maximum score. The score is computed based on Eq. (4), with S being the set of selected features and $\mathcal{X}_i \notin S$. As shown in [112], DISR is suitable for dealing with datasets that have a high feature-to-sample ratio [111].

$$\mathcal{I}_{\text{DISR}}(\mathcal{X}_i) = \sum_{\mathcal{X}_j \in S} \frac{\mathcal{I}(\mathcal{X}_i, \mathcal{X}_j; \mathcal{Y})}{\mathcal{H}(\mathcal{X}_i, \mathcal{X}_j; \mathcal{Y})} \tag{4}$$

Features were then ranked based on the selection metrics. For each method, we selected the set of features that lead to the highest accuracy using the lightweight classifier. We further compared the performance and evaluated the efficacy of feature selection methods.

Support Vector Machine (SVM) SVM is a widely used supervised learning algorithm that is known for its effectiveness and memory efficiency. We therefore utilize SVM for lightweight stress classification on computational resource-restricted devices. It aims at finding a hyperplane that separates data from different classes. It finds the hyperplane by maximizing the margin, i.e., the distance of the closest data point to the hyperplane. Formally, the distance from data point x_i to the hyperplane is defined as $f(x_i)$, as indicated in Eq. (5), where w represents the weight and b represents the bias. The objective of SVM is to maximize the margin, as described by Eq. (6), where $y_i \in \{-1, +1\}$ represents the label of the test point. The ideal constraint on margin maximization for the hyperplane to perfectly separate all the data points is $y_i f(x_i) \geq 1$. However, most real-world datasets are non-separable. In order to solve this problem, SVM loosens the constraint by introducing a penalty term ϵ, where $\epsilon \geq 0$ as shown in Eq. (7), adding tolerance on data misclassification.

$$f(x_i) = w^T x_i + b \tag{5}$$

$$\arg\max_{w,b} \frac{1}{\|w\|} \min_i y_i f(x_i) \tag{6}$$

$$y_i f(x_i) \geq 1 - \epsilon, \forall x_i \tag{7}$$

Spiking Neural Networks (SNNs) Neuromorphic computing mimics the learning patterns of neurons and synapses in the human brain [114]. It possesses an architecture with collocation of processor and memory, which enables faster processing and lower power consumption compared to the classical von Neumann architecture [114]. Spiking neural networks (SNNs) are novel models that utilize neuromorphic computing building blocks.

In contrast to neurons in artificial neural networks, neurons in SNNs output time-dependent binary values called spikes. The behavior of a spiking neuron mimics a biological neuron, where the output of neuron j at time t (Out^t_j) is threshold-based, as shown in Eq. (8). When the potential accumulated in the neuron (V^t_j) is above the threshold V_{thres}, the neuron is activated to generate a spike. It restores the resting membrane potential V_{rest} after activation. However, when the potential is less than V_{thres}, the neuron remains inactivated and outputs a zero. There are many types of spiking neurons. One of the most common types of spiking neurons is leaky integrate-and-fire (LIF) neurons. The dynamics of the potential accumulated in a LIF neuron is characterized by the differential equation presented in Eq. (9), where V^t_j is the potential of neuron j at time t, V_{rest} is the resting membrane potential, ζ and α are constants, w_{ij} is the synaptic weight between neuron i and neuron j, and x_i is the output from neuron i. A single spiking neuron has been shown to have higher computational power than a canonical neuron with the sigmoid activation function [115].

$$\mathrm{Out}^t_j = \begin{cases} 1, & \text{if } V^t_j > V_{\mathrm{thres}} \\ 0, & \text{otherwise} \end{cases} \tag{8}$$

$$\zeta \cdot \frac{d}{dt} V^t_j = -(V^t_j - V_{\mathrm{rest}}) + \alpha \sum_{i=1}^{n} w_{ij} \cdot x_i \tag{9}$$

Similar to canonical artificial neural network architectures such as fully connected neural networks, CNNs, and RNNs, architectures composed of spiking neurons have been developed and have been shown to achieve high performance in handling physiological signals such as ECG and EMG [116–118]. To handle the problem of non-differentiability inherent in Eq. (9), algorithms for approximating the gradient descent mechanism were developed [119, 120].

Inspired by the architectures presented in previous work [117, 118], we develop an SNN architecture with one recurrent cell unit and one fully connected layer utilizing LIF neurons. We encode input data based on Eqs. (8) and (9), where x_i is the input.

Conditionally Parameterized Convolutions (CondConv) CNNs are widely used for various computer vision tasks [121, 122]. The performance of a model can be optimized by increasing the capacity of the network, such as adding more convolutional layers. However, this approach increases the computational cost. The convolutional layer applies convolution using the same set of filters to all input data. Yang et al. [123] proposed conditionally parameterized convolutions (CondConv) using customized convolutional filters for each input sample to enhance model performance with low computational cost. The output from a CondConv layer is calculated based on Eq. (10), where x is the input data, $W_1, \ldots, W_N$ are convolutional filters, $\omega_1, \ldots, \omega_N$ are weights learned via gradient descent, and ς is a nonlinear activation function [123].

$$\text{Output}(x) = \varsigma((\omega_1 \cdot W_1 + \ldots + \omega_N \cdot W_N) * x) \quad (10)$$

We utilize CondConv layers to construct an advanced neural network at the edge. We perform design-space exploration by varying the number of CondConv layers followed by one fully connected layer as the architecture. The details of the evaluated architectures are explained in Sect. 3.3.1. The description of the conventional CNNs is provided in Sect. 3.2.2.

3.2.2 Stress Recognition in the Cloud

To ensure overall high performance, samples selected by our decision criteria are delivered to the cloud and converted to images for advanced analysis. We utilized a CNN and a matching network that are known for their high performance on image classification.

Convolutional Neural Network (CNNs) CNNs are a category of neural networks that consist of combinations of convolutional layers, pooling layers, and fully connected layers; it is typically used for image classification. The convolutional layer applies convolution between a set of filters and the image, converting the image to feature maps. The pooling layer subsamples values in nonoverlapping neighbors in the maps by either taking the maximum or the average in the neighborhood. The fully connected layer applies linear transforms to the input data. The activation function introduces nonlinearity to the output of the fully connected layer.

All parameters are adjusted through back-propagation to minimize a predefined loss function based on the output and the ground truth. Inspired by the CNN architecture explained in [124], we design a CNN with the architecture detailed in Table 3.

Matching Networks Matching networks are used for few-shot learning, which is aimed at achieving high classification accuracy with very few training samples in each class. These networks are typically characterized in the form of k-way n-shot learning, where k is the number of classes in the dataset and n is the number of samples in each class. In the scenario of limited sample size in the publicly available datasets for stress assessment, matching networks have the potential to achieve higher performance than typical deep learning models.

The aim of supervised learning algorithms is to learn a hypothesis that predicts the optimal approximation of the ground truth. The objective is shown in Eq. (11), where $R(\cdot)$ is the expected risk, i.e., how well an algorithm performs, $\tilde{H}$ is the

Table 3 Details of our CNN architecture

Layer	Layer type	Kernel size	Stride	Input dimensions	Output dimensions
1	Conv2d	3 × 3	1	128 × 128 × 3	126 × 126 × 32
2	MaxPool	2 × 2	2	126 × 126 × 32	63 × 63 × 32
3	Conv2d	3 × 3	1	63 × 63 × 32	61 × 61 × 64
4	MaxPool	2 × 2	2	61 × 61 × 64	30 × 30 × 64
5	Conv2d	3 × 3	1	30 × 30 × 64	28 × 28× 128
6	Conv2d	3 × 3	1	28 × 28 × 128	26 × 26 × 256
7	Conv2d	3 × 3	1	26 × 26 × 256	24 × 24 × 256
8	Linear	–	–	24 × 24 × 256	2048
9	Linear	–	–	2048	2

optimal hypothesis that minimizes the risk, H^* is the best approximation of $\tilde{H}$ in the hypothesis space $\mathbf{H}$, and $H_N \in \mathbf{H}$ is the best hypothesis that can be obtained with N samples and that minimizes the risk [125]. As shown in Eq. (11), the error is composed of two parts, with the first portion being the approximation error, and the latter the estimation error. When N is small, $R(H_N)$ grows, increasing the estimation error. Metric-based learning models such as matching networks reduce the estimation error by constraining $\mathbf{H}$ using prior knowledge learned by their embedding functions [125].

$$\mathbb{E}[R(H_N) - R(\tilde{H})] = \mathbb{E}[R(H^*) - R(\tilde{H})] + \mathbb{E}[R(H_N) - R(H^*)] \tag{11}$$

As shown in [126], metric-based learning models can strengthen relevant information, exclude redundant information, and prevent model overfitting. Koch et al. used twin neural networks that share the same weights to learn the best embedding function that reflects the similarity between within-class and between-class sample pairs in the embedded space [127]. Their model suggests the robustness of metric-based learning methods on few-shot image recognition tasks. Matching networks utilize the attention mechanism and LSTM to learn the embedding function. LSTM is a specialized RNN cell that takes the form presented in Eq. (12). Given an input sequence x, at each time step t, it generates a hidden state h_t as the output of the cell, and a cell state c_t, which contains the information memorized from previous time steps.

$$h_t, c_t = \text{LSTM}(x, h_{t-1}, c_{t-1}) \tag{12}$$

We use the matching networks approach described in [128] for one-shot learning. It first embeds all samples (labeled support set and unlabeled samples) via a convolutional encoder. Vinyals et al. [128] modified the original embedding method via full context embeddings $f(x', S)$ as presented in Eq. (13), where x' is the test sample, S is the support set, $f^{'}(\cdot)$ is a function that denotes the CNN for the original embedding, $g(\cdot)$ is a bidirectional LSTM, and T is the number of unrolling steps in LSTM. att-LSTM is an attention-based LSTM model, which is further explained in Eqs. (14)–(17). The attention mechanism att is shown in Eq. (14), where $x_i \in S$, h_{t-1} is the hidden state at step $t-1 \in T$. It is applied to compute the attention score $s_{\text{att},\, t-1}$ as shown in Eq. (15). The quantity $s_{\text{att},\, t-1}$ is then fed into a conventional LSTM model presented in Eq. (16). The final output h_t at step t is computed as shown in Eq. (17). Vinyals et al. [128] shows that this re-embedding is especially useful when the two samples from different classes are similar.

$$f(x', S) = \text{att-LSTM}(f'(x'), g(S), T) \tag{13}$$

$$\text{att}(h_{t-1}, g(x_i)) = \text{Softmax}\left(h_{t-1}^T \cdot g(x_i)\right) \tag{14}$$

$$s_{\text{att},\, t-1} = \sum_{i=1}^{|S|} \text{att}(h_{t-1}, g(x_i)) \cdot g(x_i) \tag{15}$$

$$h_t', c_t = \text{LSTM}(f'(x'), [h_{t-1}, s_{\text{att},\, t-1}], c_{t-1}) \tag{16}$$

$$h_k = h_k' + f'(x') \tag{17}$$

After the embedding step, pairwise distances between support and unlabeled samples are calculated based on the given matching function. The Softmax function is applied to pairwise distances and the results are multiplied with true labels from the support sets to generate predicted outputs for unlabeled samples.

We use the matching network architecture described in Fig. 6. Built upon the architecture depicted in [128], we constructed an encoder with five convolutional layers with batch normalization and five max-pooling layers for the respiratory signal and ECG. The encoder for EDA was assembled using four convolutional layers with batch normalization and four max-pooling layers.

3.2.3 Edge-Cloud Interplay

As introduced in Sect. 3.1, the performance at the edge layer is constrained by low computational power and limited resources. To address this problem, we utilize the cloud layer and offload a subset of data to the cloud for analysis, and then acquire results from the cloud. Through this interplay between the edge and the

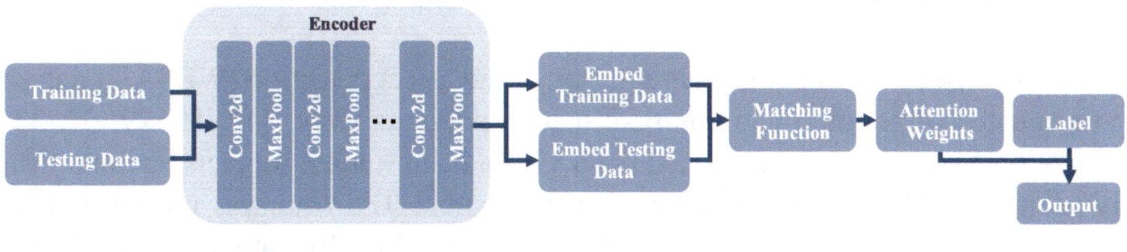

Fig. 6 Description of the matching network architecture [95]

cloud, we achieve better overall performance. However, time and energy costs for data offloading from the edge to the cloud have to be considered. We therefore investigate the criteria for data offloading to ensure low transmission cost in terms of communication delay and energy consumption.

We next present our criteria for dispatching samples from the edge to the cloud. We handle the output from the SVM at the edge as follows. Instead of outputting the predicted labels, we output the probability of a sample being in each of the classes as shown in Eq. (18), where $p_k(x_i)$ is the probability of the sample x_i being in class k, and K is the number of classes. We use the maximum probability $\max\{p_1(x_i), \ldots, p_K(x_i)\}$ as a score for the sample to make the data transmission decision. We establish a decision threshold c, so that samples with scores below the threshold are offloaded to the cloud, and the rest of the samples remain at the edge, as shown in Eq. (19). The samples left at the edge are assigned to the class where they have the maximum output probability.

$$p_k(x_i) = p(k|x_i),\ 1 \le k \le K \tag{18}$$

$$\begin{cases} \max\{p_1(x_i), \ldots, p_K(x_i)\} < c,\ \frac{1}{K} \le c \le 1, & \text{cloud sample} \\ \max\{p_1(x_i), \ldots, p_K(x_i)\} \ge c,\ \frac{1}{K} \le c \le 1, & \text{edge sample} \end{cases} \tag{19}$$

Because we perform binary classification, the threshold $c \ge 0.5$. Based on empirical evidence from experimentation, we sampled 15 thresholds spaced uniformly from 0.55 to 1 to ensure that the number of samples dispatched to the cloud at each threshold is reasonable. For each threshold, we estimated the communication latency and energy consumption during data transmission. We examined the trade-off between classification accuracy and data transmission cost at the edge.

Our work provides a data offloading scheme between a single endpoint edge device and the cloud, with an emphasis on maintaining high accuracy, for stress-level evaluation with low latency and energy cost during data transmission. Offloading schemes that consider operations among multiple endpoint edge devices, edge device battery consumption, queuing models, and communication networks while not in the scope of this work will be explored in future research.

3.3 *Experimental Results*

3.3.1 Experimental Setup

Stress Dataset We used the wearable stress and affect detection (WESAD) dataset to evaluate our solution and compare it with previous work [17, 19, 29, 31, 32]. The WESAD dataset was collected from 17 graduate student subjects but data from two of the subjects was excluded after evaluation. The data was recorded using RespiBAN, a wearable device placed on the chest, and Empatica E4 wristbands. RespiBAN measures ECG, EDA, EMG, temperature, accelerometer data, and respiratory signal at a 700 Hz sampling rate. Empatica E4 records accelerometer data sampled at 32 Hz, temperature sampled at 4 Hz, BVP sampled at 64 Hz, and EDA sampled at 4 Hz.

The participants were measured under three conditions: baseline, stress, and amusement. The baseline setting includes 20 minutes of sitting/standing/reading magazines. The stress setting was based on the Trier Social Stress Test (TSST), which includes public speaking and mental arithmetic tasks [129]. The amusement setting includes watching a set of funny video clips. A baseline was first measured and followed by a stress setting and an amusement setting with a meditation interval in between. The stress and amusement settings were interchanged for each participant. Another meditation period was measured at the end of the experiment. The devices recorded a total of 100 minutes for each participant.

In accordance with a considerable amount of work using the WESAD dataset, we selected the baseline and stress conditions for binary classification [17, 19, 32].

Computation Platform We performed signal preprocessing, feature extraction, missing value imputation, feature selection, and lightweight classification on a MacBook Pro with a 2.6 GHz Dual-Core Intel Core i5 processor and 8 GB 1600 MHz DDR3 memory. Image classification in the cloud was performed on the Google Colab Pro platform with a Tesla V100 GPU.

Training Setup As described in Sect. 3.1, all signals recorded with RespiBAN sampled at 700 Hz were downsampled to 100 Hz with normalization. Signal filtering, segmentation, and feature computation were performed using the Scipy and pyHRV packages [100]. Multiple imputation for recovering missing values was performed using Bayesian ridge regression implemented in the scikit-learn package. According to [130], we imputed $n = 5$ and $n = 9$ distinct datasets after introducing missing values at rates of 5% and 10%, respectively. Due to a limited number of samples, the synthetic minority oversampling technique was applied. Feature selection was performed based on [131].

In order to determine the best edge model, design-space exploration was performed on a recurrent SNN, a CondConv neural network, and an SVM. All the extracted features were used and shaped into a size of 10×10 as the input for the recurrent SNN and the CondConv neural network. The selected hyperparameters and values are listed in Table 4, where C is the regularization strength and γ is the parameter for RBF kernel. The range of the hyperparameters was determined based on [132] and [133]. The recurrent SNN and the CondConv neural network were trained using 70% of the samples with a batch size of 32 for 30 epochs. The number of training epochs was increased to 100 for the CondConv neural network with a larger architecture. The negative log-likelihood loss (l_{NLL}) shown in Eq. (20) was applied to both models, where $\hat{y}_i$ is the predicted value, y_i is the ground truth, and B is the batch size. Adam optimization [134, 135] with a learning rate of 10^{-3} was applied to both models. Hyperparameters for training these models were chosen based on the random search. The recurrent SNN was implemented using the Norse library [136]. We trained a leave-one-out SVM using the radial basis

Table 4 Design-space exploration on models with selected hyperparameter setting

Model	Hyperparameter	Test points
SVM	C	10 points linearly spaced in $[1, 10^4]$
	γ	20 points linearly spaced in $[10^{-9}, 10^{-8}]$
Recurrent SNN	No. of hidden neurons	10, 28, 46, 64, 82, 100
	No. of time steps	10, 15, 20, 25, 30, 35
CondConv neural network	No. of experts	1, 2, 3
	No. of CondConv layers	1, 2, 3
	No. of channels	3, 8, 16, 32, 64
	Kernel size	3, 5

function (RBF) kernel with tuned hyperparameters chosen via grid search using the scikit-learn package.

$$l_{\mathrm{NLL}} = -\frac{1}{B}\sum_{i=1}^{B} \log(\hat{y}_i) \cdot y_i \tag{20}$$

The CNN was trained using 70% of the samples with a batch size of 128 for 200 epochs. These hyperparameters were selected using random search. Data augmentation was performed with random affine transformation and random erasing to increase data diversity [137]. As a typical choice for binary classification, binary cross entropy (l_{CE}), as shown in Eq. (21), was selected as the loss function, where $\mathbb{P}$ is the true distribution of the data and $\mathbb{Q}$ is the estimated distribution.

$$l_{CE} = -\mathbb{P}\log_2 \mathbb{Q} - (1-\mathbb{P})\log_2(1-\mathbb{Q}) \tag{21}$$

Based on [128] and random search, the matching network was trained using 70% of the samples with a batch size of 128 for 400 epochs. Cosine similarity (l_{CS}), as shown in Eq. (22), was selected as the matching function to measure the distance between the prediction $\hat{y}$ and the ground truth y.

$$l_{CS} = \frac{\hat{y} \cdot y}{\|\hat{y}\|_2 \cdot \|y\|_2} \tag{22}$$

Adam optimization was applied to the CNN with learning rates of 1×10^{-3} for the respiratory signal and 1×10^{-4} for EDA and ECG. A weight decay of 4×10^{-2} was applied to the ECG model. The same optimization was applied to the matching network with learning rates of 1×10^{-4} for the respiratory signal, 3×10^{-4} for EDA, and 1×10^{-5} for ECG. These hyperparameters were tuned using random search, and all other hyperparameters were maintained at default settings from PyTorch.

Baseline Techniques/Architectures We select six state-of-the-art methods for stress-level classification as the baselines using the WESAD dataset for comparison. The baseline models include conventional CNNs [31, 48], a neural process model [32], a deep fusion network [29], a KNN [19], and an LDA [17]. The neural process model is a neural network that learns a distribution over functions and adapts to the given context. The deep fusion network is a neural network composed of multiple subnetworks, including CNNs and LSTMs.

For real-time simulation, we compared our work with models executed on only edge-based and cloud-based architectures. As stated in Sect. 3.1, we aim to show that our hierarchical model achieves a better trade-off between the overall accuracy and the cost than nonhierarchical models utilizing only edge or the cloud. Comparisons with other existing IoMT architectures are beyond the scope of this work.

3.3.2 Evaluation Metrics

By convention, we assess the performance of classifiers using the metrics of classification accuracy, precision, recall, and F1 score. Precision measures the percentage of correct predictions in the positive class and recall calculates the percentage of correct predictions in samples that are all predicted to be positive, as described in Eq. (23). The F1 score is a combined measurement of precision and recall values as shown in Eq. (24). The goal is to ensure high performance of the classifier.

$$\text{Precision} = \frac{TP}{TP + FP}, \ \text{Recall} = \frac{TP}{TP + FN} \tag{23}$$

$$\text{F1 score} = 2 \times \frac{\text{Precision} \times \text{Recall}}{\text{Precision} + \text{Recall}} \tag{24}$$

In addition, for the purpose of edge model comparison, we used the number of floating-point operations per inference sample (FLOPs) as the metric for model computational complexity. Based on [138], FLOPs for the SVM is $O(P \cdot N_s)$, where P is the dimensionality of input data and N_s is the number of support vectors utilized by the model. FLOPs for the recurrent SNN at layer l ($\text{FLOPs}^l_{\text{SNN}}$) is computed in Eq. (25) [139], where $\text{FLOPs}^l_{\text{ANN}}$ is FLOPs computed using the same layer architecture with non-spiking neurons and S^l_A is the average fraction of activated neurons at layer l in one time step. The FLOPs computation for the CondConv layer is based on Eq. (10).

$$\text{FLOPs}^l_{\text{SNN}} = \text{FLOPs}^l_{\text{ANN}} \cdot S^l_A \tag{25}$$

3.3.3 Performance Evaluation

Feature Selection As described in Sect. 3.1, a total of 100 features were computed. To test the robustness of our model, we introduced missing values at rates of 5% and 10% by convention [140, 141] and performed multiple imputation accordingly. Features were ranked based on the Fisher score, JMI, and DISR, accordingly. Considering the computational resource restriction at the edge layer, we evaluated the impact of feature size on the classification accuracy, in order to obtain the minimum number of features that yields the optimum accuracy. Figure 7 shows how the classification accuracy varies based on the top p features chosen using three selection methods, where p ranges from 3 to 15 to capture the growth in accuracy and ensure that convergence is reached. As the number of chosen features increases, the accuracy of the SVM increases and finally converges. We chose top ten features selected by each of these three methods to ensure that the SVM reaches its optimum accuracy under all three conditions. Selected features using different methods are listed in Table 5 in the order of their rankings.

Figure 7 shows that SVM based on only ten features can achieve an accuracy of around 85%. Although the accuracy is slightly worse compared to previous work [17, 29], we must also consider the limited computational resource available at the edge layer. Our model requires fewer features and thus is more feasible for implementation at the edge. The percentage contributions of each signal are presented in Fig. 8. We see that EDA and the respiratory signal are major contributors to the classification accuracy for all three feature selection methods. This observation is consistent with [17] and [29]. According to Table 5, we see

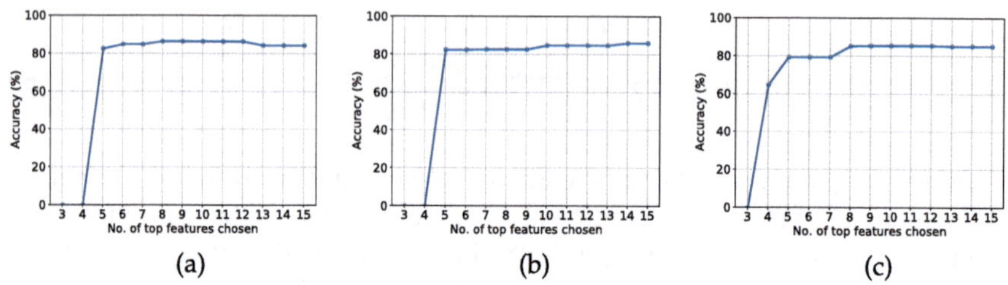

Fig. 7 Accuracy of SVM using features selected by different methods. (**a**) Feature selection using Fisher score. (**b**) Feature selection using JMI. (**c**) Feature selection using DISR

Table 5 Selected features using different selection methods

Methods	Selected features
Fisher score	Mean respiratory rate, inhalation std, SCR std, EDA AUC, SCR mean, BVP sum, ECG kurtosis, BVP mean, BVP skewness, SCR amplitude
JMI	Mean respiratory rate, inhalation std, SCR std, SCR mean, EDA AUC, Acc_x mean, BVP mean, ECG skewness, Acc_z std, respiratory signal skewness
DISR	mean respiratory rate, relative power, inhalation std, BVP sum, SCR std, ECG kurtosis, SCR mean, EDA AUC, normalized power, respiratory signal skewness

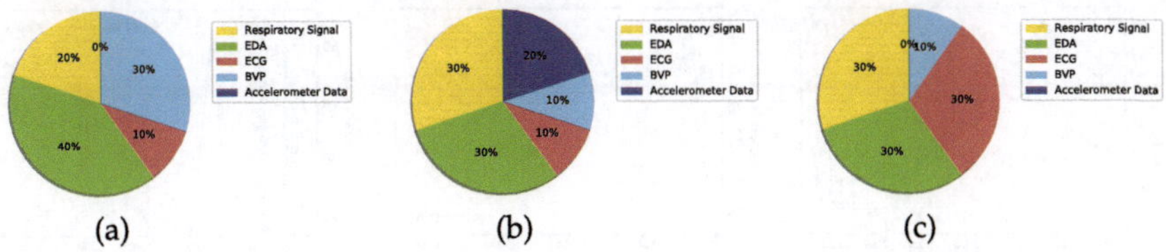

Fig. 8 Feature category contributions to the SVM performance [95]. (**a**) Feature selection using Fisher score. (**b**) Feature selection using JMI. (**c**) Feature selection using DISR

Table 6 Performance of SVM at edge before and after imputation

		Before imputation		After imputation	
Missing rate (%)	Method	Accuracy (%)	F1 score	Accuracy (%)	F1 score
0.22	Fisher	85.2041	0.8549	86.7347	0.8725
	JMI	82.3979	0.8253	81.6326	0.8172
	DISR	85.7143	0.8613	86.7347	0.8725
5	Fisher	71.9387	0.6646	76.2755	0.7223
	JMI	52.0408	0.1296	84.4388	0.8508
	DISR	80.6122	0.8041	85.4592	0.8613
10	Fisher	71.6836	0.6498	76.7857	0.7331
	JMI	58.9286	0.566	83.1632	0.8366
	DISR	80.8673	0.812	85.4592	0.8585

some key features, namely, mean respiratory rate, standard deviation of inhalation, the mean and standard deviation of SCR, and EDA AUC, selected by these three methods. This highlights the consistency of the importance of the top features between the different methods for stress-level classification.

The Fisher score is efficient for computation, but one major disadvantage associated with it is that it computes each feature independently without considering feature redundancy. We can also see from Table 6 that the Fisher score has the worst performance compared to the other two methods, when noise was introduced by multiple imputation at higher missing rates. By examining the distributions of the features, we found that the feature with similar intra-class and inter-class mean and variance can lead to a relatively high Fisher score, even though the feature introduces significant noise to the whole dataset, degrading the performance of the classifier.

JMI solves the problem of feature redundancy by measuring the amount of mutual information obtained by adding a new feature. DISR is considered to be a normalized form of JMI that measures the relative reduction of uncertainty when a new feature is added. Among these two information-based feature selection methods, we see from Table 6 that DISR is less impaired by missing values, and it generates a better result compared to that from JMI. We therefore conclude that DISR is a better method for physiological feature selection.

Design-Space Exploration on the Edge Models To optimize the performance of the edge model, we constructed a recurrent SNN and a CondConv neural

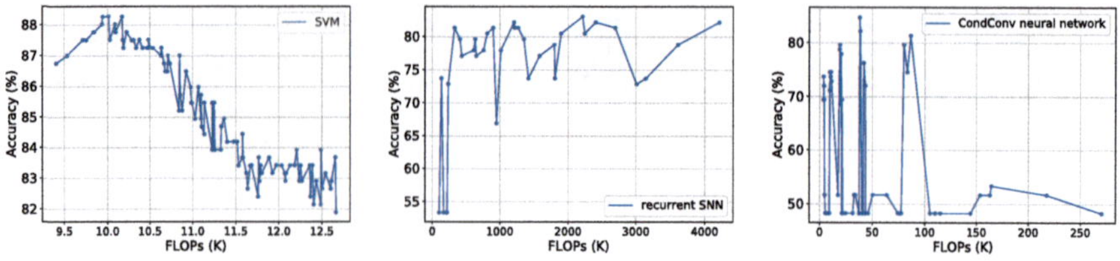

(a) (b) (c)

Fig. 9 Computational complexity and classification accuracy comparison between edge models [95]. (**a**) SVM performance. (**b**) Recurrent SNN performance. (**c**) CondConv neural network performance

network and compared them with the SVM model. We evaluate a model in terms of classification accuracy and model computational complexity as shown in Fig. 9. Points with high FLOPs and low accuracy, which indicate severe model overfitting were truncated. In Fig. 9a, we see that SVM can achieve a maximum accuracy of 88% with approximately 10,000 FLOPs. Utilizing fewer or more support vectors degrades classification performance. A reduced number of support vectors leads to a narrower margin, resulting in model overfitting. An increased number of support vectors result in a wider margin, increasing the number of misclassifications.

We present Fig. 9b for the recurrent SNN model. To identify the upper bound on the parameter S_A^l in Eq. (25) for the FLOPs associated with the recurrent SNN, we trained the model with each combination of the hyperparameters listed in Table 4 for a total of ten times. Based on the experiments, the upper bound on S_A^l is determined to be 0.6 for recurrent SNN architectures explored in our study, which is consistent with [139]. We see that for the recurrent SNN, increased computational complexity leads to increased accuracy, but large fluctuations in accuracy occur at low model complexity. We observe that the performance of the recurrent SNN with a small number of hidden neurons is affected by the high variance of the input spikes induced during data encoding. The effect of input data variance on model performance is reduced when the model complexity increases.

We also observe fluctuations in the accuracy for the CondConv neural network in Fig. 9c. This observation can be attributed to the fact that both the width and depth of the network are increasing during design-space exploration. An increase in the width or depth of a network increases model complexity, but the resulting predictions from the network show large variations for each individual sample [142]. Furthermore, as the number of model parameters increases, given the limited amount of training data, the model is not able to learn most of the parameters, resulting in low accuracy. When we compare the three models on the basis of computational complexity and accuracy, we find that SVM achieves the highest accuracy with the lowest computational complexity. Given a limited amount of training data, SVM obtains better performance compared to the other two data-hungry neural networks. Note that design-space exploration has to be conducted to re-evaluate the quality of a model when larger datasets are available.

Stress Classification at the Edge Based on the results of design-space exploration shown in Fig. 9 and the discussion in section "Design-Space Exploration on the Edge Models", we chose SVM as the edge model. In Table 6, we present the performance of the SVM classifier at the edge layer before and after missing value imputation using the top ten features selected by the Fisher score, JMI, and DISR. Signals are segmented in 90-second windows; some features cannot be calculated due to the lack of available patterns in the windows, resulting in a missing rate of 0.22%. According to Table 6, at the original missing rate, the features selected by the Fisher score yield the best accuracy of 86.7347% and an F1 score of 0.8725. At higher missing rates, the accuracy and the F1 score after imputation lead to at least a 5% increase compared to the scores obtained before imputation using all three methods. By comparing the performance at all the missing rates, we conclude that DISR outperforms the other two methods.

Stress Classification in the Cloud As indicated in Sect. 3.2, we established two neural network models for image classification using the respiratory signal, EDA, and ECG individually. Since ECG was segmented by R-peaks, a considerable number of images were generated in one 90-second window. To maintain consistency of the outputs from all three signals, we pooled ECG predictions based on the 90-second windows using a majority vote. Table 7 presents the evaluation results for the CNN and the matching network model. We see that the performance of the matching network using all three signals is superior to that of the CNN model. The CNN achieves comparable performance using ECG with an accuracy of 98.9247% and an F1 score of 0.9873. Additionally, the overall performance in the cloud significantly exceeds the results obtained at the edge.

We next compare the performance of our models in the cloud with six prior methods that utilize selected signals and labels from the WESAD dataset; the results are shown in Table 8. Considering the performance obtained from prior work that used the same evaluation metrics, our models in the cloud outperform prior methods with at least a 2% increase in the accuracy and the F1 score. Note that [31] utilized ECG signals from four datasets to train the model, therefore resulting in a higher accuracy compared to other prior work. Without comparing with [31], our models can achieve at least a 5% increase in accuracy and a 7% increase in the F1 score.

Table 7 Performance of CNN and matching network using respiratory signal, EDA, and ECG

	Models	Accuracy (%)	Precision	Recall	F1 score
Respiratory signal	CNN	94.6236	1	0.875	0.9333
	Matching network	96.4286	1	0.9091	0.9524
EDA	CNN	90.3226	0.9697	0.8	0.8767
	Matching network	96.4286	1	0.9091	0.9524
ECG	CNN	98.9247	1	0.975	0.9873
	Matching network	98.9247	0.9756	1	0.9876

Table 8 Performance comparison with prior work using WESAD

Reference	Signals	No. of classes	Accuracy (%)	F1 score	AUC	Precision
[31][a]	ECG	4	96.9	0.963	–	–
[29]	All	3	85	0.86	–	–
[48]	All	3	87.7	0.83	–	–
[32]	EDA, ECG	2	–	–	0.984	0.957
[19]	EDA	2	91.6	–	–	–
[17]	All chest	2	93.12	0.9147	–	–
Our work	Respiratory signal	2	96.4286	0.9524	–	–
	EDA	2	96.4286	0.9524	–	–
	ECG	2	98.9247	0.9876	–	–

[a]This work utilized ECG signals from four datasets (including WESAD) to train the model

Based on the performance of the models using different signals, we conclude that ECG serves as the best indicator for stress level using the matching network in the cloud. Table 7 shows that the matching network outperforms the CNN model for all three signals, indicating the robustness of the few-shot learning models under the scenario of a limited number of samples in each class. As mentioned earlier in this section, the number of ECG image samples is much greater than the number of EDA or respiratory signal image samples. A conventional CNN that requires large amounts of data for training can obtain comparable performance to the matching network model using the ECG signal, but the performance is worse when only the EDA and respiratory signals are used. This experimental finding is also supported by our theoretical explanation in Sect. 3.2.2.

In addition, we compare our work with prior work using different stress datasets. As introduced in Sect. 2, [49] and [50] achieved an accuracy of 92.05% and 94%, respectively, on binary stress classification. Our work outperforms the previous models with at least a 4% increase in accuracy and a 6% increase in the F1 score.

Trade-Off Between Overall Accuracy and Cost The results in Tables 6 and 7 show that inferencing using samples at the edge results in inadequate accuracy due to limited computational power; therefore, a subset of samples were sent to the cloud to obtain higher overall accuracy in our hierarchical architecture. In order to investigate the trade-off between the overall accuracy and the energy and communication costs during data transmission from the edge to the cloud, we sampled a series of decision thresholds to obtain the corresponding partitioned samples and the performance at the edge and the cloud, respectively. Here we use the CNN model with ECG input in the cloud as an example.

Based on empirical evidence from experimentation, we fed 30% of the data to the edge for inference. For each decision threshold, using the criteria defined in Sect. 3.2, samples with a score below the threshold were offloaded to the cloud for advanced classification. We performed lightweight classification using the rest of the samples that remained at the edge. Confusion matrices were computed at

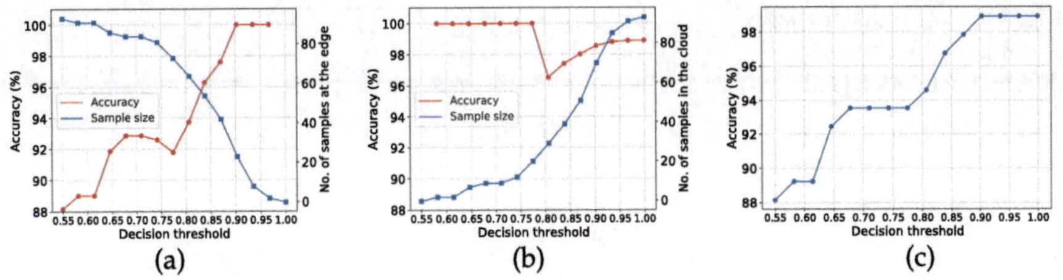

Fig. 10 Accuracy and sample partitioning obtained using ECG in the overall system and at the edge and in the cloud [95]. (a) ECG accuracy at the edge. (b) ECG accuracy in the cloud using CNN. (c) Overall ECG accuracy

Table 9 Upload bandwidth and data rate of different networks

	Bandwidth	Data rate
Wi-Fi	6.1 Mbps	759 KB/s
LTE	830 Kbps	104 KB/s
3G	55 Kbps	7 KB/s

both the edge and the cloud, and both these matrices were utilized for the overall performance assessment. The number of samples and the accuracy obtained at the edge and in the cloud, and their combined accuracy are presented in Fig. 10. We note that the accuracy drops slightly as the number of samples increases at both the edge and the cloud. As the decision threshold increases, more samples are sent to the cloud, resulting in an increase in the overall accuracy.

To compute the communication latency and the energy consumption during data offloading to the cloud, we first measured the upload bandwidth of 3G, Long-Term Evolution (LTE), and Wi-Fi using TestMy.net [143] on a mobile phone; see Table 9. For each decision threshold, we calculated the total size of the data offloaded to the cloud in KB by multiplying the number of samples by the size of a 90-second window signal sample (72 KB). We then divided the total cloud sample size by the upload Internet speed to obtain the communication latency. We computed the energy consumption based on the power model presented in [144]. Equation (26) describes the upload power model, where P_{upload} is the power level for uploading, α_u (mW/Mbps) and β_u (mW) are parameters with the values provided in [144], and t_u is the upload bandwidth in Mbps. The power value was then converted to J/s and multiplied with the communication latency at each decision threshold to determine the energy consumption during data transmission.

$$P_{\text{Upload}} = \alpha_u \cdot t_u + \beta_u \tag{26}$$

Figures 11 and 12 present the energy consumption and the communication latency during data transmission. As the decision threshold increases, the cost of using 3G increases rapidly, and the cost escalates at higher thresholds using LTE. Wi-Fi incurs the minimum energy and time cost at all thresholds compared to LTE

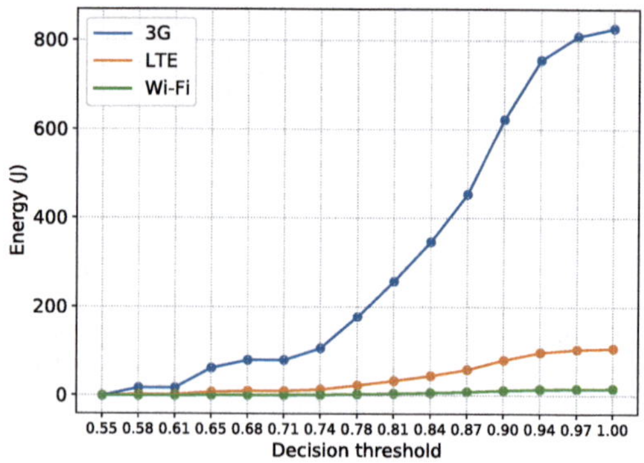

Fig. 11 Data transfer energy at different thresholds for different networks [95]

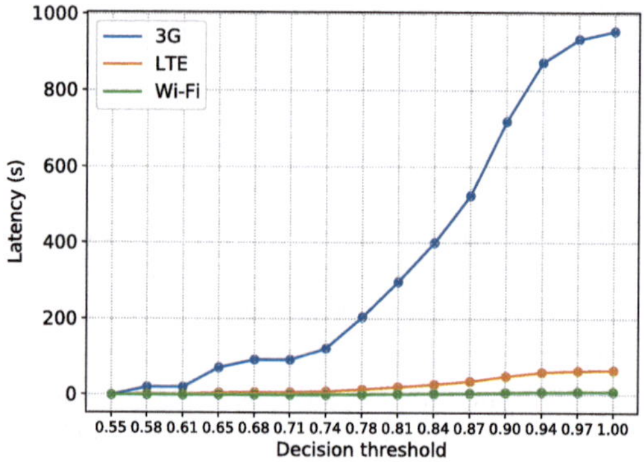

Fig. 12 Communication latency at different thresholds for different networks [95]

and 3G. We therefore assess the trade-off between the overall performance and the data transmission cost using only Wi-Fi.

We illustrate the trade-off between the overall accuracy, the communication latency, and the energy consumption during data transmission from the edge to the cloud layer. As shown in Fig. 13, the time and energy costs grow slowly at lower thresholds, but they grow faster at intermediate thresholds. The escalation of the cost declines at higher thresholds because only a few new samples are offloaded to the cloud at each of the high thresholds, according to Fig. 10b. The overall accuracy shares a similar trend with the cost as indicated in Fig. 10c. The decision threshold is determined by the point with the minimal communication latency and energy consumption given the overall system accuracy. To illustrate this selection, suppose our target is an overall accuracy of 93.55%; in this case, a threshold of either 0.68 or 0.71 should be chosen to achieve the minimum communication latency and energy consumption for data transmission.

Real-Time Stress Monitoring Simulation We have shown in section "Stress Classification in the Cloud" that the accuracy of our cloud models is higher than that of existing state-of-the-art methods. We also assessed the trade-off between the overall accuracy of our hierarchical model and the cost during data transmission between the edge and the cloud. To further highlight the merit of our solution

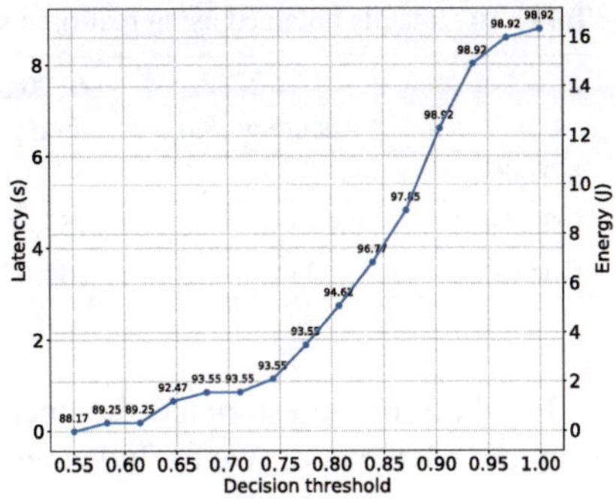

Fig. 13 Communication latency and energy costs at different thresholds with overall accuracy (%) labeled [95]

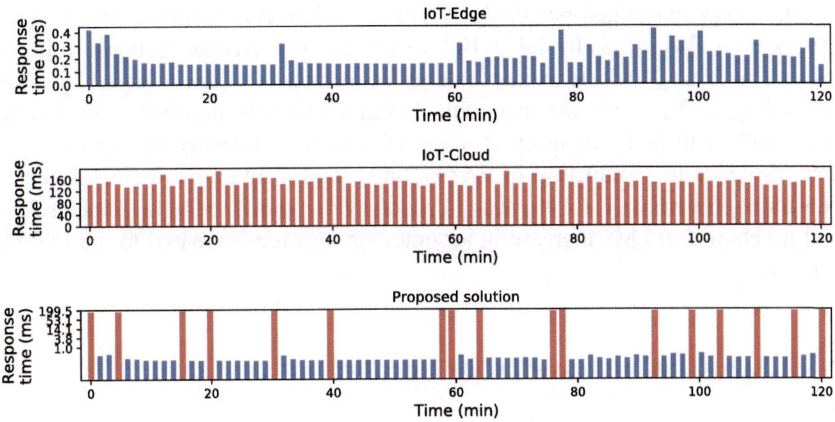

Fig. 14 Response time for different schemes based on 2-hour real-time simulation [95]

in terms of classification accuracy, response time of data inference, and energy consumption, we performed a real-time stress-level monitoring simulation in a 2-hour window. We next compare our hierarchical solution with the use of only the edge-based or the cloud-based models. We randomly selected 80 samples with each sample being a 90-second window containing stress information from the test set described in Sect. 3.3.1. Inference using the samples was conducted under IoT-edge (edge-based model using SVM), IoT-cloud (cloud-based model using CNN), and our hierarchical scheme with an offloading threshold of 0.68 selected as described in section "Trade-Off Between Overall Accuracy and Cost" based on Fig. 13. We measured the response time of each sample in the 2-hour window for each scheme as shown in Fig. 14. The response time for each sample inferred in the cloud includes data transmission latency using Wi-Fi computed from Eq. (26). Fluctuations in the response time in IoT-edge are caused by other task executions on the CPU. To minimize this effect, we measured the response time ten times and recorded the lowest time for each sample.

Table 10 Results obtained using real-time simulation under different IoT architectures

	Accuracy (%)	Average response time per sample (ms)	Average energy consumption per sample (μJ)
IoT-edge	86.25	0.21	0.43
IoT-cloud	98.75	152.46	176.08
Our work	91.25	33.65	37.75

In addition, we estimated the energy consumption based on FLOPs computed from the edge model. We excluded energy consumption in the cloud but considered energy consumed during data transmission for the IoT-cloud scheme. Based on [145], energy consumed per double-precision arithmetic operation on an x86-64 processor is 387 pJ/op. In Table 10, we present the average response time, the average accuracy, and the average energy consumption for all three architectures. According to Table 10, the hierarchical edge-cloud scheme obtains an accuracy of 91.25% with high efficiency in terms of latency and energy consumption. Our solution leads to a reduction of 77.89% in response time and 78.56% in energy consumption with only a 7.6% drop in accuracy compared to the IoT-cloud scheme and it achieves a 5.8% increase in accuracy on average compared to the IoT-edge scheme.

3.3.4 Open Challenges and Future Directions

There still exist some unresolved challenges in current solutions to remote stress monitoring. In this section, we discuss potential future directions/aspects to advance the development of next-generation stress monitoring.

Data and Label Acquisition It is widely acknowledged that wearable data can contain a lot of artifacts. Sweat and intense body movement can lead to low-quality data or missing data. Improvement in biomarker sensors for more precise data acquisition is therefore crucial. Instead of sensing conventional biomarkers such as ECG and EDA, which are indirect causes of the stress response, Samson and Koh [146] introduced biosensors that detect saliva and sweat cortisol for stress monitoring for more precise sensing. Ates et al. [147] presented an electronic skin that can sense physiochemical signals from sweat with an extended drift-free measurement period. Additionally, the corresponding stress labels used in the literature are mostly binary and can be inaccurate [148]. Creating a higher level of granularity in data or using other precise measurements, such as the cortisol level [149], as the ground truth is necessary for better indication of stress levels.

Real-Time Stress Monitoring Most of the work did not consider stress monitoring in the IoT context. Nath et al. [150], in their survey, introduced limited existing literature on real-time stress monitoring leveraging edge computing with hardware implementation. Although most of the work reported latency, only one study reported power consumption. Similarly, Gedam and Paul [151] also indicated the lack of real-time stress monitoring literature. Developing a more comprehensive model evaluation under the IoT framework is thus essential for the deployment of stress monitoring in practice.

Additional Aspects Besides the challenges described above, there are many other open issues on model generalizability [148], availability of large-scale public datasets [148, 151], stress monitoring in different environment settings outside of controlled lab experiments [152], wearable data heterogeneity (inter-user and intra-user variances) [153], multimodal data fusion [153], and clinical evaluation [153]. Further investigations into the challenges mentioned above are required to establish a fully functional and automated real-time stress monitoring system.

4 Low-Overhead Clustered Federated Learning for Personalized Stress Monitoring

Since data is collected and trained locally on each device, and each person has distinct physiological patterns, FL has to address the problem of data heterogeneity caused by nonindependent identically distributed (non-i.i.d.) data. Previous FL methods that rely on aggregation over all client model parameters lead to significant performance degradation in some of the clients [45]. To address this problem, researchers have proposed many approaches. One such method groups clients based on their similarity [83]. The similarity can be measured either based on the data or the model. Duan et al. [154] suggested the clustering of clients based on distances between the gradients of local models using a novel measure. Ouyang et al. [155] proposed to use an indicator matrix and update it via alternating optimization during training to cluster clients. Kim et al. [84] also developed an algorithm to estimate the optimum number of clusters required for FL. However, existing clustering methods require predefining the number of clusters to form (i.e., set a threshold), of which we have no prior knowledge. Furthermore, there does not exist an optimal way of defining similarities between the clients. We thus present an algorithm that heuristically estimates the optimum number of clusters and optimizes clustering utilizing ClusterGAN.

In addition to data heterogeneity, high overhead in terms of overall training time is another limitation of FL. Frequent model parameter updates between clients and the central server and large numbers of model parameters are the major cause of high overhead. To mitigate this limitation, Liang et al. [156] proposed to first learn local representations of the raw data via a deep learning model and update a subset of the model parameters to improve individual performance and to reduce overhead.

However, training overhead for cluster-based FL is still significant. Furthermore, despite the reduction in the impact of data heterogeneity, existing clustered FL algorithms do not provide personalized models. Studies on healthcare applications based on FL suffer from the same deficiency. To the best of our knowledge, cluster-based FL algorithms for stress-level evaluation have yet to be studied. In addition, the cold-start problem, i.e., when a new client joins an existing system, has not received much attention. In this section, we describe a novel dynamic clustering FL algorithm for stress monitoring with the following highlights:

- We utilize heuristic data-based clustering techniques and transfer learning for developing personalized stress-level classification models.
- We present a novel cold-start algorithm based on the two-sample test that achieves significant performance with tolerable runtime.
- We show a personalized, communication-efficient dynamic clustered FL algorithm that significantly improves the accuracy and F1 score in the stress classification task by over 7% compared to two state-of-the-art baseline methods.
- We obtain a reduction of 37.5% and 9.6% in the training runtime compared to the state-of-the-art dynamic clustering algorithm and DF_FL algorithm, respectively.

The remainder of the section is organized as follows: Sect. 4.1 introduces related work on generative adversarial networks and two-sample hypothesis testing. Section 4.2 describes our dynamic clustered FL algorithm for stress classification; Sect. 4.3 presents the analysis of the task performance, overhead, and comparison with baselines.

4.1 Related Prior Work

4.1.1 Cluster Generated Adversarial Network (ClusterGAN)

Mukherjee et al. [157] suggested a novel generative adversarial network (GAN) architecture named ClusterGAN (see Fig. 15) that learns a better clustering in the latent space. A conventional GAN is composed of a generator **G** and a discriminator **D**. The generator generates artificial samples x_g using latent variables z sampled from the latent space, and the discriminator distinguishes between generated samples and real samples x_r. The goal of ClusterGAN is to obtain a projection of synthetic data on the latent space $\hat{z}$ so that $\hat{z}$ reveals a cluster structure. Figure 15 illustrates the overall ClusterGAN architecture.

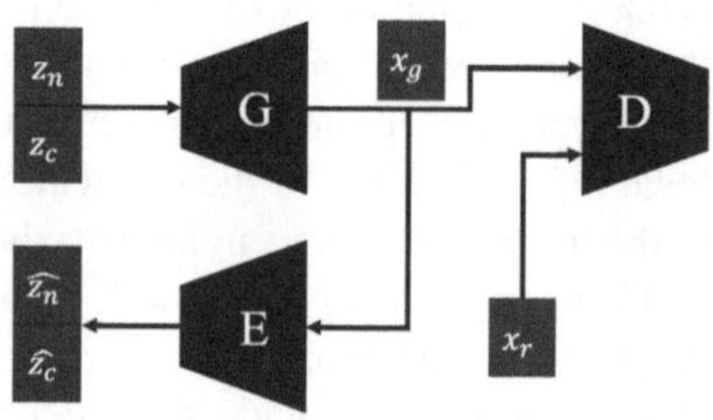

Fig. 15 ClusterGAN architecture [81]

The latent variables z are composed of a continuous variable $z_n \sim \mathcal{N}(0, \sigma^2 I)$ and a discrete, one-hot variable $z_c = e_k, k \in \{1, \ldots, K\}$, where σ^2 is the variance, I is the identity matrix, e_k is the kth one-hot vector, and K is the number of clusters. In addition to conventional GAN training, they introduced an encoder $\mathbf{E}$ to project synthetic data on the latent space. We express the new objective formally in Eq. (27), where $l(\cdot)$ is the cross-entropy loss, and β_n and β_c are regularization coefficients.

$$
\begin{aligned}
\min_{\Theta_{\mathbf{G}}, \Theta_{\mathbf{E}}} \max_{\Theta_{\mathbf{D}}} \; & \mathop{\mathbb{E}}_{x \sim \mathbb{P}_x^r} \mathbf{D}(x) + \mathop{\mathbb{E}}_{z \sim \mathbb{P}_z} (1 - \mathbf{D}(\mathbf{G}(z))) \\
& + \beta_n \mathop{\mathbb{E}}_{z \sim \mathbb{P}_z} ||z_n - \mathbf{E}(\mathbf{G}(z_n))||_2^2 \\
& + \beta_c \mathop{\mathbb{E}}_{z \sim \mathbb{P}_z} l(z_c, \mathbf{E}(\mathbf{G}(z_c)))
\end{aligned} \tag{27}
$$

4.1.2 Nonparametric Two-Sample Tests

In statistics, people often use nonparametric two-sample tests to verify if the given two samples come from the same distribution. One advantage of such tests is that they require no assumption on the distributions of samples. Rabanser et al. [158] suggested identifying data distribution shifts by introducing nonparametric two-sample tests. They proposed to embed the given source and target data samples into a latent space via neural networks. In addition to the traditional Kolmogorov-Smirnov (KS) and chi-squared tests, they utilized maximum mean discrepancy (MMD) as a test statistic for the embedded data. They performed extensive experiments in identifying data shifts on MNIST and CIFAR-10 datasets.

Given x_1 and x_2 sampled independently from distribution $\mathbb{P}$, and x_1' and x_2' sampled independently from distribution $\mathbb{Q}$, we formally define MMD^2 in Eq. (28), where $k(\cdot)$ is the kernel function [159]. Liu et al. [160] proposed a kernel-based two-sample test built upon deep learning models. By extending the work of [159], they optimized the simple Gaussian kernel and provided a theoretical guarantee, enhancing the test power using MMD.

$$
\text{MMD}^2[\mathbb{P}, \mathbb{Q}] = \mathop{\mathbb{E}}_{x_1, x_2 \sim \mathbb{P}} [k(x_1, x_2)] - 2 \mathop{\mathbb{E}}_{x_1, x_1' \sim \mathbb{P}, \mathbb{Q}} [k(x_1, x_1')] + \mathop{\mathbb{E}}_{x_1', x_2' \sim \mathbb{Q}} [k(x_1', x_2')] \tag{28}
$$

4.2 *Our Method*

4.2.1 Overall Clustered Federated Learning Framework

As discussed at the beginning of Sects. 4 and 4.1, existing cluster-based FL algorithms can reduce performance degradation caused by data heterogeneity, but

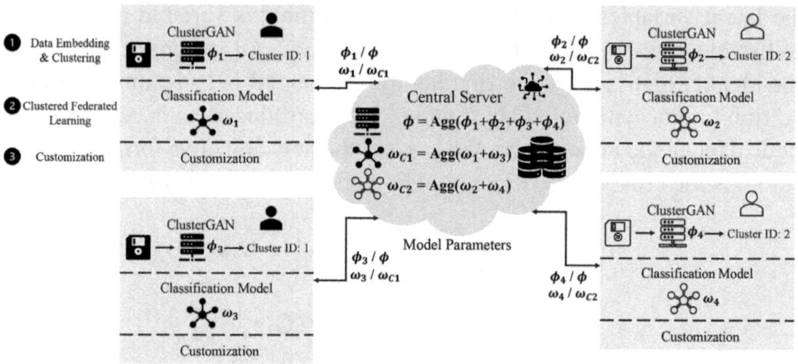

Fig. 16 Overall architecture for customized clustered federated learning [81]

they suffer from high training overhead due to the need for extensive communication between local clients and the central server. Additionally, previously introduced methods do not provide personalized models and they do not focus on stress monitoring. We provide a personalized solution to stress-level monitoring with low training overhead. We explain a data-driven clustered federated learning algorithm, as shown in Fig. 16. Note that $\boldsymbol{\phi}$ represents the aggregated global ClusterGAN model parameters, and ω_{C1} and ω_{C2} represent aggregated model parameters within Clusters 1 and 2, respectively. It consists of three modules: cluster group estimation, clustered federated learning, and personalization module. The cluster group estimation utilizes ClusterGAN and K-means clustering. ClusterGAN provides remarkable interpretability in the latent space compared to conventional dimension reduction techniques [157]. In addition, ClusterGAN yields high data separability in the latent space and is thus promising for clustering [157]. We first train federated ClusterGAN and assign a cluster ID to each client based on the embedded data from the ClusterGAN. The assigned cluster ID is obtained by estimating the best-quality cluster groups. We next perform clustered federated learning based on the cluster ID, conducting FedAvg [36] for clients within each cluster. After the training is complete, we execute the personalization module. We freeze the parameters of the bulk layers of the client model and train on the weights of the last few layers. We provide explanations for abbreviations used in the presented algorithm in Table 11.

To explore the power of our algorithm, we develop different variations based on clustering criteria and the level of personalization. We next define each variation and present detailed descriptions for each step in the following sections.

Table 11 Abbreviations used in the presented algorithm

Abbreviation	Definition
$NumCluster$	Number of clusters to form
$NumRound$	Number of communication rounds
$i.data$	Data of client i
$i.encoder$	Encoder from ClusterGAN of client i
$i.data_{enc}$	Encoded data of client i
$i.model$	Classification model of client i
$BestIndex$	Current best cluster ID assignments
C^{id}	List of selected clients C with a given cluster ID
μ	List of means within each cluster
σ	List of standard deviations within each cluster
$Update$	Indicator of change of $NumCluster$ and corresponding cluster ID assignments
$N.loss$	List of client loss for all N clients
$SavedNModels.endLayer$	Last layer of each of N saved client classification models

Algorithm 1: Data-driven cluster number estimation

```
Input: N clients
Output: ClusterID for all N clients
initialize(MaxScore, BestIndex, NumCluster, Count, Patience);
while NumCluster ≤ N do
    ClusterID = [ ];
    for i ∈ N clients do
        i.encoder ← TrainClusterGAN(i.data);
        i.data_enc. ← i.encoder(i.data);
        labels ← KMeans(i.data_enc., NumCluster);
        ClusterID ← Mode(labels);
    s ← SilhouetteScore([i.data_enc., i ∈ N], ClusterID);
    if s > MaxScore then
        MaxScore ← s;
        BestIndex ← ClusterID;
    else
        Count += 1;
    if Count ≥ Patience then
        break;
    NumCluster += 1
```

Algorithm 2: Clustered federated learning

```
Input: N clients, Server, ClusterID, NumCluster
Output: Trained N clients, NumRound
initialize(i.model for i ∈ N clients);
synchronize(Server.model);
initialize(loss_log, δ, Count, Patience, FLAG);
// Execute only for model-driven and hybrid clustering
initialize(μ_min, σ_min, Update);
while FLAG == False do
    if Update then
        NumCluster += 1;
        ClusterID ← ClusterUpdate(N, NumCluster);
    C ← SelectClients(N);
    for c ∈ C clients do
        LocalTraining(c.model);
    for id in Unique(ClusterID) do
        C^id ← clients from C belong to cluster id;
        AggregateAndUpdate(Server.model, C^id.model);
    μ, σ ← ObtainClusterStat(N, ClusterID);
    if Sum(μ) > μ_min & Sum(σ) > σ_min & NumCluster < N then
        Update = True;
    else
        μ_min, σ_min ← Sum(μ), Sum(σ);
        Update = False;
    loss_client ← N.loss;
    if loss_log − loss_client > δ then
        loss_log ← loss_client;
        SaveClientModel(N);
    else
        Count += 1;
        if Count ≥ Patience then
            FLAG ← True;
    NumRound += 1;
/* Client Customization */
FreezeFrontLayers(SavedNModels);
LocalTraining(SavedNModels.endLayer);
SaveClientModel(N);
```

4.2.2 Variations on Clustered Federated Learning

To highlight the robustness of our method, we present variations of the algorithm introduced early in the section. We explore three different clustering criteria for the algorithm: data-driven, model-driven, and hybrid. We also investigate the impact of the level of customization on model performance.

Clustering Criteria

Data-Driven Clustering Data-driven clustering clusters clients purely based on the similarity between encoded data generated from the encoder of ClusterGAN at each local client. We first find the optimum number of clusters as described in Algorithm 1. We then perform clustered federated learning shown in Algorithm 2 with a fixed number of clusters. Details of the two algorithms are described in Sect. 4.2.3.

Model-Driven Clustering Model-driven clustering clusters clients purely based on the similarity between client model weights. As shown in Algorithm 3, we first generate a distance matrix where the distance between the models is the sum of cosine similarity between model weights at each layer. The formula for cosine similarity ($\text{Dist}_{\text{Cos.}}$) is shown in Eq. (29), where ω_1 and ω_2 are layer-wise model weights in our context. After obtaining the distance matrix, we perform spectral clustering according to [161, 162]. We next assign a cluster index to each client based on the initial client models.

$$\text{Dist}_{\text{Cos.}} = \frac{\omega_1 \cdot \omega_2}{\|\omega_1\|_2 \cdot \|\omega_2\|_2} \tag{29}$$

Algorithm 3: Model-driven clustering

```
Input: N clients, NumCluster
Output: ClusterID for all N clients
/* Model-driven Clustering                                                  */
distMatrix ← ComputeSimilarity(N.model);
/* Compute model similarity based on the sum of cosine similarity
   between client models at each layer                                      */
ClusterID ← SpectralClustering(NumCluster, distMatrix);
initialize(MaxMu, MaxSigma, Update);
/* Execute after clustered FL update in each communication round           */
for id in Unique(ClusterID) do
    μ_loss ← i^id.loss, i ∈ N^id;
    σ_loss ← i^id.loss, i ∈ N^id;
    if Sum(μ_loss) > MaxMu and Sum(σ_loss) > MaxSigma and NumCluster < N then
        Update ← True;
        NumCluster += 1;
        /* Update ClusterID based on model-driven clustering                */
    else
        MaxMu ← Sum(μ_loss);
        MaxSigma ← Sum(σ_loss);
        Update ← False;
```

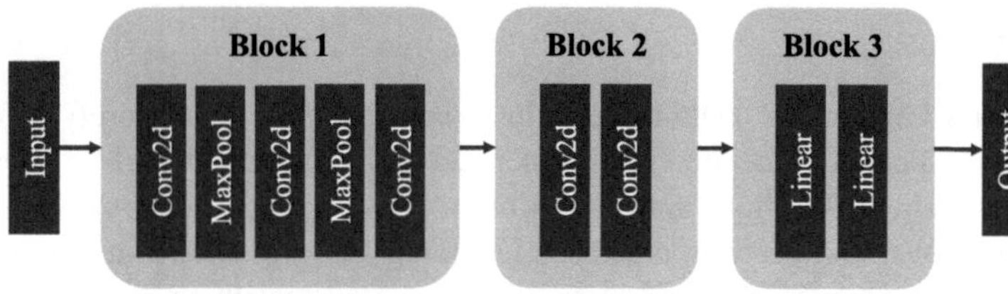

Fig. 17 Division of the client classification model into three blocks [81]

Hybrid Clustering Data-model-hybrid clustering assigns an initial cluster index to each client using data-driven clustering; see Algorithm 1. During clustered federated learning training, we dynamically update cluster numbers and assignments based on the performance of client models shown in Algorithm 2. We assign cluster indices during each communication round based on the output from the ClusterGAN encoder and K-means clustering.

Level of Personalization To investigate the impact of model customization on model performance, we evaluate three schemes. We split the classification model into three blocks as shown in Fig. 17. We define $Block1$ as the first five layers of the model with three convolutional layers and two max-pooling layers. We define $Block2$ as the rest of the two convolutional layers. $Block3$ is defined as two fully connected layers at the end of the model. Note that the batch normalization and activation layers after each main layer mentioned above are included in each block.

Scheme 1 Weights from $Block1$ and $Block2$ are averaged over all clients. Weights from $Block3$ are averaged cluster-wise.

Scheme 2 Weights from $Block1$ are averaged over all clients. Weights from $Block2$ are averaged cluster-wise. Weights from $Block3$ are trained client-wise.

Scheme 3 Weights from $Block1$ and $Block2$ are averaged cluster-wise. Weights from $Block3$ are trained client-wise.

4.2.3 Clustered Federated Learning

Building upon [157], we estimate the optimum number of clusters based on data from all clients. In Algorithm 1, we iteratively search for the best cluster number. We first train a ClusterGAN at each client, aggregate and update the GAN model for all clients, and collect encoded data via the trained encoder of the ClusterGAN. We then perform K-means clustering to determine the cluster ID for each client based on the majority of labels assigned to the encoded data. We evaluate the quality of the clusters utilizing the mean silhouette coefficient [163]. We record the cluster number that yields the best silhouette coefficient. In addition, we stop the search if the silhouette coefficient does not improve for $Patience$ rounds (i.e., the number of iterations to wait before stopping the search). The overall time complexity of Algorithm 1 is $O(N \cdot (T_{NN}(\text{GAN}) + M \cdot n_{\text{iter}} \cdot K))$, where N is the number of clients, $T_{NN}(\text{GAN})$ is the time complexity of training a ClusterGAN, M is the data size, and n_{iter} and K are the number of iterations and the number of clusters for K-means clustering, respectively. Details for computation of $T_{NN}(\cdot)$ for a given neural network architecture are explained in [164].

After obtaining the cluster ID for each client from Algorithm 1, we execute Algorithm 2 to conduct clustered FL with client customization. We randomly select C clients, a portion of all N clients, and perform local model training. We adopt the CNN model architecture for stress-level classification from our previous work [95]. According to the assigned cluster ID, we perform FedAvg on client models that belong to the same cluster. We evaluate model convergence by introducing an early stopping mechanism. When the loss from all clients stops to decrease by the amount δ for $Patience$ rounds, we claim the models reach convergence and record the number of communication rounds $NumRound$. We log the best models during training. The time complexity of Algorithm 2 is $O(n_{\text{iter}} \cdot T_{NN}(\text{CNN}) + N \cdot P)$, where n_{iter} is the number of communication rounds, $T_{NN}(\text{CNN})$ is the time complexity of training a client classification model, and P is the number of parameters in the classification model.

Note that for model-driven and hybrid clustered FL, additional steps are required to update the number of clusters, highlighted in brown in Algorithm 2. During each communication round of clustered FL, we evaluate the need to update cluster numbers and assignments based on within-cluster mean and variance of current training loss obtained from each client based on assigned cluster ID from Algorithm 3 and Algorithm 1 for model-driven and hybrid methods, respectively. The time complexity of executing model-driven clustering is $O(N^3 \cdot P)$.

In addition, to enable client personalization, we retrieve the best client models saved in the earlier steps and freeze the majority layers in the client model. We initialize the last two layers of the model and retrain the layers for a few epochs. We then return the personalized client model at the end of this procedure.

4.2.4 Cold-Start Problem

We have explained the training process for all clients earlier in this section. However, we need to consider the situation when a new user joins the trained system. This situation is referred to as the cold-start problem. One naive approach is to train the new client in the same way as existing clients, but the training overhead for this approach is considerable. This problem has not received much attention in the literature. We present a low-overhead novel cold-start algorithm based on a two-sample test; see Algorithm 4.

We first send the trained ClusterGAN to the new client and extract the encoded data $(N + 1).data_{enc.}$ from the encoder. Encoded data $i.data_{enc.}$ is also collected from each existing i client. For each existing cluster, we perform a two-sample test between $(N + 1).data_{enc.}$ and $i.data^{id}_{enc.}$ that belongs to the current cluster using

Algorithm 4: Cold-start algorithm

Input: Trained N clients, Trained ClusterGAN, $ClusterID, NumRound, \alpha$
Output: Trained $(N+1)^{th}$ client
initialize($(N+1)^{th}$ client);
$(N+1)^{th} client \leftarrow Send(ClusterGAN)$;
for $i \in N+1$ *clients* **do**
  $i.data_{enc.} \leftarrow ClusterGAN(i.data)$;
for *id in Unique(ClusterID)* **do**
  $i.data^{id}_{enc.} \leftarrow$ randomly pick encoded data from N^{id};
  p-value $\leftarrow MMDTest(i.data^{id}_{enc.}, (N+1).data_{enc.})$;
  if p-value $\leq \alpha$ **then**
    $(N+1).id \leftarrow id$;
    $(N+1).model \leftarrow Copy(i.model^{id})$;
    $ClientCustomization((N+1).model)$;
  break;
if *all* p-value $> \alpha$ **then**
  $LocalTraining((N+1).model)$;

MMD as test statistic defined in Eq. (28). If the p-value is significant, we assign the new client to the current cluster and load the cluster model to the new client. Client customization is performed at the end. The time complexity of Algorithm 4 is $O(T_{NN}(Enc))$, where $T_{NN}(Enc)$ is the time complexity of inference using the encoder of the ClusterGAN.

If none of the p-values from all clusters is significant, it suggests data from the new client has a different distribution from all other existing clients. We, therefore, assign the new client to a new cluster and train the model locally.

4.3 Results

4.3.1 Experimental Setting

Dataset Following our prior work [95], we used the wearable stress and affect detection (WESAD) dataset [165]. This dataset includes wearable data recorded from participants each with a chest-worn device called RespiBAN and a wristband called Empatica E4. A total of 15 participants were involved in the study and were asked to perform a series of tasks that are known to be associated with different levels of stress: amusement, stress, and baseline [129]. These stress levels were used as labels for the data. The RespiBAN records electrocardiogram (ECG), electrical dermal activities (EDA), electromyogram (EMG), respiratory signal, temperature, and accelerometer data with a sampling rate of 700 Hz. In this study, we used the ECG signal sampled from RespiBAN and labeled with stress and baseline only.

Evaluation Metrics We report accuracy and F1 score in two cases as follows:

- Client accuracy/F1 score: accuracy/F1 score of a single client
- Overall accuracy/F1 score: mean accuracy/F1 score across all clients

In addition, we report the total training time for our algorithm. The total training time is composed of the following:

- Client training time: the maximum time required to train a single client
- Communication time: the maximum time needed to upload and download parameters of a single client model to the server
- Model aggregation time: time needed to compute the global model parameters
- Cluster update time: total time needed to assign and update a cluster ID to each client

Hyperparameter Selection We utilized an R-peak detection algorithm [100] to detect ECG cycles based on the locations of R-peaks. We then converted each ECG cycle to an image of size 128 × 128. We set the train/test split for data in each client to be 70%/30%.

We present the employed ClusterGAN architecture in Table 12 with the generator, the discriminator, and the encoder listed, respectively. For convolutional layers and transposed convolutional layers, *c_out* is the number of output channels, *k* is the kernel size, and *s* is the stride. Values in brackets of the linear layers indicate output dimensions. z_n and z_c represent dimensions of sampled continuous and discrete latent space data, respectively, as introduced in Sect. 4.1.1.

Table 12 Details of the ClusterGAN architecture

Generator
Linear (1024)
Linear (32×15×15)
ConvTranspose2d (c_out = 32, k = 2, s = 2)
ConvTranspose2d (c_out = 32, k = 2, s = 2)
ConvTranspose2d (c_out = 1, k = 2, s = 2)
Discriminator
Conv2d (c_out = 32, k = 4, s = 2)
Conv2d (c_out = 32, k = 5, s = 2)
Conv2d (c_out = 32, k = 4, s = 2)
Linear (1024)
Linear (1)
Encoder
Conv2d (c_out = 32, k = 3, s = 2)
Conv2d (c_out = 32, k = 3, s = 2)
MaxPool2d (k = 2)
Linear (1024)
Linear ($z_n + z_c$)

We searched z_n from 5 to 40, based on [157], with a step size of 5 and set $z_n = 10$ for optimized performance. We initialized $z_c = 2$ as the smallest possible number of clusters to begin with for the iterative search in Algorithm 1. Regularization coefficients β_n and β_c are set to 10, following prior work [157]. We trained the ClusterGAN at each client with a batch size of 32 for 5 epochs utilizing stochastic gradient descent (SGD) with a learning rate of 10^{-5}, where these hyperparameters are optimized via random search.

We trained the classification model for each client with a batch size of 32 for 1 epoch at each communication round with a learning rate of 10^{-4} using SGD based on our prior work [95]. As explained in Algorithm 2, we terminate clustered FL via early stopping when we achieve model convergence. We chose $Patience = 10$ and $\delta = 0.1$, which are optimized via random search to determine model convergence. For the cold-start experiment, we set significance level $\alpha = 0.05$ by convention.

All experiments were executed using the Scikit-learn [166] and the PyTorch package [167].

Platform All experiments were executed using two NVIDIA GeForce RTX 2080Ti each with 11 GB RAM and two NVIDIA TITAN V each with 12 GB RAM.

Baselines To demonstrate that our algorithm is robust and efficient, we compare our method with two baseline methods from prior work.

- Federated Dynamic Clustering [84]: A clustered federated learning scheme that utilizes ClusterGAN to assign cluster indices to clients, performs cluster calibration, and executes cluster division to update the number of clusters.
- Dynamic-Fusion Federated Learning (DF_FL) [168]: An extension on the FedAvg algorithm that selects participating clients based on local training time at each communication round and performs aggregation based on client model performance and server waiting time.

4.3.2 Comparison with FL Baselines

In Fig. 18, we present the performance comparison between our data-driven clustered FL with personalization and the baseline algorithms described in section "Baselines." Our algorithm achieves over a 7% increase in both the accuracy and the F1 score compared to the dynamic clustering algorithm. Our method outperforms the dynamic-fusion federated learning algorithm (DF_FL) with over 12% increase in both the accuracy and the F1 score. The dynamic clustering algorithm calibrates and updates cluster groups during federated training, but our algorithm first finds an optimum number of clusters and then performs clustered FL. Our strategy thus achieves better performance than dynamic clustering. Since DF_FL does not involve client clustering, averaging parameters across all client models would lead to performance degradation on clients with shifted data distributions.

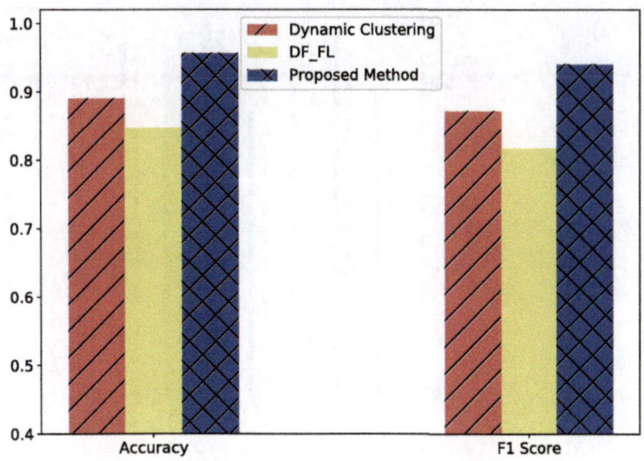

Fig. 18 Global classification accuracy and F1 score comparison with baseline methods [81]

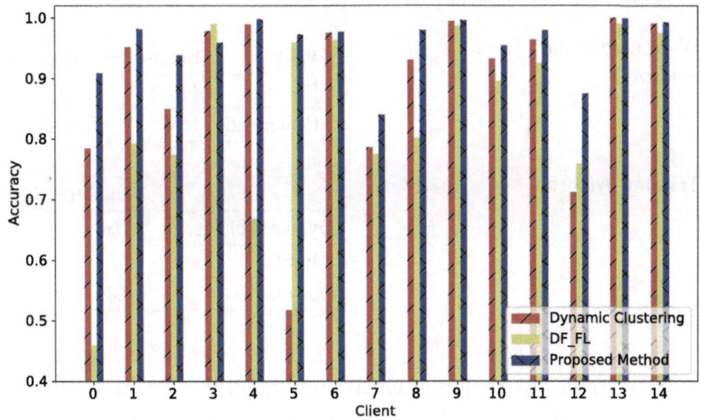

Fig. 19 Individual accuracy of dynamic clustering, DF_FL, and our method [81]

We further explore the performance comparison between dynamic clustering, DF_FL, and our method by investigating individual accuracy and F1 score presented in Figs. 19 and 20. We observe that the accuracy and F1 score of five clients (clients 0, 2, 5, 7, and 12) using the explained method are notably higher than that for dynamic clustering. The performance of six clients (clients 0, 1, 2, 4, 8, 12) using our method significantly exceeds the results obtained using DF_FL. Our results therefore show that our method improves individual performance compared to existing state-of-the-art methods. We thus conclude that estimating optimal cluster groups before federated training leads to superior performance.

We next compare the total training time needed for our method and the baselines. As described in section "Evaluation Metrics", we recorded the time needed for each step of the algorithm. We estimated the communication time based on our prior method in [95]. We measured the upload and download bandwidth of Wi-Fi via TestMy.net [143] and present these numbers in Table 13. We estimated the size of classification model parameters by assuming all parameters are 32-bit floats and we then computed the communication latency.

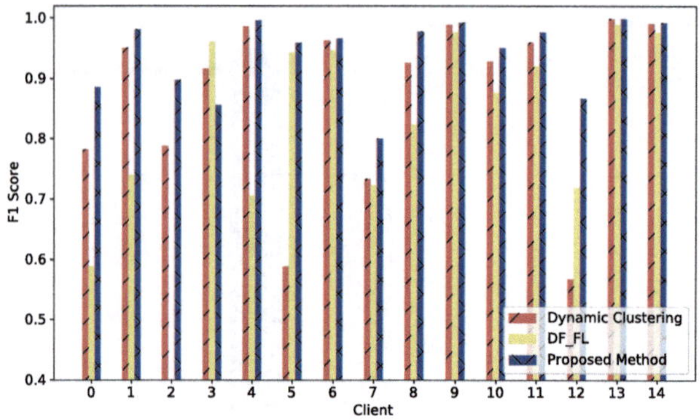

Fig. 20 Individual F1 score of dynamic clustering, DF_FL, and our method [81]

Table 13 Wi-Fi upload and download bandwidth

Wi-Fi	Bandwidth	Data rate (MB/s)
Upload	114.4	14.3
Download	164	20.5

Table 14 Training overhead comparison

	Total training time (hr)
Dynamic clustering	1.36
DF_FL	0.94
Our method	0.85

We see from Table 14 that our work significantly reduces training overhead (by 37.5%) compared to the dynamic clustering algorithm. The reduction in overhead is due to the fact that the dynamic clustering algorithm contains a cluster calibration step that reassigns each client to existing cluster groups. This step causes extra overhead and does not improve classification performance according to Fig. 18. In addition, the cluster division step in dynamic clustering is also time-consuming since it involves classification model parameter transferring. Although we performed a similar procedure in our work to estimate the optimum number of clusters, we transmitted ClusterGAN parameters with a much smaller size than the classification model.

We also observe a 9.6% decrease in the total training time of the presented method compared to the DF_FL algorithm, which does not involve client clustering and is an extension of the FedAvg algorithm with the addition of aggregation criteria as mentioned in section "Baselines." The decrease in training time can be attributed to the faster convergence of our method. Furthermore, since the classification performance of DF_FL degrades significantly compared to our approach, we can therefore claim that our method outperforms DF_FL.

Table 15 Performance comparison on clustering criteria

Clustering criterion	Accuracy	F1 score	Total training time (hr)
Data-driven	0.957	0.941	0.85
Model-driven	0.9552	0.9393	1.79
Hybrid	0.9564	0.94	0.98

4.3.3 Selection on Clustering Criteria

To demonstrate that the described method is effective across variations, we explored the clustering criteria, namely, data-driven, model-driven, and hybrid clustering, as defined in section "Clustering Criteria." We present evaluation metrics in terms of accuracy, F1 score, and total training time in Table 15. There is no noticeable difference between the three different criteria in terms of classification accuracy and the F1 score but the total training time for data-driven clustering is the minimum among these clustering criteria. The total training time of model-driven clustering is doubled compared to data-driven clustering due to the large size of the classification model. Estimating cluster groups based on model parameters is computationally expensive when the model size is large since this procedure involves transferring classification model parameters. We conclude that data-driven clustering results in significantly less communication overhead compared to the model-driven approach at the same level of task performance, and is thus superior.

We observe that the training time for hybrid clustering is slightly higher than that for data-driven clustering. The additional time is due to the cluster division update step. At each communication round, we reexamined criteria to update the cluster group based on average within-cluster performance. We note that the algorithm updated the cluster group four times during the complete training stage. The cluster groups do not update frequently and thus consume a small amount of additional time.

We present the breakdown of total training time for different clustering criteria in Fig. 21. We see that the communication time and the cluster update time are two major components of the total training time. Since model-driven clustering requires transferring a large classification model for cluster updates during federated learning, it results in a much higher overhead than the algorithm using the other two clustering criteria. There is an increase in the cluster update time in hybrid clustering compared to data-driven clustering. This increase is due to cluster updates during federated learning.

Significant amounts of communication time and the cluster update time that we observe are due to the large size of the ClusterGAN and the client classification model. To further reduce the overhead, we plan to apply selective model parameter sharing or model compression techniques in our future work [169, 170].

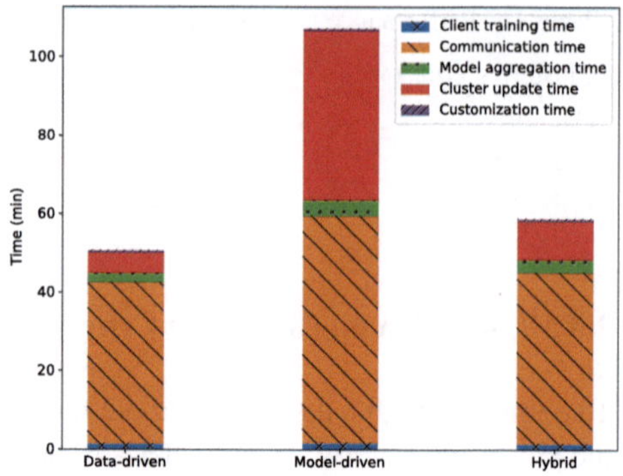

Fig. 21 Overall training-time breakdown for different clustering criteria [81]

Table 16 Performance comparison on the degree of customization

Customization level	Accuracy	F1 score	Total training time (hr)
Scheme 1	0.8522	0.8536	0.98
Scheme 2	0.9549	0.9383	0.99
Scheme 3	0.957	0.941	0.85

4.3.4 Impact of Personalization

We next address the importance of customization on the client model based on the three different customization schemes described in section "Level of Personalization." We present evaluation results in Table 16. We see that Scheme 2 and Scheme 3 with personalized layers obtain a significant 10% increase in the global accuracy and F1 score compared to Scheme 1 without personalized layers. The results suggest that personalization is promising in enhancing classification performance. Scheme 3 achieves slightly better global accuracy and F1 score than Scheme 2. Since Scheme 2 averages model parameters in $Block1$ across all clients, such an averaging method would lead to performance degradation in clients with shifted data distributions.

We also notice that the total training time for Scheme 3 is less than that for Scheme 1 and Scheme 2. Aggregating $Block1$ model parameters across all clients may result in additional time needed for Schemes 1–2.

We next compare individual performance in terms of accuracy and F1 score for the different schemes; the results are presented in Figs. 22 and 23. We observe a significant increase in accuracy and F1 score in Clients 0, 1, 2, 4, 7, 8, and 12 in schemes with personalized layers compared to Scheme 1. Except for Client 3, the performance of the rest of the clients in Scheme 3 achieves a moderate increase in accuracy and F1 score compared to Scheme 1. We also see that individual performance in Scheme 3 is moderately better than in Scheme 2 for the majority of the clients. The reason for the performance degradation of Scheme 2 can be attributed to the averaging of model parameters across all clients with different data distributions, as discussed earlier in this section.

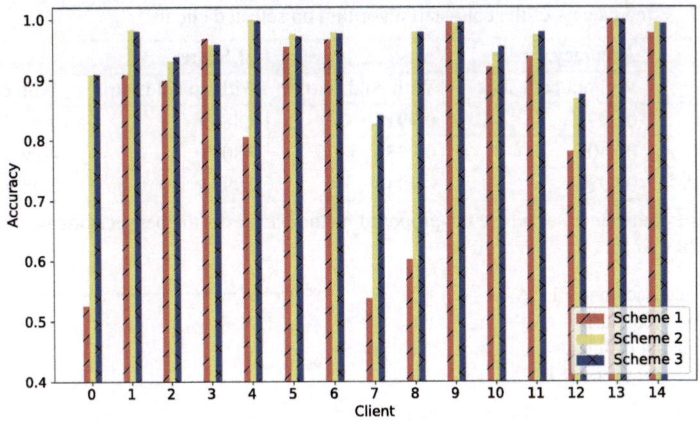

Fig. 22 Individual accuracy at different customization schemes [81]

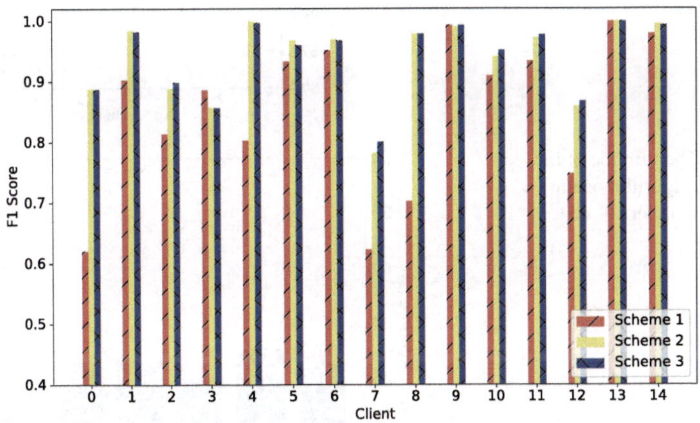

Fig. 23 Individual F1 scores for different customization schemes [81]

4.3.5 Cold-Start Results

To illustrate the effectiveness of the presented cold-start algorithm, we compare the accuracy and the F1 score using Algorithm 2 without cold start and with cold start. Note that for the cold-start case, we assume that all clients in the system are trained except the new incoming client. We executed the cold-start algorithm three times by selecting a different client each time. We chose clients based on the individual performance achieved during data-driven clustered FL without the cold start. We selected clients with high, moderate, and low accuracy and F1 scores.

We present the results in Table 17. The results obtained using the cold start that are significantly better than those without the cold start are marked in bold. We see that Client 1 and Client 8 achieve remarkable increases in accuracy and F1 score with the cold-start algorithm. Client 15 experienced a slight decrease in accuracy

Table 17 Effectiveness of the cold-start algorithm on selected clients

	Accuracy		F1 Score	
	Without cold start	With cold start	Without cold start	With cold start
Client 1	0.9095	**0.9917**	0.8869	**0.9897**
Client 8	0.8405	**0.9959**	0.8011	**0.9948**
Client 15	0.9929	0.9847	0.9934	0.9838

Bold values indicate cases where the proposed method achieves the best performance compared with the baselines

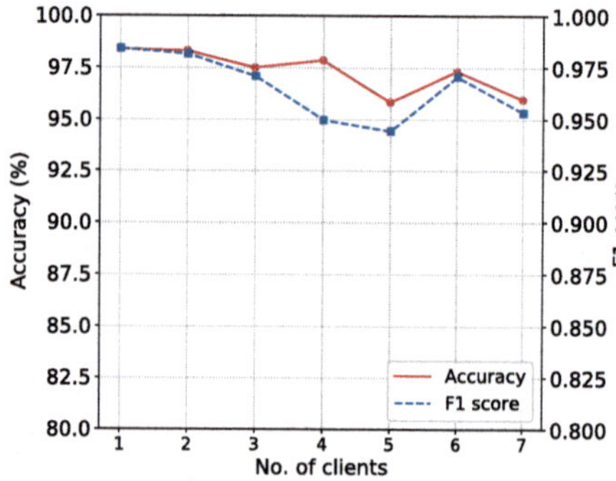

Fig. 24 Accuracy and F1 score of the cold-start algorithm using different numbers of new users [81]

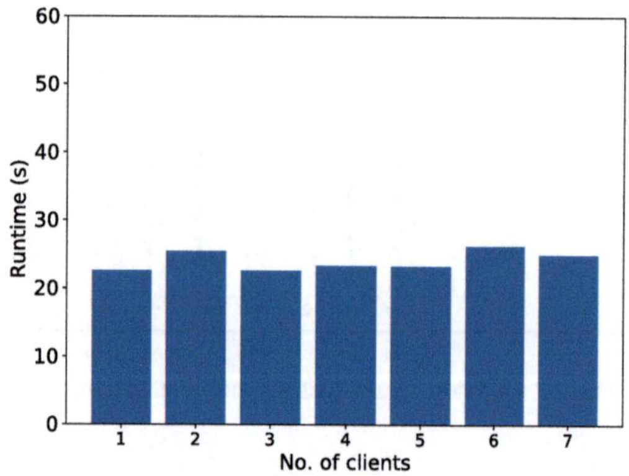

Fig. 25 The runtime of the cold-start algorithm using different numbers of new users [81]

and F1 score after the cold start compared to the results without the cold start. In experiments with the cold start, all the selected new clients were successfully assigned to an existing cluster group, eliminating cluster update time and reducing communication time, i.e., no iterative model parameter updates.

To further investigate the effectiveness of our cold-start algorithm, we record the average accuracy, F1 score, and the runtime of the algorithm when different numbers of new users join the trained system. We randomly selected new users from one up to seven to ensure that the number of new users is smaller than the existing users in the system for training. We present the average accuracy and F1 score of the new users in Fig. 24 and the runtime in Fig. 25, respectively. We notice in Fig. 24 that as the number of new users increases, the average accuracy and F1 score of the new users experience a slight decrease. This observation can be attributed to the decrease in the number of existing clients in the trained system. Fewer existing clients may result in worse cluster assignments of the new users, leading to performance degradation.

We see in Fig. 21 that cluster update time and communication time are two major factors for high training overhead. Based on Fig. 25, our method significantly reduces the training overhead (to less than 0.5 min) compared to clustering and training the new client from zero knowledge. Results show that handling intra-subject variations, i.e., data distribution shifts between subjects, via the two-sample test is promising. Note that this technique can also be applied to detect inter-subject variations, i.e., within-subject data distribution shifts over time.

4.3.6 Limitations and Future Work

As described in Sect. 4.2.3, the time complexity of the presented algorithm depends on the size of the classification model P, the amount of client data N, and the number of clients K. The increase in these variables will raise computation and communication overhead during client model aggregation, affecting the scalability of our algorithm. Advanced approaches such as pruning and quantization [170] and asynchronous aggregation [171] will be investigated to improve the scalability of our algorithm.

In addition, the focus of the presented algorithm is on developing a personalized healthcare model. Our algorithm does not consider cases when clients become stragglers due to device heterogeneity or drop out of the system because this is out of the scope of this work. Improvements in the resilience of our algorithm will be investigated in future work.

Besides, note that our algorithm only considers the condition when data heterogeneity exists between clients. Alleviating data heterogeneity within a client is beyond the scope of this work, and we are addressing this problem in our ongoing work.

Finally, with the advances in AI technology, discussions about the ethics of using AI have been brought to the attention of researchers and the general public. Yang et al. advocated research that considers fairness/equity of federated learning algorithms with clients that contain sensitive information such as gender and race [172]. Zhang et al. [173] and Ezzeldin et al. [174] proposed federated learning algorithms that reduce group bias due to distribution differences in demographics in the database. Fairness in federated learning has not received much attention, and it will be investigated and incorporated in future work.

5 Federated Clustered Multi-domain Learning for Health Monitoring

Data heterogeneity occurs in the following two scenarios in the FL scheme:

- Inter-client data heterogeneity: It refers to the case when there exist domain differences across clients. For example, as illustrated in Fig. 26a, images from different clients have diverse backgrounds, viewpoints, and image resolutions, indicating that they belong to different domains. Inter-client data heterogeneity deteriorates the performance of the conventional one-size-fits-all FL algorithms such as FedAvg [175] since the aggregated models obscure individual differences.
- Intra-client data heterogeneity: It refers to the case when there exist domain differences within a client. As illustrated in Fig. 26b, data from a single client belongs to multiple domains, having different data distributions. Intra-client data heterogeneity is a common issue in the field of healthcare. Long et al. [176] investigated cardiorespiratory activities and electrophysiological signals at different stages of sleep and found that the same patient's body signals vary by stage. In addition, Karimian et al. [177] indicate that the impact of the intra-client variation of electrocardiogram (ECG) signals cannot be neglected. The aforementioned examples illustrate how a single individual/client can encompass multiple data distributions (domains), stemming from the different patterns involved in a physiological event over time. Note that patterns of the physiological attributes (domains) for a disease of interest over time are usually unknown ahead of time [178].

There exists literature on alleviating inter-client data heterogeneity in FL. For example, Ghosh et al. [180] and Briggs et al. [181] proposed to divide clients into clusters according to their underlying similarities and train a model for each cluster. Fallah et al. [182] suggested that model-agnostic meta-learning (MAML)

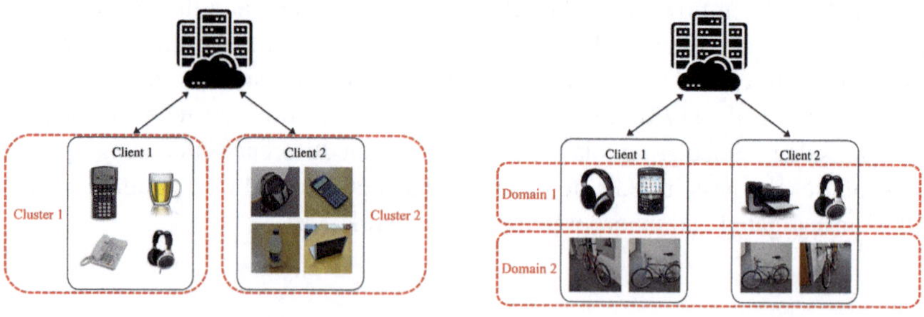

(a) Inter-client data heterogeneity (b) Intra-client data heterogeneity

Fig. 26 Illustration of inter-client and intra-client data heterogeneity [179]. (**a**) Inter-client data heterogeneity. (**b**) Intra-client data heterogeneity

[183] is a promising framework for building a personalized model for each client. Smith et al. [184] integrated multitask learning into FL, which considers each client having a model for a different task and trains those correlated models simultaneously [185]. However, previous approaches on building customized models for a subset/an individual client fail to capture the data heterogeneity within each client.

Intra-client heterogeneity has not been studied much, and there is only a limited body of literature addressing this issue. In this context, Caldarola et al. [186] proposed to model data heterogeneity by identifying domains for each client via a combination of knowledge distillation, domain-specific learning, and graph representation learning techniques. Shenaj et al. [187] further extended the idea of learning across domains to tackle data heterogeneity via style transfer and server-side pretraining. However, their assumptions include prior knowledge of the domains and their relations, which is not applicable to our context. Consequently, we develop an algorithm that addresses intra-client data heterogeneity with unknown domains in FL.

Our work is inspired by multi-domain learning. Multi-domain learning learns correlated domains simultaneously by training a set of models and then adapting them to specific domains [188]. There are two mainstream categories of methods for multi-domain learning. One set of approaches utilizes adaptors, which are small neural network modules attached to a large pretrained neural network model [189, 190]. The pretrained portion of the model learns shared information across domains (i.e., domain-agnostic), and the model obtains domain-specific knowledge by training the adaptors. The other set of approaches utilizes model parameter sharing. Studies [191, 192] show that retraining later layers of the neural network models can effectively capture domain-specific information.

There is only a limited amount of prior work on incorporating multi-domain learning into the context of FL. Parekh et al. [193] performed FL in the multi-domain, multitask setting for medical image object detection and segmentation. Li et al. [194] presented FedH2L, which utilizes mutual learning to tackle inter-client data and model architecture heterogeneity. Sun et al. [195] utilized partial model parameter sharing of the global model to mitigate inter-client data heterogeneity due to the cross-domain effect. Elvebakken et al. [196] suggested introducing adaptors in the federated learning system to reduce communication overhead. However, none of the existing federated multi-domain learning literature addresses the problem of intra-client data heterogeneity in FL.

We thus propose to integrate multi-domain learning into FL to identify domain differences within individual clients. We perform federated clustering to learn domain categories within each client. We then construct a neural network architecture shared across all domains. We enable parameters at selected layers of the client model to be domain-specific and the rest of the parameters to be domain-agnostic. In addition, we introduce an auxiliary graph neural network to fine-tune the domain-agnostic and domain-specific knowledge. In this section, we list three features of our work as follows:

- We identify a crucial source of data heterogeneity, i.e., within-client data heterogeneity, and propose a novel federated clustered multi-domain learning algorithm to overcome this problem.
- We employ a graph attention network as an auxiliary representation to connect and capture the implicit relationships between different domains.
- We validate the effectiveness of our algorithm using a case study on stress-level prediction. The presented method improves the overall accuracy and the F1 score by over 4.41% and over 7.8%, respectively, compared to two state-of-the-art methods. The performance of our model is also robust for all individual clients.

The organization of the rest of the section is as follows: We introduce our method in the Method section. In the Results section, we specify the dataset and experimental setup used for the case study; we demonstrate the effectiveness of our algorithm by comparing it with state-of-the-art approaches and analyzing different modules of our method.

5.1 Method

As explained in the Introduction, existing literature alleviates only inter-client data heterogeneity; it fails to capture intra-client data heterogeneity. This section presents a clustered multi-domain FL algorithm to mitigate the ill effects of intra-client data heterogeneity. As depicted in Fig. 27, the presented algorithm comprises three modules: a federated clustering module, a multi-domain learning module, and a graph-based domain fine-tuning module. The federated clustering module assigns domain labels to data samples at each client in a shared domain space. The multi-domain learning module enables the learning of domain-specific knowledge across all clients. A graph attention network is utilized at each client to facilitate fine-tuning of the domain-agnostic and domain-specific knowledge in the graph-based domain fine-tuning module.

Specifically, given a dataset $D = \{(X_i, Y_i), i = 1, \ldots, N\}$ where X_i is the set of images from client i, Y_i is the set of corresponding labels for X_i, and

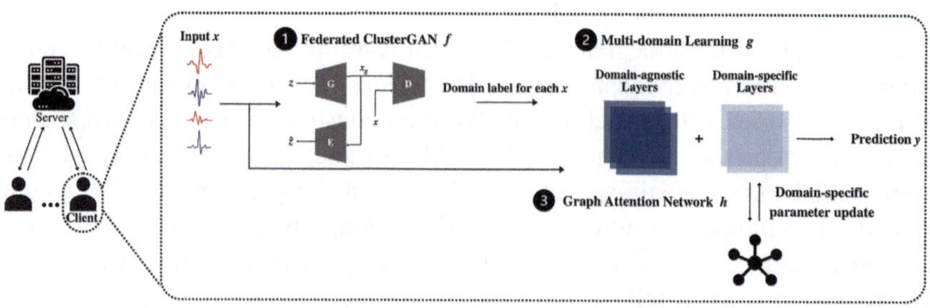

Fig. 27 Overall framework of our federated clustered multi-domain learning algorithm [179]

N is the number of clients, the objective of our algorithm is to learn a function $F(w;\theta;\phi;\omega) = \{f(w), g(\theta;\phi), h(\omega)\} : X_i \rightarrow Y_i$. The federated clustering module $f(w) : X_i \rightarrow \mathcal{D}_i = \{d, d \in \{1,\ldots,K\}\}$ assigns domain labels $\mathcal{D}_i$ to images X_i, where w is the module parameter, d is the domain label for a single image sample in X_i, and K is the number of domains obtained from the estimation algorithm from our prior work [81]. The multi-domain learning module $g(\theta;\phi) : X_i^d \rightarrow Y_i$ maps the data labeled with domain tags X_i^d to the set of labels Y_i, where θ is the domain-agnostic parameter and ϕ is the domain-specific parameter. In the last module, we fine-tune parameter ϕ by applying a graph attention network $h(\omega) : \phi \rightarrow \phi'$. We update $g(\theta;\phi')$ with the fine-tuned ϕ' and obtain the final classification output Y_i. We next explain each module in detail.

5.1.1 Federated ClusterGAN

The federated ClusterGAN groups each set of X_i into K domains, where these domains are shared across all clients. It is known that clustering in the latent space utilizing ClusterGAN provides more stable clustering results compared to clustering in the data space [197]. Unlike conventional generative adversarial networks, ClusterGAN is composed of three deep neural networks: a generator **G**, a discriminator **D**, and an encoder **E**, as shown in Fig. 28 parameterized by $\Theta_\mathbf{G}$, $\Theta_\mathbf{D}$, and $\Theta_\mathbf{E}$, respectively.

The generator **G** maps the latent space to the data space. It samples from the latent space z, which consists of a continuous normal distribution z_n and a discrete one-hot vector z_c, and generates artificial sample x_g. Formally, $z = (z_n, z_c)$, $z_n \sim \mathcal{N}(0, \sigma^2 I)$, $z_c = e_k$, $k \in \{1,\ldots,K\}$, where e_k is the kth one-hot vector and K is the number of clusters. The discriminator **D** projects the artificial sample x_g and the real sample x_r in the data space to a real value that indicates the probability of the sample being real. The encoder **E** generates the discriminative latent space variables $\hat{z_n}$ and $\hat{z_c}$ using x_g. The objective of ClusterGAN training is to minimize the loss function presented in Eq. (30) [197], where $\mathbb{P}_x^r$ is the distribution of real data samples, $\mathbb{P}_z$ is the prior distribution in the latent space. $q(\cdot)$ is the quality function

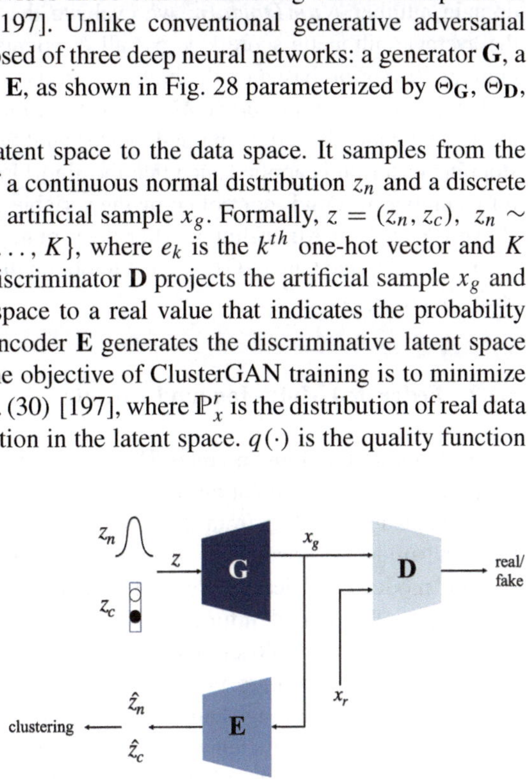

Fig. 28 ClusterGAN architecture [179]

which is $\log(x)$ for conventional GAN and x for Wasserstein GAN. $l(\cdot)$ is the cross-entropy loss, and β_n and β_c are regularization coefficients.

$$\min_{\Theta_{\mathbf{G}},\Theta_{\mathbf{E}}} \max_{\Theta_{\mathbf{D}}} \mathop{\mathbb{E}}_{x\sim\mathbb{P}_x^r} q(\mathbf{D}(x)) + \mathop{\mathbb{E}}_{z\sim\mathbb{P}_z} q(1-\mathbf{D}(\mathbf{G}(z))) + \beta_n \mathop{\mathbb{E}}_{z\sim\mathbb{P}_z} ||z_n - \mathbf{E}(\mathbf{G}(z_n))||_2^2$$
$$+ \beta_c \mathop{\mathbb{E}}_{z\sim\mathbb{P}_z} l(z_c, \mathbf{E}(\mathbf{G}(z_c))) \tag{30}$$

We first train ClusterGAN at each client. The central server aggregates local model parameters and updates the ClusterGAN from each client with the global model. Once the ClusterGAN converges during this iterative process, we apply K-means clustering to the latent space data encoded by the global encoder to identify the cluster (domain) membership for each data sample based on Euclidean distance. As mentioned earlier in this section, ClusterGAN provides a better cluster separation in the latent space than in the original data space. Therefore, we apply K-means to the latent space data. Since we have no prior knowledge of the number of domains K, we adapt an iterative search algorithm from our prior work [81]: We search for the optimum number of clusters (domains) incrementally. Given a specific domain number $NumDomain$, we evaluate the clustering quality by computing the average silhouette score s across all N clients. We record the optimal domain number $BestNum$ based on the highest average silhouette score $MaxScore$. If the current score does not improve $MaxScore$, we increase $NumDomain$ and proceed to the next round of clustering and cluster quality evaluation. In addition, an early stopping criterion is applied such that the algorithm stops when s does not improve for a number of rounds controlled by the variable $Patience$. The process described above is shown in Algorithm 5. After determining the optimal domain number K, we apply federated ClusterGAN again to obtain the final domain labels $\mathcal{D}_i$ for X_i in each client.

5.1.2 Federated Multi-domain Learning

After obtaining the domain labels $\mathcal{D}_i$ for each X_i from federated ClusterGAN, we learn customized classification models for all the domains. A previously reported study indicates that data from different domains can still share a large amount of low- and mid-level visual information such that establishing individual domain-specific models will lead to loss of shared information across domains [190]. To preserve the shared information while learning domain-specific knowledge, we apply federated multi-domain learning by developing a convolutional neural network (CNN) g that is composed of domain-agnostic layers with a set of parameters θ for obtaining shared information across domains and domain-specific layers with a set of parameters ϕ. We define the nth layer of the CNN as the domain-specific layer and the rest of the layers to be domain-agnostic layers. We know from prior research that the end layer of a CNN has lower representation capacity compared to other layers in the architecture and is thus more sensitive to domain-

Algorithm 5: Data-driven domain number estimation

Input: $D = \{(X_i, Y_i), i = 1, \ldots, N\}$, *Round*
Output: *BestNum*
Initialize(*NumDomain, MaxScore, BestNum, Count, Patience*);
Initialize(Generator $G(\Theta_\mathbf{G})$, Discriminator $D(\Theta_\mathbf{D})$, Encoder $E(\Theta_\mathbf{E})$);
while $NumDomain \leq N$ **do**
  for *r = 1:Round* **do**
    for $X_i \in$ *n clients sampled from N* **do**
      $i.encoder \leftarrow TrainClusterGAN(X_i)$;
      $X_i^{enc.} \leftarrow i.encoder(X_i)$;
      $\mathcal{D}_i \leftarrow KMeans(X_i^{enc.}, NumDomain)$;
    $\Theta_\mathbf{G}^r \leftarrow \sum_{c=1}^{n} \frac{i_c}{N} \Theta_\mathbf{G}^r$;
    $\Theta_\mathbf{D}^r \leftarrow \sum_{c=1}^{n} \frac{i_c}{N} \Theta_\mathbf{D}^r$;
    $\Theta_\mathbf{E}^r \leftarrow \sum_{c=1}^{n} \frac{i_c}{N} \Theta_\mathbf{E}^r$;
  $s \leftarrow SilhouetteScore(\{X_i^{enc.}, \mathcal{D}_i, i = 1, \ldots, N\})$;
  if $s > MaxScore$ **then**
    $MaxScore \leftarrow s$;
    $BestNum \leftarrow NumDomain$;
  else
    $Count += 1$;
  if $Count \geq Patience$ **then**
    break;
  $NumDomain += 1$

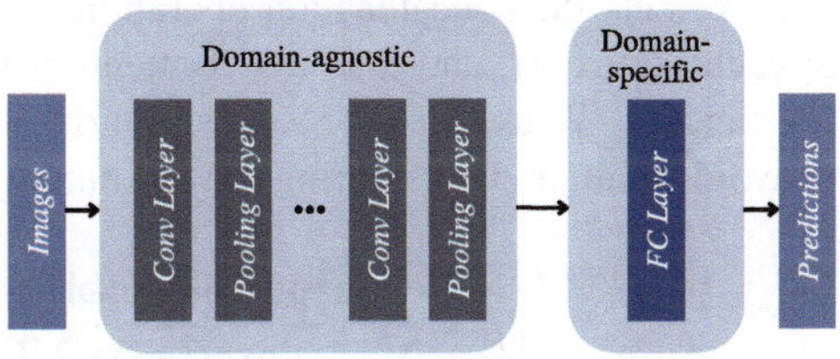

Fig. 29 Federated multi-domain learning architecture [179]

specific information [198]. We, therefore, choose n to be the last fully connected layer of the CNN architecture shown in Fig. 29.

5.1.3 Graph-Based Domain Knowledge Fine-Tuning

We obtain customized classification models for data from different domains via federated multi-domain learning. To further fine-tune the domain-specific parameter ϕ, we adopt a graph attention network (GAT) [199] to model relationships between domains. We fine-tune ϕ by treating each domain-specific ϕ as a node in a graph. Due to the intrinsic nature of the graph architecture, a graph neural network (GNN) can effectively learn relationships between the nodes. This makes GNNs an excellent choice for modeling domain relationships. GAT overcomes

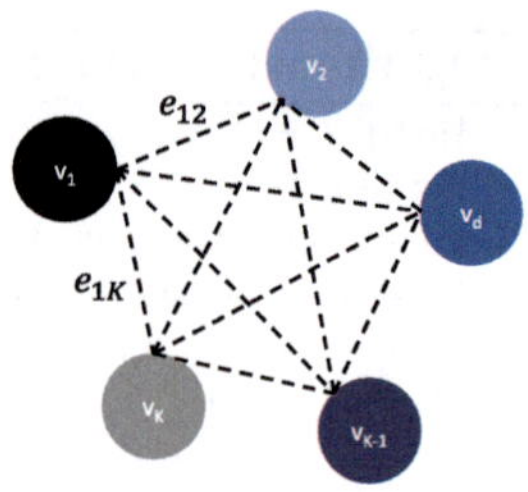

Fig. 30 Illustration of GAT architecture [179]

several limitations of prior GNNs, such as the GCN: it allows different weightings to different neighboring vertices (domains) in the graph, thereby enhancing the interpretability of domain relationships [200]. Moreover, it does not make any assumptions on edge connections between the nodes (domains) [201], aligning well with our scenario where domain relationships are unknown. Additionally, GAT is computationally efficient [199].

We define the GAT as $h = (\mathcal{V}, \mathcal{E})$, where $\mathcal{V}$ is the set of vertices that represent domains and $\mathcal{E}$ is the set of edges that connects domains. As shown in Fig. 30, $\mathcal{V} = \{v_d,\ v_d \in \mathbb{R}^M,\ d = 1, \ldots, K\}$, where v_d is the set of parameters from the domain-specific layer of the classification model that belongs to domain d, $M = |v_d|$ represents the number of v_d, and K is the number of vertices (domains). We denote the output of the GAT as $\hat{\mathbf{V}} = \{\hat{v}_d,\ d = 1, \ldots, K\}$. Each output $\hat{v}_d$ is a linear combination of its neighboring vertices $\mathcal{N}_d = \{v_j,\ j \in \mathcal{V} \setminus d\}$ weighted by normalized self-attention coefficients e_{dj}, as presented in Eq. (31) [199], where $\oplus$ represents the concatenation operation, $\mathbf{W}$ is a learnable weight matrix applied to each vertex v_d, and α is a self-attention mapping that computes self-attention coefficients. We use scaled cosine similarity instead of the conventional dot product to avoid small gradients during model training for the case study [202].

$$e_{dj} = \frac{\exp(\text{LeakyReLU}(\alpha\left[\mathbf{W}v_d \oplus \mathbf{W}v_j\right]))}{\sum_{k \in \mathcal{N}_d} \exp(\text{LeakyReLU}(\alpha\left[\mathbf{W}v_d \oplus \mathbf{W}v_k\right]))} \tag{31}$$

We obtain $\hat{\mathbf{V}} = \{\hat{v}_d,\ d = 1, \ldots, K\}$ based on Eq. (32) [199], where $\varsigma(\cdot)$ is a nonlinear function, $\mathcal{L}$ is the number of multi-head attention mechanisms utilized to stabilize the learning process, e^l_{dj} is the normalized self-attention coefficients computed from the lth attention mechanism, and W^l is the corresponding weight matrix.

$$\hat{v}_d = \varsigma\left(\frac{1}{\mathcal{L}}\sum_{l=1}^{\mathcal{L}}\sum_{j \in \mathcal{N}_d} e^l_{dj}\mathbf{W}^l v_j\right) \tag{32}$$

5.2 *Results*

5.2.1 Dataset

In this study, we used the wearable stress and affect detection (WESAD) dataset [203]. The data was collected from 15 participants using 2 devices, i.e., a chest-worn device (RespiBAN) and a wrist-worn device (Empatica E4). The RespiBAN measures accelerometer data, electrocardiogram (ECG), electrodermal activities (EDA), electromyogram (EMG), respiratory signal, and temperature at a sampling rate of 700 Hz. Empatica E4 records accelerometer data, blood volume pressure, EDA, and temperature at different sampling rates. All signals were labeled with baseline, stress, and amusement tags [204]. Following the prior work [205], we utilized the ECG signal sampled from RespiBAN with baseline and stress tags only. Note that we consider each subject with the individual's data as a client for federated learning with non-i.i.d. data due to intrinsic physiological differences between individuals.

5.2.2 Experimental Setup

Evaluation Metrics We evaluated our framework using the following criteria for all the experiments:

- Overall accuracy/F1 score: the accuracy/F1 score across all clients in the test dataset
- Client accuracy/F1 score: the accuracy/F1 score of the individual client in the test dataset

Implementation Details To preprocess the raw ECG signals, we used an R-peak detection algorithm [206] to detect ECG cycles according to the locations of R-peaks and then converted each ECG cycle to an image of size 128×128. Since the WESAD dataset contains valid data from 15 clients, we randomly selected data from 11 clients for training and the rest of the 4 clients for testing. We fixed the train/test data split for all the experiments.

Following our prior work [81], we utilized stochastic gradient descent (SGD) as the training optimizer with a weight decay of 10^{-4} and a momentum of 0.9. The maximum communication round is 50, with an early stopping criterion of stopping training after having 5 consecutive epochs without improved accuracy. We implemented the federated clustering module using the model from our prior work [81]. We trained the ClusterGAN at each client with a batch size of 32 and a learning rate of 5×10^{-4} for 5 epochs. We chose the standard ResNet50 [207] as the CNN backbone architecture for multi-domain learning due to its remarkable performance on image classification tasks and training efficiency. We froze domain-agnostic parameters from all convolutional layers of the ResNet50 and enabled parameters from its last fully connected layer to be adaptable to each domain. We trained the multi-domain learning module using a batch size of 32 and a learning rate of 10^{-3} for 20 epochs. Random search is employed to optimize the hyperparameters. We applied a five-layer GAT model with all layers having $K = 5$ attention heads followed by an exponential linear unit (ELU) activation layer. We applied a dropout rate of 0.2 to avoid overfitting. The GAT was initialized using Xavier initialization [208] and trained to minimize the cross-entropy loss with a learning rate of 5×10^{-4}.

Baselines To demonstrate the effectiveness of our method, we compare our model with the following two baselines:

- Dynamic-Fusion Federated Learning (DF_FL) [209]: A modified version of the FedAvg algorithm that dynamically selects the participating clients based on local model performance and performs model fusion according to participating clients' training time. It yields excellent model performance with non-i.i.d. health data.
- Cluster-Driven Graph Federated Learning (FedCG) [186]: A state-of-the-art federated learning algorithm that addresses intra-client data heterogeneity by utilizing a teacher-student model, cluster-specific models, and a graph convolutional network (GCN) that connects the cluster-specific models.

5.2.3 Effect of Intra-client Data Heterogeneity

As described in the Introduction, there exist two types of data heterogeneity, i.e., inter-client and intra-client data heterogeneity. In this section, we demonstrate the effect of data heterogeneity on the client classification model performance by comparing our method with FedAvg [175] and cluster-based FL. We present the corresponding accuracy and F1 score in Fig. 31. FedAvg aggregates all client models to form a global model and sends back the shared global model to each client without considering inter-client/intra-client data heterogeneity. We observe

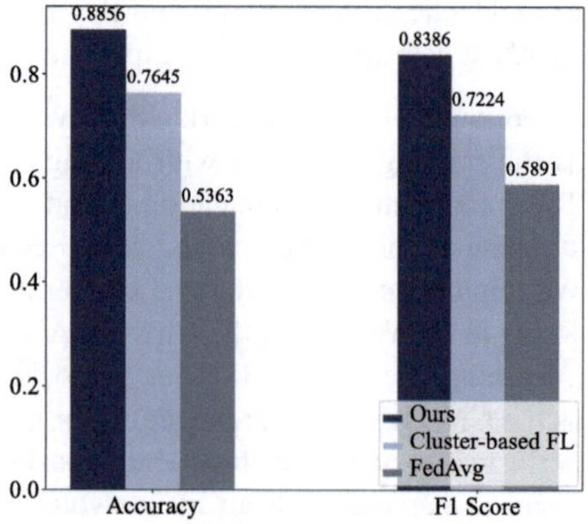

Fig. 31 Overall accuracy and F1 score comparison with FedAvg, cluster-based FL, and our algorithm on stress-level prediction [179]

that FedAvg yields a prediction accuracy of 53.63% and an F1 score of 0.5891, indicating that the global model fails to provide a good generalization across all clients when there exists data heterogeneity.

Cluster-based FL aims at reducing inter-client data heterogeneity by grouping similar patients such that clients within a cluster share similar data distributions and those across different clusters have distinct data distributions. Following our prior work [81], we performed cluster number estimation, trained federated ClusterGAN, and assigned a cluster ID to each client based on the latent embedding from the ClusterGAN. We then applied FedAvg for clients in the same cluster. The cluster-based FL achieves an accuracy of 76.45% and an F1 score of 0.7224, which is approximately 12% lower in accuracy and 0.12 lower in the F1 score compared to our model. The result suggests that considering inter-client data heterogeneity in the algorithm helps improve classification performance. However, the performance is adversely affected by the effect of intra-client data heterogeneity. Our algorithm considers intra-client data heterogeneity and thus significantly improves the accuracy and the F1 score.

Note that identifying outliers, i.e., clients with data distributions significantly diverging from other clients, is beyond the scope of this section. Prior work on mitigating FL data heterogeneity has not addressed the problem of outliers. This is an area we aim to explore in future research. Additionally, outlier clients in federated learning are often considered malicious attackers, and there exists literature that utilizes anomaly detection techniques to identify and remove such attackers [210–212]. Techniques for detecting outliers to improve the privacy of the FL will be investigated in future work.

5.2.4 Comparison with Baselines

We present the performance comparison between the described federated clustered multi-domain learning algorithm and the selected baselines. As shown in Table 18, our approach achieves an overall accuracy of 88.56% and an overall F1 score of 0.8386, outperforming the DF_FL and FedCG baselines. Since the DF_FL model does not differentiate heterogeneous data distributions within each client, its mediocre performance demonstrates that the one-size-fits-all federated learning model does not generalize well. The knowledge learned from other clients does not apply to new clients with different data distributions. Our approach is significantly superior since it identifies intra-client heterogeneous data distributions (i.e., domains).

The presented method obtains around a 4.4% increase in accuracy and an increase of 0.06 in F1 score compared to FedCG. The FedCG utilizes a GCN that defines an adjacency matrix based on predetermined parameter similarity between vertices/domains. The GAT does not make assumptions about the connections between vertices/domains and assigns different weights to the connecting edges via the attention mechanism [199]. Our results suggest the GAT is a better graph representation learning model with learnable weight coefficients between domains.

Table 18 Overall prediction accuracy and F1 score of our model and baselines

Method	Overall accuracy (%)	Overall F1 score
DF_FL [209]	48.04	0.6479
FedCG [186]	84.15	0.7776
Our method	**88.56**	**0.8386**

Bold values indicate cases where the proposed method achieves the best performance compared with the baselines

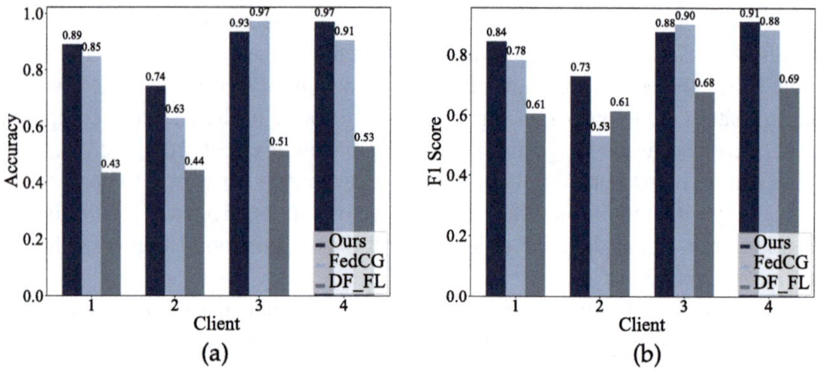

Fig. 32 Client accuracy and F1 score on stress-level prediction [179]. (**a**) Accuracy of individual clients. (**b**) F1 score of individual clients

We next compare the client accuracy and F1 score obtained from our method and the baselines shown in Fig. 32a, b, respectively. The results manifest the robustness of our approach since it outperforms DF_FL and FedCG on three out of the four clients in the test dataset. The DF_FL yields low accuracy and F1 score on all clients. This low performance can be attributed to significant inter-client domain differences between clients in the training and test datasets. By comparing our method with FedCG, we observe an increase of at least over 4% in accuracy and an increase of at least 0.03 in the F1 score in Clients 1, 2, and 4, while FedCG achieves slightly better accuracy and F1 score in Client 3. Both methods did not perform well in Client 2, indicating that Client 2 may contain different data distributions from the other three clients in the test dataset. To conclude, our model outperforms FedCG in terms of both client accuracy and F1 score.

Additionally, to ensure a more comprehensive comparison between the presented method and the baselines, we determine their model size by computing the number of model parameters (assuming all of them are 32-bit floats), computation cost in terms of mega floating point operations per second (MFLOPS), and communication cost in terms of latency in minutes; the results are shown in Table 19. Note that we provide a breakdown of our method by each module. Upon observation, we find that our method significantly outperforms DF_FL, featuring both a smaller model size and lower computation and communication costs. Our method exhibits a similar model size, computation cost, and communication cost to FedCG, yet it achieves

Table 19 Computation and communication cost comparison with baselines

Method	Model size (MB)	Computation cost (MFLOPS)	Communication cost (min)
DF_FL [209]	1212	977.47	120
FedCG [186]	175.15	144.72	51.4
Our method	Total: 172.88	147.21	50.87
	ClusterGAN: 81.18	68.54	
	CNN Backbone: 89.67	78.46	
	GAT: 2.03	0.21	

Table 20 Comparison of overall accuracy and F1 score on federated clustering methods

Federated clustering module	Overall accuracy (%)	Overall F1 score
Teacher-student	84.22	0.8093
ClusterGAN	88.56	0.8386

higher accuracy and F1 score. We thus conclude that the improvement in prediction results is attributed to the presented algorithm itself, rather than other factors, such as increased model size.

5.2.5 Effect of Federated ClusterGAN

In this section, we further demonstrate the effectiveness of our federated ClusterGAN module in identifying domains across all clients. To illustrate the superiority of latent space clustering via ClusterGAN, we replace the federated ClusterGAN module in our algorithm with a teacher-student model [186], which is effective for domain adaptation, and compare it with the algorithm without replacement. The teacher-student model consists of a teacher network and a student network, where the teacher classifier provides domain pseudo-labels as targets to guide student network training. Note that the work of Caldarola et al. [186], which minimizes the cross-entropy loss between the domain labels obtained from the teacher and the student model, is inspired by the work of Asano et al. [213]. In their research, they proposed an alternating minimization algorithm for self-labeling. We used the same number of clusters as the hyperparameter for a fair comparison. Table 20 presents the accuracy and the F1 score using the ClusterGAN and the teacher-student model. Results show that our federated ClusterGAN outperforms the teacher-student model by 4.34% in accuracy and 0.0293 in F1 score, highlighting the advantage of utilizing the federated ClusterGAN.

Table 21 Comparison of overall accuracy and F1 score on auxiliary domain fine-tuning methods

Auxiliary domain fine-tuning module	Overall accuracy (%)	Overall F1 score
None	84.73	0.8109
GCN	87.89	0.8320
GAT	**88.56**	**0.8386**

5.2.6 Effect of Graph Attention Layer

We next investigate the effectiveness of the auxiliary GAT in the graph-based domain fine-tuning module. To demonstrate that the auxiliary network enhances classification model performance, we conducted an ablation study by removing the GAT. We also compare the GAT with GCN, which has no attention mechanism. We present the overall accuracy and F1 score in Table 21. We observe that the model without a graph neural network yields an overall accuracy of 84.73% and an F1 score of 0.8109, and using a graph-based auxiliary network improves the classification performance. The model with GAT as the auxiliary network outperforms the model with GCN by a 0.67% increase in accuracy and a 0.0066 increase in the F1 score. This increase may be due to the fact that the GAT learns the weights between vertices via the attention mechanism rather than using predefined weights as in GCN. Based on the results in Table 21, we conclude that the GAT as an auxiliary domain fine-tuning module improves overall classification performance.

6 Final Remarks

Stress is becoming a significant concern with the potential to cause mental and physical health impairment in the long term. IoT and AI/ML techniques are enabling significant advances in stress monitoring. We have presented a resilient hierarchical IoT system for stress-level evaluation. The trade-off between communication latency, data transfer energy, and overall accuracy was evaluated when making offloading decisions. We also conducted a real-time stress monitoring simulation to demonstrate the benefits of our solution.

Conventional AI/ML models require combining data from all participants and training in a centralized server. However, such an approach gives rise to privacy concerns. Federated learning is thus employed to achieve a trade-off between privacy and AI/ML model utility. We have described customized, low-overhead clustered federated learning for stress monitoring. In addition, we have described a cold-start algorithm that yields high performance with low overhead when a new user joins the system. Additionally, we have introduced a novel federated clustered multi-domain learning algorithm to overcome intra-client data heterogeneity while preserving privacy using stress monitoring as a case study.

Future work for stress monitoring includes investigating other privacy-preserving techniques such as differential privacy and data encryption/compression, exploring personalized federated learning to develop customized models for each domain of the clients, incorporating feedback from clinical experts for model adjustment, and establishing a recommendation system for the subjects.

References

1. S. Jiang, F. Firouzi, K. Chakrabarty, and E. B. Elbogen, "A resilient and hierarchical IoT-based solution for stress monitoring in everyday settings," *IEEE Internet of Things Journal*, vol. 9, no. 12, pp. 10224–10243, June 15, 2022, https://doi.org/10.1109/JIOT.2021.3122015.
2. S. Jiang, F. Firouzi, and K. Chakrabarty, "Low-overhead clustered federated learning for personalized stress monitoring," *IEEE Internet of Things Journal*, vol. 11, no. 3, pp. 4335–4347, Feb. 1, 2024, https://doi.org/10.1109/JIOT.2023.3299736.
3. S. Jiang, Y. Li, F. Firouzi, et al., "Federated clustered multi-domain learning for health monitoring," Scientific Reports, vol. 14, article number: 903, 2024, https://doi.org/10.1038/s41598-024-51344-9.
4. C. Hammen, "Stress and depression," *Annual Review of Clinical Psychology*, vol. 1, pp. 293–319, 2005.
5. H. Yaribeygi, Y. Panahi, H. Sahraei, T. Johnston, and A. Sahebkar, "The impact of stress on body function: A review," *EXCLI Journal*, vol. 16, pp. 1057–1072, 2017.
6. J. Kim and K. Yoon, "Stress: metaplastic effects in the hippocampus," *Trends in neurosciences*, vol. 21, no. 12, pp. 505–509, 1998.
7. S. Lupien, B. McEwen, M. Gunnar, and C. Heim, "Effects of stress throughout the lifespan on the brain, behaviour and cognition," *Nature Reviews Neuroscience*, vol. 10, pp. 434–445, 2009.
8. L. Greco, G. Percannella, P. Ritrovato, F. Tortorella, and M. Vento, "Trends in IoT based solutions for health care: Moving AI to the edge," *Pattern Recognition Letters*, vol. 135, pp. 346–353, 2020.
9. B. Farahani, F. Firouzi, V. I. C. Chang, M. Badaroglu, N. Constant, and K. Mankodiya, "Towards fog-driven IoT eHealth: Promises and challenges of IoT in medicine and healthcare," *Future Gener. Comput. Syst.*, vol. 78, pp. 659–676, 2018.
10. P. Pace, G. Aloi, R. Gravina, G. Caliciuri, G. Fortino, and A. Liotta, "An Edge-Based Architecture to Support Efficient Applications for Healthcare Industry 4.0," *IEEE Transactions on Industrial Informatics*, vol. 15, pp. 481–489, 2019.
11. I. Azimi, J. Takalo-Mattila, A. Anzanpour, A. Rahmani, J. Soininen, and P. Liljeberg, "Empowering Healthcare IoT Systems with Hierarchical Edge-Based Deep Learning," *2018 IEEE/ACM International Conference on Connected Health: Applications, Systems and Engineering Technologies (CHASE)*, pp. 63–68, 2018.
12. M. Mozafari, F. Firouzi, and B. Farahani, "Towards IoT-enabled Multimodal Mental Stress Monitoring," in *2020 International Conference on Omni-layer Intelligent Systems (COINS)*, pp. 1–8, IEEE, 2020.
13. K. Plarre, A. Raij, and M. Scott, "Continuous inference of psychological stress from sensory measurements collected in the natural environment," in *2011 10th International Conference on Information Processing in Sensor Networks (IPSN)*, pp. 97–108, 2011.
14. K. Hovsepian, M. al'Absi, and S. Kumar, "cStress: Towards a gold standard for continuous stress assessment in the mobile environment," in *2015 UbiComp*, pp. 493–504, 2015.
15. M. Gjoreski, H. Gjoreski, and M. Gams, "Continuous stress detection using a wrist device: In laboratory and real life," in *2016 UbiComp*, pp. 1185–1193, 2016.

16. J. Kim and E. André, "Emotion recognition based on physiological changes in music listening," *IEEE Transactions on Pattern Analysis and Machine Intelligence*, vol. 30, no. 12, pp. 2067–2083, 2008.
17. P. Schmidt, A. Reiss, R. Duerichen, C. Marberger, and K. Van Laerhoven, "Introducing WESAD, a Multimodal Dataset for Wearable Stress and Affect Detection," in *2018 20th ACM International Conference on Multimodal Interaction (ICMI)*, pp. 400–408, 2018.
18. P. Paredes, R. Gilad-Bachrach, M. Czerwinski, A. Roseway, K. Rowan, and J. Hernandez, "PopTherapy: Coping with Stress through Pop-Culture," in *2014 8th International Conference on Pervasive Computing Technologies for Healthcare (PervasiveHealth)*, pp. 109–117, 2014.
19. S. A. H. Aqajari, E. K. Naeini, M. A. Mehrabadi, S. Labbaf, A. Rahmani, and N. Dutt, "GSR Analysis for Stress: Development and Validation of an Open Source Tool for Noisy Naturalistic GSR Data." arXiv:2005.01834, 2020.
20. A. Oskooei, S. M. Chau, J. R. M. Weiss, A. Sridhar, M. R. Martínez, and B. Michel, "DeStress: Deep Learning for Unsupervised Identification of Mental Stress in Firefighters from Heart-rate Variability (HRV) Data." arXiv:1911.13213, 2019.
21. S. Tripathi and H. Beigi, "Multi-Modal Emotion recognition on IEMOCAP Dataset using Deep Learning." arXiv:1804.05788, 2018.
22. C. Busso, S. Parthasarathy, A. Burmania, M. AbdelWahab, N. Sadoughi, and E. M. Provost, "MSP-IMPROV: An Acted Corpus of Dyadic Interactions to Study Emotion Perception," *IEEE Transactions on Affective Computing*, vol. 8, no. 1, pp. 67–80, 2017.
23. M. Jaiswal, C.-P. Bara, Y. Luo, M. Burzo, R. Mihalcea, and E. M. Provost, "MuSE: a Multimodal Dataset of Stressed Emotion," in *2020 12th Conference on Language Resources and Evaluation (LREC)*, pp. 1499–1510, 2020.
24. S. C. Guntuku, A. Buffone, K. Jaidka, J. C. Eichstaedt, and L. H. Ungar, "Understanding and measuring psychological stress using social media," in *2019 the Thirteenth International AAAI Conference on Web and Social Media (ICWSM)*, pp. 214–225, 2019.
25. H. Lin *et al.*, "Detecting stress based on social interactions in social networks," *IEEE Transactions on Knowledge and Data Engineering*, vol. 29, no. 9, pp. 1820–1833, 2017.
26. E. Turcan and K. McKeown, "Dreaddit: A Reddit Dataset for Stress Analysis in Social Media." arXiv:1911.00133, 2019.
27. R. Alharthi, R. Alharthi, B. Guthier, and A. E. Saddik, "CASP: context-aware stress prediction system," *Multimed Tools Appl*, vol. 78, pp. 9011–9031, 2019.
28. K. Nkurikiyeyezu, A. Yokokubo, and G. Lopez, "The effect of person-specific biometrics in improving generic stress predictive models." arXiv:1910.01770, 2019.
29. J. Lin, S. Pan, C. S. Lee, and S. Oviatt, "An explainable deep fusion network for affect recognition using physiological signals," in *28th ACM International Conference on Information and Knowledge Management (CIKM '19)*, pp. 2069–2072, 2019.
30. M. Mozafari, E. Khazaei, S. Bagheri, N. Hasanzadeh, B. TaghiBeyglou, and F. Ghassemi, "Mental stress level detection using physiological signals and evolutionary algorithm," in *2019 26th National and 4th International Iranian Conference on Biomedical Engineering (ICBME)*, pp. 75–81, 2019.
31. P. Sarkar and A. Etemad, "Self-supervised ECG Representation Learning for Emotion Recognition," in *2020 IEEE International Conference on Acoustics, Speech and Signal Processing (ICASSP)*, pp. 3217–3221, 2020.
32. C. Stewart, A. Folarin, and R. Dobson, "Personalized acute stress classification from physiological signals with neural processes." arXiv:2002.04176, 2020.
33. P. Mohassel and Y. Zhang, "SecureML: A System for Scalable Privacy-Preserving Machine Learning," *2017 IEEE Symposium on Security and Privacy (SP)*, pp. 19–38, 2017.
34. Z. Ji, Z. C. Lipton, and C. P. Elkan, "Differential Privacy and Machine Learning: a Survey and Review," *arXiv:1412.7584*, 2014.

35. N. Dowlin, R. Gilad-Bachrach, K. Laine, K. E. Lauter, M. Naehrig, and J. R. Wernsing, "CryptoNets: applying neural networks to encrypted data with high throughput and accuracy," in *Proceedings of The 33rd International Conference on Machine Learning*, vol. 48, pp. 201–210, 2016.
36. H. B. McMahan, E. Moore, D. Ramage, S. Hampson, and B. A. y Arcas, "Communication-Efficient Learning of Deep Networks from Decentralized Data," *ARTIFICIAL INTELLIGENCE AND STATISTICS*, vol. 54, pp. 1273–1282, 2017.
37. S. Abdulrahman, H. Tout, H. Ould-Slimane, A. Mourad, C. Talhi, and M. Guizani, "A Survey on Federated Learning: The Journey From Centralized to Distributed On-Site Learning and Beyond," *IEEE Internet of Things Journal*, vol. 8, no. 7, pp. 5476–5497, 2021.
38. J. Xu and F. Wang, "Federated Learning for Healthcare Informatics," *Journal of Healthcare Informatics Research*, vol. 5, pp. 1–19, 2021.
39. Q. Wu, K. He, and X. Chen, "Personalized Federated Learning for Intelligent IoT Applications: A Cloud-Edge Based Framework," *IEEE Open Journal of the Computer Society*, vol. 1, pp. 35–44, 2020.
40. D. C. Nguyen, M. Ding, P. N. Pathirana, A. Seneviratne, J. Li, and H. Vincent Poor, "Federated Learning for Internet of Things: A Comprehensive Survey," *IEEE Communications Surveys & Tutorials*, vol. 23, no. 3, pp. 1622–1658, 2021.
41. L. U. Khan, W. Saad, Z. Han, E. Hossain, and C. S. Hong, "Federated Learning for Internet of Things: Recent Advances, Taxonomy, and Open Challenges," *IEEE Communications Surveys & Tutorials*, vol. 23, no. 3, pp. 1759–1799, 2021.
42. P. Chhikara, P. Singh, R. Tekchandani, N. Kumar, and M. Guizani, "Federated Learning Meets Human Emotions: A Decentralized Framework for Human–Computer Interaction for IoT Applications," *IEEE Internet of Things Journal*, vol. 8, no. 8, pp. 6949–6962, 2021.
43. Y. Chen, J. Wang, C. Yu, W. Gao, and X. Qin, "FedHealth: A Federated Transfer Learning Framework for Wearable Healthcare," *IEEE Intelligent Systems*, vol. 35, pp. 83–93, 2020.
44. Y. Zhao, M. Li, L. Lai, N. Suda, D. Civin, and V. Chandra, "Federated Learning with Non-IID Data," *arXiv preprint arXiv:1806.00582*, 2018.
45. Q. Li, Y. Diao, Q. Chen, and B. He, "Federated Learning on Non-IID Data Silos: An Experimental Study," *2022 IEEE 38th International Conference on Data Engineering (ICDE)*, pp. 965–978, 2021.
46. Y. Lin and G. Wei, "Speech emotion recognition based on HMM and SVM," in *2005 International Conference on Machine Learning and Cybernetics*, vol. 8, pp. 4898–4901, 2005.
47. M. Hossain, M. Ghulam, M. F. Alhamid, B. Song, and K. A. Mutib, "Audio-Visual Emotion Recognition Using Big Data Towards 5G," *Mobile Networks and Applications*, vol. 21, pp. 753–763, 2016.
48. A. Kumar, K. Sharma, and A. Sharma, "Hierarchical deep neural network for mental stress state detection using IoT based biomarkers," *Pattern Recognit. Lett.*, vol. 145, pp. 81–87, 2021.
49. S. A. Khowaja, A. G. Prabono, F. Setiawan, B. N. Yahya, and S.-L. Lee, "Toward soft real-time stress detection using wrist-worn devices for human workspaces," *Soft Comput.*, vol. 25, pp. 2793–2820, 2021.
50. R. K. Nath and H. Thapliyal, "Smart wristband-based stress detection framework for older adults with cortisol as stress biomarker," *IEEE Transactions on Consumer Electronics*, vol. 67, pp. 30–39, 2021.
51. J.-Y. Jiang, Z. Chao, A. L. Bertozzi, W. Wang, S. D. Young, and D. Needell, "Learning to Predict Human Stress Level with Incomplete Sensor Data from Wearable Devices," in *Proceedings of the 28th ACM International Conference on Information and Knowledge Management*, pp. 2773–2781, Association for Computing Machinery, 2019.
52. Y. S. Can, N. Chalabianloo, D. Ekiz, and C. Ersoy, "Continuous Stress Detection Using Wearable Sensors in Real Life: Algorithmic Programming Contest Case Study," *Sensors (Basel, Switzerland)*, vol. 19, no. 8, p. 1849, 2019.

53. Y. S. Can, B. Arnrich, and C. Ersoy, "Stress detection in daily life scenarios using smart phones and wearable sensors: A survey," *Journal of biomedical informatics*, vol. 92, p. 103139, 2019.
54. K. M. Dalmeida and G. L. Masala, "HRV Features as Viable Physiological Markers for Stress Detection Using Wearable Devices," *Sensors (Basel, Switzerland)*, vol. 21, no. 8, p. 2873, 2021.
55. L. Coutts, D. Plans, A. W. Brown, and J. P. Collomosse, "Deep learning with wearable based heart rate variability for prediction of mental and general health," *Journal of biomedical informatics*, vol. 112, p. 103610, 2020.
56. E. Smets *et al.*, "Large-scale wearable data reveal digital phenotypes for daily-life stress detection," *NPJ Digital Medicine*, vol. 1, p. 67, 2018.
57. G. Vos, K. Trinh, Z. Sarnyai, and M. R. Azghadi, "Ensemble Machine Learning Model Trained on a New Synthesized Dataset Generalizes Well for Stress Prediction Using Wearable Devices," *Journal of biomedical informatics*, vol. 148, p. 104556, 2023.
58. M. Usak, M. Kubiatko, M. Shabbir, O. Dudnik, K. Jermsittiparsert, and L. Rajabion, "Health care service delivery based on the Internet of things: A systematic and comprehensive study," *Int. J. Commun. Syst.*, vol. 33, 2020.
59. Y. A. Qadri, A. Nauman, Y. B. Zikria, A. Vasilakos, and S. Kim, "The Future of Healthcare Internet of Things: A Survey of Emerging Technologies," *IEEE Communications Surveys & Tutorials*, vol. 22, pp. 1121–1167, 2020.
60. F. Alshehri and G. Muhammad, "A Comprehensive Survey of the Internet of Things (IoT) and AI-Based Smart Healthcare," *IEEE Access*, vol. 9, pp. 3660–3678, 2021.
61. F. Firouzi, B. Farahani, M. Ibrahim, and K. Chakrabarty, "Keynote Paper: From EDA to IoT eHealth: Promises, Challenges, and Solutions," *IEEE Transactions on Computer-Aided Design of Integrated Circuits and Systems*, vol. 37, pp. 2965–2978, 2018.
62. V. Bianchi, M. Bassoli, G. Lombardo, P. Fornacciari, M. Mordonini, and I. D. Munari, "IoT Wearable Sensor and Deep Learning: An Integrated Approach for Personalized Human Activity Recognition in a Smart Home Environment," *IEEE Internet of Things Journal*, vol. 6, pp. 8553–8562, 2019.
63. Y. Qiu, H. Zhang, and K. Long, "Computation Offloading and Wireless Resource Management for Healthcare Monitoring in Fog-Computing based Internet of Medical Things," *IEEE Internet of Things Journal*, 2021. Early Access.
64. Z. Ning, P. Dong, X. Wang, X. Hu, L. Guo, B. Hu, Y. Guo, T. Qiu, and R. Y. K. Kwok, "Mobile Edge Computing Enabled 5G Health Monitoring for Internet of Medical Things: A Decentralized Game Theoretic Approach," *IEEE Journal on Selected Areas in Communications*, vol. 39, pp. 463–478, 2021.
65. A. Rahmani *et al.*, "Exploiting smart e-Health gateways at the edge of healthcare Internet-of-Things: A fog computing approach," *Future Gener. Comput. Syst.*, vol. 78, pp. 641–658, 2018.
66. X. Zhou, W. Liang, K. Wang, H. Wang, L. Yang, and Q. Jin, "Deep-Learning-Enhanced Human Activity Recognition for Internet of Healthcare Things," *IEEE Internet of Things Journal*, vol. 7, pp. 6429–6438, 2020.
67. F. Samie, L. Bauer, and J. Henkel, "Hierarchical Classification for Constrained IoT Devices: A Case Study on Human Activity Recognition," *IEEE Internet of Things Journal*, vol. 7, no. 9, pp. 8287–8295, 2020.
68. F. Samie, S. Paul, L. Bauer, and J. Henkel, "Highly Efficient and Accurate Seizure Prediction on Constrained IoT Devices," in *2018 Design, Automation and Test in Europe Conference and Exhibition (DATE)*, pp. 955–960, 2018.
69. B. Farahani, M. Barzegari, F. S. Aliee, and K. A. Shaik, "Towards collaborative intelligent IoT eHealth: From device to fog, and cloud," *Microprocess. Microsystems*, vol. 72, 2020.
70. M. A. Rahman and M. S. Hossain, "An Internet of Medical Things-Enabled Edge Computing Framework for Tackling COVID-19," *IEEE Internet of Things Journal*, 2021. Early Access.
71. M. Aazam, S. Zeadally, and K. A. Harras, "Offloading in fog computing for IoT: Review, enabling technologies, and research opportunities," *Future Gener. Comput. Syst.*, vol. 87, pp. 278–289, 2018.

72. M. Min, X. Wan, L. Xiao, Y. Chen, M. Xia, D. Wu, and H. Dai, "Learning-Based Privacy-Aware Offloading for Healthcare IoT With Energy Harvesting," *IEEE Internet of Things Journal*, vol. 6, pp. 4307–4316, 2019.
73. X. Liu, J. Yu, J. Wang, and Y. Gao, "Resource Allocation With Edge Computing in IoT Networks via Machine Learning," *IEEE Internet of Things Journal*, vol. 7, pp. 3415–3426, 2020.
74. X. Q. Pham and E. N. Huh, "Towards task scheduling in a cloud-fog computing system," *2016 18th Asia-Pacific Network Operations and Management Symposium (APNOMS)*, pp. 1–4, 2016.
75. Z. Zhou, S. Yu, W. Chen, and X. Chen, "CE-IoT: Cost-Effective Cloud-Edge Resource Provisioning for Heterogeneous IoT Applications," *IEEE Internet of Things Journal*, vol. 7, pp. 8600–8614, 2020.
76. Z. Zhou, Z. Wang, H. Yu, H. Liao, S. Mumtaz, L. Oliveira, and V. Frascolla, "Learning-Based URLLC-Aware Task Offloading for Internet of Health Things," *IEEE Journal on Selected Areas in Communications*, vol. 39, pp. 396–410, 2021.
77. S. Din, A. Paul, N. Guizani, S. H. Ahmed, M. Khan, and M. Rathore, "Features Selection Model for Internet of E-Health Things Using Big Data," *GLOBECOM 2017 - 2017 IEEE Global Communications Conference*, pp. 1–7, 2017.
78. S. Ghosh, A. Mukherjee, S. K. Ghosh, and R. Buyya, "Mobi-IoST: Mobility-Aware Cloud-Fog-Edge-IoT Collaborative Framework for Time-Critical Applications," *IEEE Transactions on Network Science and Engineering*, vol. 7, pp. 2271–2285, 2020.
79. M. S. Hossain and M. Ghulam, "Emotion recognition using secure edge and cloud computing," *Inf. Sci.*, vol. 504, pp. 589–601, 2019.
80. O. Krestinskaya, A. P. James, and L. Chua, "Neuromemristive Circuits for Edge Computing: A Review," *IEEE Transactions on Neural Networks and Learning Systems*, vol. 31, pp. 4–23, 2020.
81. S. Jiang, F. Firouzi, and K. Chakrabarty, "Low-overhead Clustered Federated Learning for Personalized Stress Monitoring," *IEEE Internet of Things Journal*, vol. 11, no. 3, pp. 4335–4347, 2024.
82. C. Briggs, Z. Fan, and P. András, "Federated learning with hierarchical clustering of local updates to improve training on non-IID data," *2020 International Joint Conference on Neural Networks (IJCNN)*, pp. 1–9, 2020.
83. A. Ghosh, J. Chung, D. Yin, and K. Ramchandran, "An Efficient Framework for Clustered Federated Learning," *arXiv:2006.04088*, 2020.
84. Y. Kim, E. A. Hakim, J. Haraldson, H. Eriksson, J. M. B. da Silva, and C. Fischione, "Dynamic Clustering in Federated Learning," *ICC 2021-IEEE International Conference on Communications*, pp. 1–6, 2021.
85. A. Li, J. Sun, B. Wang, L. Duan, S. Li, Y. Chen, and H. H. Li, "LotteryFL: Empower Edge Intelligence with Personalized and Communication-Efficient Federated Learning," *2021 IEEE/ACM Symposium on Edge Computing (SEC)*, pp. 68–79, 2021.
86. A. Li, J. Sun, P. Li, Y. Pu, H. H. Li, and Y. Chen, "Hermes: an efficient federated learning framework for heterogeneous mobile clients," *Proceedings of the 27th Annual International Conference on Mobile Computing and Networking*, pp. 420–437, 2021.
87. H. B. McMahan, D. Ramage, K. Talwar, and L. Zhang, "Learning Differentially Private Recurrent Language Models," in *International Conference on Learning Representations*, 2018.
88. K. Wei *et al.*, "Federated Learning With Differential Privacy: Algorithms and Performance Analysis," *IEEE Transactions on Information Forensics and Security*, vol. 15, pp. 3454–3469, 2019.
89. C. Zhang, S. Li, J. Xia, W. Wang, F. Yan, and Y. Liu, "BatchCrypt: Efficient Homomorphic Encryption for Cross-Silo Federated Learning," in *Proceedings of the 2020 USENIX Conference on Usenix Annual Technical Conference*, pp. 493–506, 2020. Article no. 33.
90. J. Xu, H. Wang, and L. Chen, "Bandwidth Allocation for Multiple Federated Learning Services in Wireless Edge Networks," *IEEE Transactions on Wireless Communications*, vol. 21, no. 4, pp. 2534–2546, 2022.

91. Z. Yang, M. Chen, W. Saad, C. S. Hong, and M. R. Shikh-Bahaei, "Energy Efficient Federated Learning Over Wireless Communication Networks," *IEEE Transactions on Wireless Communications*, vol. 20, no. 3, pp. 1935–1949, 2021.
92. M. Chen, N. Shlezinger, H. V. Poor, Y. C. Eldar, and S. Cui, "Communication-efficient federated learning," *Proceedings of the National Academy of Sciences*, vol. 118, no. 17, 2021. e2024789118.
93. V. Smith, C.-K. Chiang, M. Sanjabi, and A. S. Talwalkar, "Federated Multi-Task Learning," in *31st Conference on Neural Information Processing Systems (NIPS)*, 2017.
94. Y. Chen, Y. Ning, M. Slawski, and H. Rangwala, "Asynchronous Online Federated Learning for Edge Devices with Non-IID Data," *2020 IEEE International Conference on Big Data (Big Data)*, pp. 15–24, 2019.
95. S. Jiang, F. Firouzi, K. Chakrabarty, and E. B. Elbogen, "A Resilient and Hierarchical IoT-based Solution for Stress Monitoring in Everyday Settings," *IEEE Internet of Things Journal*, vol. 9, no. 12, pp. 10224–10243, 2022.
96. A. Emam, A. Abdellatif, A. M. Mohamed, and K. A. Harras, "EdgeHealth: An Energy-Efficient Edge-based Remote mHealth Monitoring System," *2019 IEEE Wireless Communications and Networking Conference (WCNC)*, pp. 1–7, 2019.
97. F. Firouzi, B. Farahani, and A. Marinšek, "The convergence and interplay of edge, fog, and cloud in the AI-driven Internet of Things (IoT)," *Information Systems*, p. 101840, 2021.
98. F. Firouzi and B. Farahani, "Architecting IoT Cloud," in *Intelligent Internet of Things* (F. Firouzi, K. Chakrabarty, and S. Nassif, eds.), pp. 173–241, Springer, 2020.
99. "National Institute of Standards and Technology." https://www.nist.gov. Accessed: 2021-08.
100. P. Gomes, P. Margaritoff, and H. Silva, "pyHRV: Development and evaluation of an open-source python toolbox for heart rate variability (HRV)," in *Proc. Int'l Conf. on Electrical, Electronic and Computing Engineering (IcETRAN)*, pp. 822–828, 2019.
101. I. Azimi, T. Pahikkala, A. Rahmani, H. Niela-Vilén, A. Axelin, and P. Liljeberg, "Missing data resilient decision-making for healthcare IoT through personalization: A case study on maternal health," *Future Generation Computer Systems*, vol. 96, pp. 297–308, 2019.
102. A. Donders, G. van der Heijden, T. Stijnen, and K. Moons, "Review: A gentle introduction to imputation of missing values," *Journal of Clinical Epidemiology*, vol. 59, no. 10, pp. 1087–1091, 2006.
103. S. van Buuren and K. Groothuis-Oudshoorn, "mice: Multivariate Imputation by Chained Equations in R," *Journal of Statistical Software*, vol. 45, no. 3, pp. 1–67, 2011.
104. Nurzaman, T. Siswantining, S. Soemartojo, and D. Sarwinda, "Application of sequential regression multivariate imputation method on multivariate normal missing data," in *2019 3rd International Conference on Informatics and Computational Sciences (ICICoS)*, pp. 1–6, 2019.
105. H. Gamboa, *Multi-Modal Behavioral Biometrics based on HCI and Electrophysiology*. PhD thesis, Instituto Superior Técnico, 2008.
106. A. Greco, G. Valenza, A. Lanata, E. P. Scilingo, and L. Citi, "cvxEDA: A Convex Optimization Approach to Electrodermal Activity Processing," *IEEE Transactions on Biomedical Engineering*, vol. 63, no. 4, pp. 797–804, 2016.
107. I. Christov, "Real time electrocardiogram QRS detection using combined adaptive threshold," *Biomed Eng Online*, vol. 3, no. 1, 2004. doi: 10.1186/1475-925X-3-28.
108. M. Malik and C. Terrace, "Heart rate variability. standards of measurement, physiological interpretation, and clinical use," *European Heart Journal*, vol. 17, pp. 354–381, 1996.
109. P. Welch, "The use of fast fourier transform for the estimation of power spectra: A method based on time averaging over short, modified periodograms," *IEEE Transactions on Audio and Electroacoustics*, vol. 15, pp. 70–73, 1967.
110. X. He, D. Cai, and P. Niyogi, "Laplacian score for feature selection," in *Advances in Neural Information Processing Systems 18*, NIPS'05, (Cambridge, MA, USA), pp. 507–514, MIT Press, 2006.

111. G. Brown, A. Pocock, M.-J. Zhao, and M. Luján, "Conditional likelihood maximisation: A unifying framework for information theoretic feature selection," *J. Mach. Learn. Res.*, vol. 13, pp. 27–66, 2012.
112. P. E. Meyer and G. Bontempi, "On the use of variable complementarity for feature selection in cancer classification," in *Applications of Evolutionary Computing*, vol. 3907 of *EuroGP'06*, (Berlin, Heidelberg), pp. 91–102, Springer-Verlag, 2006.
113. L. Paninski, "Estimation of entropy and mutual information," *Neural Computation*, vol. 15, no. 6, pp. 1191–1253, 2003.
114. C. D. Schuman *et al.*, "A survey of neuromorphic computing and neural networks in hardware." arXiv:1705.06963, 2017.
115. W. Maass, "Noisy Spiking Neurons with Temporal Coding have more Computational Power than Sigmoidal Neurons," in *Advances in Neural Information Processing Systems*, vol. 9, pp. 211–217, 1997.
116. A. Amirshahi and M. Hashemi, "ECG Classification Algorithm Based on STDP and R-STDP Neural Networks for Real-Time Monitoring on Ultra Low-Power Personal Wearable Devices," *IEEE Transactions on Biomedical Circuits and Systems*, vol. 13, no. 6, pp. 1483–1493, 2019.
117. Y. Ma, B. Chen, P. Ren, N. Zheng, G. Indiveri, and E. Donati, "EMG-Based Gestures Classification Using a Mixed-Signal Neuromorphic Processing System," *IEEE Journal on Emerging and Selected Topics in Circuits and Systems*, vol. 10, pp. 578–587, 2020.
118. F. Corradi *et al.*, "ECG-based Heartbeat Classification in Neuromorphic Hardware," in *2019 International Joint Conference on Neural Networks (IJCNN)*, pp. 1–8, 2019.
119. E. Neftci, H. Mostafa, and F. Zenke, "Surrogate Gradient Learning in Spiking Neural Networks." arXiv:1901.09948, 2019.
120. Y. Jin, W. Zhang, and P. Li, "Hybrid Macro/Micro Level Backpropagation for Training Deep Spiking Neural Networks," in *Proceedings of the 32nd International Conference on Neural Information Processing Systems (NIPS'18)*, pp. 7005–7015, Curran Associates Inc., 2018.
121. K. He, X. Zhang, S. Ren, and J. Sun, "Deep Residual Learning for Image Recognition," *2016 IEEE Conference on Computer Vision and Pattern Recognition (CVPR)*, pp. 770–778, 2016.
122. A. G. Howard *et al.*, "MobileNets: Efficient Convolutional Neural Networks for Mobile Vision Applications." arXiv:1704.04861, 2017.
123. B. Yang, G. Bender, Q. V. Le, and J. Ngiam, "CondConv: Conditionally Parameterized Convolutions for Efficient Inference," in *Advances in Neural Information Processing Systems*, vol. 32, Curran Associates, Inc., 2019.
124. T. J. Jun, H. M. Nguyen, D. Kang, D. Kim, D. Kim, and Y. Kim, "ECG arrhythmia classification using a 2-D convolutional neural network." arXiv:1804.06812, 2018.
125. Y. Wang, Q. Yao, J. T. Kwok, and L. M. Ni, "Generalizing from a Few Examples: A Survey on Few-Shot Learning," *ACM Comput. Surv.*, vol. 53, no. 3, 2020. article no.63, 34 pages.
126. M. Fink, "Object classification from a single example utilizing class relevance metrics," in *Proceedings of the 17th International Conference on Neural Information Processing Systems (NIPS'04)*, pp. 449–456, MIT Press, 2004.
127. G. Koch, R. Zemel, and R. Salakhutdinov, "Siamese Neural Networks for One-Shot Image Recognition," in *Proceedings of the 32nd International Conference on Machine Learning (ICML'15)*, vol. 37, 2015.
128. O. Vinyals, C. Blundell, T. P. Lillicrap, K. Kavukcuoglu, and D. Wierstra, "Matching networks for one shot learning." arXiv:1606.04080, 2016.
129. K. P. C. Kirschbaum and D. Hellhammer, "The 'Trier Social Stress Test'–a tool for investigating psychobiological stress responses in a laboratory setting," *Neuropsychobiology*, vol. 28, no. 1–2, pp. 76–81, 1993.
130. J. Graham, A. Olchowski, and T. Gilreath, "How Many Imputations are Really Needed? Some Practical Clarifications of Multiple Imputation Theory," *Prevention Science*, vol. 8, pp. 206–213, 2007.
131. J. Li *et al.*, "Feature selection: A data perspective," *ACM Computing Surveys (CSUR)*, vol. 50, no. 6, pp. 94:1–94:45, 2018.

132. C.-W. Hsu, C.-C. Chang, and C.-J. Lin, “A Practical Guide to Support Vector Classification,” tech. rep., National Taiwan University, 2003.
133. T. Elsken, J. H. Metzen, and F. Hutter, “Neural Architecture Search: A Survey.” arXiv:1808.05377, 2019.
134. D. P. Kingma and J. Ba, “Adam: A method for stochastic optimization.” arXiv:1412.6980, 2017.
135. I. Loshchilov and F. Hutter, “Decoupled weight decay regularization.” arXiv:1711.05101, 2019.
136. C. Pehle and J. E. Pedersen, “Norse-A deep learning library for spiking neural networks.” https://doi.org/10.5281/zenodo.4422025, 2021. Documentation: https://norse.ai/docs/.
137. Z. Zhong, L. Zheng, G. Kang, S. Li, and Y. Yang, “Random erasing data augmentation.” arXiv:1708.04896, 2017.
138. C. Burges, “A tutorial on support vector machines for pattern recognition,” *Data Mining and Knowledge Discovery*, vol. 2, pp. 121–167, 1998.
139. P. Panda, S. A. Aketi, and K. Roy, “Toward Scalable, Efficient, and Accurate Deep Spiking Neural Networks With Backward Residual Connections, Stochastic Softmax, and Hybridization,” *Frontiers in Neuroscience*, vol. 14, 2020.
140. J. L. Schafer, “Multiple imputation: a primer,” *Statistical Methods in Medical Research*, vol. 8, no. 1, pp. 3–15, 1999.
141. Y. Dong and C. Peng, “Principled missing data methods for researchers,” *Springer Plus*, vol. 2, 2013. doi: 10.1186/2193-1801-2-222.
142. T. Nguyen, M. Raghu, and S. Kornblith, “Do Wide and Deep Networks Learn the Same Things? Uncovering How Neural Network Representations Vary with Width and Depth.” arXiv:2010.15327, 2021.
143. “TestMy.net Internet Speed Test.” Accessed: 2022-02.
144. J. Huang, F. Qian, A. Gerber, Z. M. Mao, S. Sen, and O. Spatscheck, “A Close Examination of Performance and Power Characteristics of 4G LTE Networks,” in *Proceedings of the 10th International Conference on Mobile Systems, Applications, and Services (MobiSys'12)*, pp. 225–238, Association for Computing Machinery, 2012.
145. D. Molka, D. Hackenberg, R. Schöne, and M. S. Müller, “Characterizing the energy consumption of data transfers and arithmetic operations on x86-64 processors,” in *International Conference on Green Computing*, pp. 123–133, 2010.
146. C. Samson and A. Koh, “Stress Monitoring and Recent Advancements in Wearable Biosensors,” *Frontiers in Bioengineering and Biotechnology*, vol. 8, p. 1037, 2020.
147. H. C. Ates, C. Ates, and C. Dincer, “Stress monitoring with wearable technology and AI,” *Nature Electronics*, vol. 7, pp. 98–99, 2024.
148. G. Vos, K. Trinh, Z. Sarnyai, and M. R. Azghadi, “Generalizable machine learning for stress monitoring from wearable devices: A systematic literature review,” *International journal of medical informatics*, vol. 173, p. 105026, 2023.
149. R. K. Nath, H. Thapliyal, and A. Caban-Holt, “Machine Learning Based Stress Monitoring in Older Adults Using Wearable Sensors and Cortisol as Stress Biomarker,” *Journal of Signal Processing Systems*, vol. 94, pp. 513–525, 2022.
150. R. K. Nath, H. Thapliyal, A. Caban-Holt, and S. P. Mohanty, “Machine Learning Based Solutions for Real-Time Stress Monitoring,” *IEEE Consumer Electronics Magazine*, vol. 9, pp. 34–41, 2020.
151. S. Gedam and S. Paul, “A Review on Mental Stress Detection Using Wearable Sensors and Machine Learning Techniques,” *IEEE Access*, vol. 9, pp. 84045–84066, 2021.
152. H. J. Han, S. Labbaf, J. L. Borelli, N. D. Dutt, and A.-M. Rahmani, “Objective stress monitoring based on wearable sensors in everyday settings,” *Journal of Medical Engineering & Technology*, vol. 44, pp. 177–189, 2020.
153. E. Garcia-Ceja, M. Riegler, T. Nordgreen, P. Jakobsen, K. J. Oedegaard, and J. Tørresen, “Mental health monitoring with multimodal sensing and machine learning: A survey,” *Pervasive Mob. Comput.*, vol. 51, pp. 1–26, 2018.

154. M. Duan, D. Liu, X. Ji, Y. Wu, L. Liang, X. Chen, Y. Tan, and A. Ren, "Flexible Clustered Federated Learning for Client-Level Data Distribution Shift," *IEEE Transactions on Parallel and Distributed Systems*, pp. 1–1, 2021. Early Access.
155. X. Ouyang, Z. Xie, J. Zhou, J. Huang, and G. Xing, "ClusterFL: A Similarity-Aware Federated Learning System for Human Activity Recognition," in *Proceedings of the 19th Annual International Conference on Mobile Systems, Applications, and Services*, MobiSys'21, (New York, NY, USA), pp. 54–66, Association for Computing Machinery, 2021.
156. P. P. Liang, T. Liu, L. Ziyin, R. Salakhutdinov, and L.-P. Morency, "Think locally, act globally: Federated learning with local and global representations," *arXiv:2001.01523*, 2020.
157. S. Mukherjee, H. Asnani, E. Lin, and S. Kannan, "ClusterGAN: Latent Space Clustering in Generative Adversarial Networks," in *Proceedings of the Thirty-Third AAAI Conference on Artificial Intelligence and Thirty-First Innovative Applications of Artificial Intelligence Conference and Ninth AAAI Symposium on Educational Advances in Artificial Intelligence*, AAAI'19/IAAI'19/EAAI'19, AAAI Press, 2019. Article no. 566.
158. S. Rabanser, S. Günnemann, and Z. C. Lipton, "Failing Loudly: An Empirical Study of Methods for Detecting Dataset Shift," in *NeurIPS*, 2019.
159. A. Gretton, K. M. Borgwardt, M. J. Rasch, B. Schölkopf, and A. Smola, "A Kernel Two-Sample Test," *J. Mach. Learn. Res.*, vol. 13, pp. 723–773, 2012.
160. F. Liu, W. Xu, J. Lu, G. Zhang, A. Gretton, and D. J. Sutherland, "Learning Deep Kernels for Non-Parametric Two-Sample Tests," in *ICML*, 2020.
161. Yu and Shi, "Multiclass spectral clustering," in *Proceedings Ninth IEEE International Conference on Computer Vision*, vol. 1, pp. 313–319, 2003.
162. U. von Luxburg, "A tutorial on spectral clustering," *Statistics and Computing*, vol. 17, pp. 395–416, 2007.
163. P. Rousseeuw, "Silhouettes: a graphical aid to the interpretation and validation of cluster analysis," *Journal of Computational and Applied Mathematics*, vol. 20, pp. 53–65, 1987.
164. K. He and J. Sun, "Convolutional Neural Networks at Constrained Time Cost," in *2015 IEEE Conference on Computer Vision and Pattern Recognition (CVPR)*, pp. 5353–5360, IEEE Computer Society, 2015.
165. P. Schmidt, A. Reiss, R. Dürichen, C. Marberger, and K. V. Laerhoven, "Introducing WESAD, a Multimodal Dataset for Wearable Stress and Affect Detection," in *Proceedings of the 20th ACM International Conference on Multimodal Interaction*, pp. 400–408, 2018.
166. F. Pedregosa *et al.*, "Scikit-learn: Machine Learning in Python," *Journal of Machine Learning Research*, vol. 12, pp. 2825–2830, 2011.
167. A. Paszke, S. Gross, F. Massa, A. Lerer, J. Bradbury, G. Chanan, *et al.*, "PyTorch: An Imperative Style, High-Performance Deep Learning Library," in *Advances in Neural Information Processing Systems 32* (H. Wallach, H. Larochelle, A. Beygelzimer, F. d'Alché-Buc, E. Fox, and R. Garnett, eds.), pp. 8024–8035, Curran Associates, Inc., 2019.
168. W. Zhang, T. Zhou, Q. Lu, X. Wang, C. Zhu, H. Sun, Z. Wang, S. K. Lo, and F. Y. Wang, "Dynamic-Fusion-Based Federated Learning for COVID-19 Detection," *IEEE Internet of Things Journal*, vol. 8, no. 21, pp. 15884–15891, 2021.
169. T. Hoefler, D. Alistarh, T. Ben-Nun, N. Dryden, and A. Peste, "Sparsity in Deep Learning: Pruning and growth for efficient inference and training in neural networks," *J. Mach. Learn. Res.*, vol. 22, pp. 241:1–241:124, 2021.
170. T. Liang, C. J. Glossner, L. Wang, and S. Shi, "Pruning and Quantization for Deep Neural Network Acceleration: A Survey," *Neurocomputing*, vol. 461, pp. 370–403, 2021.
171. D. Huba *et al.*, "Papaya: Practical, private, and scalable federated learning," in *Proceedings of Machine Learning and Systems*, vol. 4, pp. 814–832, 2022.
172. S. Yang, W. Zheng, M. Xie, and X. Zhang, "Research of Federated Learning Application Methods and Social Responsibility," *IEEE Transactions on Big Data*, 2022. Early Access.
173. F. Zhang *et al.*, "Unified Group Fairness on Federated Learning," *arXiv:2111.04986*, 2021.
174. Y. H. Ezzeldin, S. Yan, C. He, E. Ferrara, and A. S. Avestimehr, "FairFed: Enabling Group Fairness in Federated Learning," *arXiv:2110.00857*, 2021.

175. B. McMahan, E. Moore, D. Ramage, S. Hampson, and B. A. y Arcas, "Communication-efficient learning of deep networks from decentralized data," in *Artificial intelligence and statistics*, pp. 1273–1282, PMLR, 2017.
176. X. Long, R. Haakma, T. R. Leufkens, P. Fonseca, and R. M. Aarts, "Effects of between- and within-subject variability on autonomic cardiorespiratory activity during sleep and their limitations on sleep staging: a multilevel analysis," *Computational intelligence and neuroscience*, vol. 2015, 2015.
177. N. Karimian, Z. Guo, M. Tehranipoor, and D. Forte, "Highly reliable key generation from electrocardiogram (ECG)," *IEEE Transactions on Biomedical Engineering*, vol. 64, no. 6, pp. 1400–1411, 2016.
178. X. Teng, S. Pei, and Y. Lin, "StoCast: Stochastic Disease Forecasting With Progression Uncertainty," *IEEE Journal of Biomedical and Health Informatics*, vol. 25, pp. 850–861, 2020.
179. S. Jiang, Y. Li, F. Firouzi, and K. Chakrabarty, "Federated clustered multi-domain learning for health monitoring," *Scientific Reports*, vol. 14, p. 903, 2024.
180. A. Ghosh, J. Chung, D. Yin, and K. Ramchandran, "An efficient framework for clustered federated learning," *Advances in Neural Information Processing Systems*, vol. 33, pp. 19586–19597, 2020.
181. C. Briggs, Z. Fan, and P. Andras, "Federated learning with hierarchical clustering of local updates to improve training on non-IID data," in *2020 International Joint Conference on Neural Networks (IJCNN)*, pp. 1–9, IEEE, 2020.
182. A. Fallah, A. Mokhtari, and A. Ozdaglar, "Personalized federated learning: A meta-learning approach," *arXiv preprint arXiv:2002.07948*, 2020.
183. C. Finn, P. Abbeel, and S. Levine, "Model-agnostic meta-learning for fast adaptation of deep networks," in *International conference on machine learning*, pp. 1126–1135, PMLR, 2017.
184. V. Smith, C.-K. Chiang, M. Sanjabi, and A. S. Talwalkar, "Federated multi-task learning," *Advances in neural information processing systems*, vol. 30, 2017.
185. Y. Deng, M. M. Kamani, and M. Mahdavi, "Adaptive personalized federated learning," *arXiv preprint arXiv:2003.13461*, 2020.
186. D. Caldarola, M. Mancini, F. Galasso, M. Ciccone, E. Rodolà, and B. Caputo, "Cluster-driven graph federated learning over multiple domains," in *Proceedings of the IEEE/CVF Conference on Computer Vision and Pattern Recognition*, pp. 2749–2758, 2021.
187. D. Shenaj *et al.*, "Learning Across Domains and Devices: Style-Driven Source-Free Domain Adaptation in Clustered Federated Learning," *2023 IEEE/CVF Winter Conference on Applications of Computer Vision (WACV)*, pp. 444–454, 2022.
188. M. Dredze, A. Kulesza, and K. Crammer, "Multi-domain learning by confidence-weighted parameter combination," *Machine Learning*, vol. 79, no. 1, pp. 123–149, 2010.
189. H. Bilen and A. Vedaldi, "Universal representations: The missing link between faces, text, planktons, and cat breeds," *arXiv preprint arXiv:1701.07275*, 2017.
190. S.-A. Rebuffi, H. Bilen, and A. Vedaldi, "Learning multiple visual domains with residual adapters," *Advances in neural information processing systems*, vol. 30, 2017.
191. S. S. Sarwar, A. Ankit, and K. Roy, "Incremental learning in deep convolutional neural networks using partial network sharing," *IEEE Access*, vol. 8, pp. 4615–4628, 2019.
192. S. He, H. Luo, W. Chen, M. Zhang, Y. Zhang, F. Wang, H. Li, and W. Jiang, "Multi-domain learning and identity mining for vehicle re-identification," in *Proceedings of the IEEE/CVF Conference on Computer Vision and Pattern Recognition Workshops*, pp. 582–583, 2020.
193. V. S. Parekh *et al.*, "Cross-Domain Federated Learning in Medical Imaging," *arXiv preprint arXiv:2112.10001*, 2021.
194. Y. Li, W. Zhou, H. Wang, H. Mi, and T. M. Hospedales, "FedH2L: Federated Learning with Model and Statistical Heterogeneity," *arXiv preprint arXiv:2101.11296*, 2021.
195. B. Sun, H. Huo, Y. Yang, and B. Bai, "PartialFed: Cross-Domain Personalized Federated Learning via Partial Initialization," in *Neural Information Processing Systems*, vol. 34, pp. 23309–23320, 2021.

196. M. F. Elvebakken, A. Iosifidis, and L. Esterle, "Adaptive Parameterization of Deep Learning Models for Federated Learning," *arXiv preprint arXiv:2302.02949*, 2023.
197. S. Mukherjee, H. Asnani, E. Lin, and S. Kannan, "ClusterGAN: Latent space clustering in generative adversarial networks," in *Proceedings of the AAAI conference on artificial intelligence*, pp. 4610–4617, 2019. Article No.: 566.
198. L. Zhang, Q. Yang, X. Liu, and H. Guan, "Rethinking Hard-Parameter Sharing in Multi-Domain Learning," in *2022 IEEE International Conference on Multimedia and Expo (ICME)*, pp. 01–06, IEEE, 2022.
199. P. Veličković, G. Cucurull, A. Casanova, A. Romero, P. Lio, and Y. Bengio, "Graph attention networks," *arXiv preprint arXiv:1710.10903*, 2017.
200. Z. Wu, S. Pan, F. Chen, G. Long, C. Zhang, and P. S. Yu, "A Comprehensive Survey on Graph Neural Networks," *IEEE Transactions on Neural Networks and Learning Systems*, vol. 32, pp. 4–24, 2019.
201. J. Zhou, G. Cui, S. Hu, Z. Zhang, C. Yang, Z. Liu, L. Wang, C. Li, and M. Sun, "Graph neural networks: A review of methods and applications," *AI Open*, vol. 1, pp. 57–81, 2020.
202. A. Vaswani *et al.*, "Attention is All you Need," in *Proceedings of the 31st International Conference on Neural Information Processing Systems*, pp. 6000–6010, 2017.
203. P. Schmidt, A. Reiss, R. Duerichen, C. Marberger, and K. Van Laerhoven, "Introducing WESAD, a Multimodal Dataset for Wearable Stress and Affect Detection," in *Proceedings of the 20th ACM international conference on multimodal interaction*, pp. 400–408, 2018.
204. C. Kirschbaum, K.-M. Pirke, and D. H. Hellhammer, "The 'Trier Social Stress Test'–a tool for investigating psychobiological stress responses in a laboratory setting," *Neuropsychobiology*, vol. 28, no. 1-2, pp. 76–81, 1993.
205. S. Jiang, F. Firouzi, K. Chakrabarty, and E. B. Elbogen, "A Resilient and Hierarchical IoT-based Solution for Stress Monitoring in Everyday Settings," *IEEE Internet of Things Journal*, vol. 9, no. 12, pp. 10224–10243, 2021.
206. P. Gomes, P. Margaritoff, and H. Silva, "pyHRV: Development and evaluation of an open-source python toolbox for heart rate variability (HRV)," in *Proc. Int'l conf. On electrical, electronic and computing engineering (icetran)*, pp. 822–828, 2019.
207. K. He, X. Zhang, S. Ren, and J. Sun, "Deep residual learning for image recognition," in *Proceedings of the IEEE conference on computer vision and pattern recognition*, pp. 770–778, 2016.
208. X. Glorot and Y. Bengio, "Understanding the difficulty of training deep feedforward neural networks," in *Proceedings of the thirteenth international conference on artificial intelligence and statistics*, pp. 249–256, JMLR Workshop and Conference Proceedings, 2010.
209. W. Zhang, T. Zhou, Q. Lu, X. Wang, C. Zhu, H. Sun, Z. Wang, S. K. Lo, and F.-Y. Wang, "Dynamic-fusion-based federated learning for COVID-19 detection," *IEEE Internet of Things Journal*, vol. 8, no. 21, pp. 15884–15891, 2021.
210. S. Li, Y. Cheng, W. Wang, Y. Liu, and T. Chen, "Learning to Detect Malicious Clients for Robust Federated Learning," *arXiv preprint arXiv:2002.00211*, 2020.
211. J. So, B. Guler, and A. S. Avestimehr, "Byzantine-Resilient Secure Federated Learning," *IEEE Journal on Selected Areas in Communications*, vol. 39, pp. 2168–2181, 2020.
212. W. Liu *et al.*, "D2MIF: A Malicious Model Detection Mechanism for Federated-Learning-Empowered Artificial Intelligence of Things," *IEEE Internet of Things Journal*, vol. 10, pp. 2141–2151, 2021.
213. Y. M. Asano, C. Rupprecht, and A. Vedaldi, "Self-labelling via simultaneous clustering and representation learning," *arXiv preprint arXiv:1911.05371*, 2019.

Stool Image Analysis for Digital Health Monitoring by Smart Toilets

Jin Zhou, Jackson McNabb, Nick DeCapite, Jose R. Ruiz, Deborah A. Fisher, Sonia Grego, and Krishnendu Chakrabarty

> *"Every system is perfectly designed to get the result that it does."*
>
> – *W. Edwards Deming*

Abstract Gastrointestinal (GI) conditions are widespread and significantly impact quality of life and healthcare. Stool appearance is a valuable GI diagnostic indicator, and with the advent of the Internet of Things (IoT), daily monitoring of excreta from a toilet is emerging as a promising digital health tool. This chapter describes a stool image analysis approach that classifies two physiologically relevant stool characteristics, form and color, for an IoT-based smart toilet. We present the design of hierarchical convolutional neural network (CNN) architectures for training a stool

This chapter is adapted from Zhou et al. [1].

J. Zhou (✉) · J. McNabb · N. DeCapite
Department of Electrical and Computer Engineering, Duke University, Durham, NC, USA
e-mail: jin.zhou@duke.edu

J. R. Ruiz
Division of Gastroenterology, School of Medicine, Duke University, Durham, NC, USA

Gastro Health, Miami, FL, USA

D. A. Fisher
Division of Gastroenterology, School of Medicine, Duke University, Durham, NC, USA

Eli Lilly and Company, Indianapolis, IN, USA

S. Grego
Department of Electrical and Computer Engineering, Duke University, Durham, NC, USA

Center for Water, Sanitation, Hygiene and Infectious Diseases (WaSH-AID), Durham, NC, USA

K. Chakrabarty
Department of Electrical and Computer Engineering, Duke University, Durham, NC, USA

Arizona State University, Tempe, AZ, USA

B. Farahani et al. (eds.), *Smart and Connected Health: Applications and Real-World Use Cases*, https://doi.org/10.1007/978-3-032-09253-3_5

form classifier. Additionally, we describe a pipeline that employs perceptual color quantization coupled with machine learning techniques to optimize the color feature space for the classification of stool color. Furthermore, we present an edge-cloud approach to pursue an optimal balance between accuracy and latency and for the classification of stool form.

Keywords Smart toilets · Digital health monitoring · Gastrointestinal diagnostics · Stool image classification · Edge-cloud computing · Cancer screening

1 Introduction

The impact of gastrointestinal (GI) diseases on patients and the healthcare system is substantial, affecting approximately one in ten people in industrialized countries [2]. For example, in the USA, GI healthcare cost is higher than the cost associated with heart disease [3]. Stool physical characteristics such as form (i.e., consistency) and color contribute to the diagnosis and management of many acute and chronic GI conditions. Stool appearance is an early indicator of colitis, malabsorption syndromes, and upper and lower GI bleeding [4]. Stool morphology is used as part of the diagnostic criteria for irritable bowel syndrome (IBS) (as much as 10–15% of the world population is estimated to suffer from IBS) [5].

In clinical practice, patient self-reporting of bowel movement is limited by subjectivity, poor recall accuracy, and the burden of regular tracking [5]. Discrepancies have been documented between patient self-reports and standardized stool descriptors of color and frequency [4] and between patient report and clinician assessment of diarrhea [6], despite the development of paper- and digital-based diaries [5].

Recent years have witnessed a rapid growth in the Internet of Things (IoT), which has led to a paradigm shift in a number of areas associated with human-machine interaction, especially in healthcare [7–10]. The global IoT healthcare market is expected to reach USD 534.3 billion by 2025 [11]. In addition, artificial intelligence (AI) has advanced significantly in recent years in terms of algorithms, software, hardware design, and many applications and has profoundly changed our lives. A variety of machine learning algorithms, especially deep learning, have been utilized for decision-makings in healthcare applications [12–15]. With the convergence of mobile technology, IoT, and AI, the monitoring of physiological functions from a toilet during bathroom visits is emerging as an active area of research for digital health monitoring [16–21].

"Smart toilet" health monitoring holds promise for objective and long-term data collection. As opposed to a snapshot of the visit to a doctor's office, such at-home health monitoring can provide individualized risk assessment and early warning of disease onset that can be used to trigger interventions. Research on "smart toilet" has thus far mainly focused on urine analysis [18, 19]. To collect data related to

GI conditions, image capture of stool from the content of the toilet bowl either by the user [22] or without user intervention [16] has been proposed. Our group is developing a smart toilet that enables discreet imaging of stool in the toilet plumbing, after flushing and outside the purview of the user, thereby not changing the user experience nor engendering any discomfort.

Stool image analysis is a key enabler of smart toilets for the monitoring of GI health and disease. In this chapter, we present two functions for a smart toilet stool data collection system that provide clinically relevant information: stool color and stool form classifications. We use white balance, color quantization, and a logistic regression (LR) classifier framework for stool color classification. For stool form classification, we use a hierarchical convolutional neural network (CNN) architecture and class-balanced softmax cross-entropy loss for model training. We further provide an edge-assisted IoT-based architecture and investigate the trade-offs between edge processing and cloud computing for both functions.

The rest of this chapter is organized as follows: Sect. 2 describes related prior work and provides further motivation for this research. Section 3 describes the background of the smart toilet and our framework for stool analysis. Section 4 describes the methods for stool form classification and color analysis. Section 5 presents experimental results. Section 6 describes the limitation of this approach and highlights directions for future research. Finally, Sect. 7 concludes the chapter.

2 Related Prior Work

In this section, we describe some representative prior work on stool form and color classification and then focus on recent work on using machine learning (ML) for stool image assessment.

2.1 Stool Assessment

2.1.1 Bristol Stool Form Classification

The BSFS scale [23] is a standard medical diagnostic tool for categorizing adult stool based on its physical appearance. It is an ordinal 7-point scale ranging from hard to soft stool and uses both written and pictorial descriptor (a chart is shown in Fig. 1). Normal stool consistency is defined as BSFS type 3, 4, and 5 [24]. Constipation is defined as type 1 (separate hard lumps, like nuts) or type 2 (sausage-like but lumpy). Diarrhea is defined as a minimum of three loose stools (type 6 and type 7) per day. The BSFS has demonstrated substantial validity and reliability overall; however, difficulties arise around clinical decision points (types 2 and 3 and 5 and 6) and when using two-dimensional photographs of stool [25].

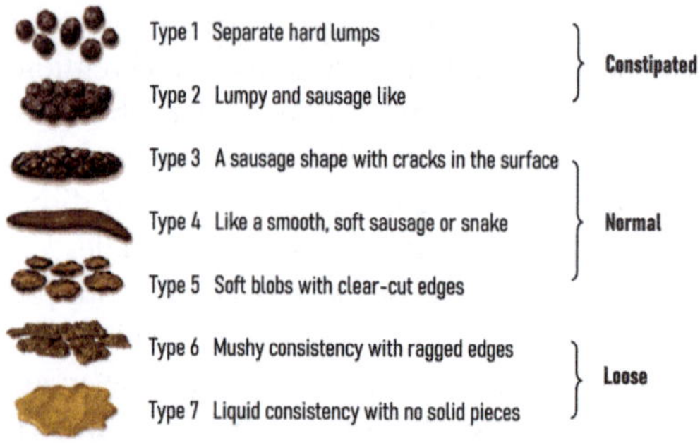

Fig. 1 Illustration of the BSFS chart (Adapted from http://cdn.intechopen.com/pdfs-wm/46082.pdf)

Fig. 2 List of stool colors and possible associated causes, either pathological or dietary/medications [26]

Color	Associated Pathology or Substance Intake
Brown	Healthy condition
Green	May be a sign of infectious or non-infectious enteritis, causing a rapid passing of unabsorbed bile in loose stools Dietary: leafy green vegetables, dye in foods Meds: iron supplementation
Grey	Not enough bile: bile duct obstruction such as with cancer or choledocholithiasis, or decreased liver production of bile such as with cirrhosis, hepatitis, primary sclerosis cholangitis and primary biliary cholangitis Medication intake: bismuth subsalicylate (sold as Pepto-bismol)
Yellow	Steatorrhea: oily stool, such as from malabsorption, celiac disease, exocrine pancreatic insufficiency
Black	Melena: Upper GI bleeding such as from peptic ulcer disease, cancer, angioectasias Dietary: licorice Medications: iron supplements, bismuth subsalicylate
Red	Hematochezia: Lower GI bleeding such as from a diverticular bleed, cancer, IBD, ischemic colitis; brisk upper GI bleeding Dietary: beets, red dye in foods

A 2019 study validated the use of the BSFS by having participants use a printed card tool with graphics to assess the properties of their bowel movements [26].

2.1.2 Stool Color Analysis

The color of healthy stool is brown from secretion of digestive fluids such as bile and from breakdown of red blood cells. Stool color is also affected by diet and medication intake, and, importantly, deviations from brown color can occur due to pathological conditions (Fig. 2) resulting in color such as gray, yellow, and black. The color of gross (i.e., visible) blood in stool, a symptom associated with GI bleeding, ranges from bright red (usually distal intestinal bleeding) to dark red/maroon (colonic bleeding or rapid upper GI bleeding) to black, tarry stool, termed melena (usually gastric or proximal small intestinal bleeding).

Stool color cards have been developed for screening for biliary atresia, a life-threatening liver condition in early childhood resulting in pale stool [27–29], and for describing the color of gross (i.e., visible) blood in stool [4, 30]. For healthy stool, a six-point color scale ranging from yellow/light brown to dark olive gray was proposed and, using a printed card, was evaluated with participants rating specimens on a whole tray with moderate validity; however, accuracy is expected to decrease for observation of stool in a toilet bowl [26].

2.2 Machine Learning Approaches for Stool Image Analysis

Yang et al. [31] introduced StoolNet, which combines the region of interest (ROI) detection and a shallow CNN for color classification of stool images. Park et al. [16] utilized transfer learning to train a classifier on top of a trained deep learning architecture. While these studies provide key insights into automated stool classification, two major challenges have yet to be addressed, namely, uneven separability and class imbalance.

Uneven Separability Visual separability between different BSFS categories is uneven, for example, it is difficult to distinguish type 3 from type 4, while it is easy to tell a type 1 from type 3. The traditional CNNs [32, 33] use the flat structure to train a multi-way classifier and do not consider such uneven separability, which often leads to sub-optimal performance in the task of fine-grained classification. A common strategy to address this problem is to predefine a hierarchy or taxonomy of classifiers so that a given testing image can be first evaluated by a coarse classifier and use the corresponding fine classifier to make the fine prediction [34, 35].

Class Imbalance Medical diagnostic data may have a normal distribution (bell-shaped curve) or a skewed distribution, for instance, in the stool image dataset collected by Park et al. [16], only a few images report constipated stool, while most images indicate normal stool. A number of solutions have been proposed in the literature to address the problem of class imbalance. The first approach is resampling, which aims to alter the training data distribution, usually by random under-sampling and over-sampling techniques [36, 37]. Another common resampling approach is based on data augmentation. Data augmentation synthesizes class-specific samples by applying random transformations, such as image rotation. However, augmentation fails to provide enough data variations in extreme imbalance cases [38]. Recently, generative adversarial networks (GANs) have been exploited to address the problem of class imbalance [39, 40]. However, a major hurdle for GAN-based methods is that the training a high-quality GAN requires an abundance of real data, which is not always available in practice [41]. The second approach is cost-sensitive learning, which assigns higher misclassification costs to the minority classes compared to the majority classes [42–44].

2.3 Internet of Things (IoT) for Health Monitoring

The Internet of Things (IoT) is a network that integrates a variety of elements such as sensors, computing devices, software, and physical objects, wherein the elements are able to interact, communicate, collect, and exchange data [45, 46]. IoT-based solutions have been developed for a variety of healthcare applications. Bianchi et al. present a human activity recognition system based on wearable sensors and deep learning for long-term personalized monitoring of daily activities at home [47]. Rahman et al. develop an edge IoT system that uses deep learning techniques to detect different types of health-related COVID-19 symptoms [48].

With the emergence of IoT, microcontrollers provide low-cost computation and are widely used for intelligence at the edge. However, these endpoint IoT devices have limitations associated with battery life and computational capability [49]. Therefore, commercial smart IoT systems often offload intelligence to the cloud, which has nearly limitless computing power and storage [50]. However, data exchanged between the cloud and the edge devices may be inaccurate due to the high latency in data transmission [51]. Therefore, in order to strike the right balance between the energy consumption, battery lifetime, latency, as well as the reliability of the ML model, efficient data offloading schemes in this interplay between the edge and the cloud are critical. Min et al. propose a reinforcement learning (RL)-based offloading scheme to help healthcare IoT devices protect both the user location privacy and the usage pattern privacy [52]. Ning et al. formulate the bandwidth scheduling problem as a bargaining game, and a game theory solution is utilized to compute the optimal allocation decision [53].

3 IoT-Based Smart Toilet System

In this section, we first provide an overview of the smart toilet system and then formulate the stool analysis problem in terms of two levels of hierarchy, at the edge and in the cloud, respectively.

3.1 Smart Toilet

Smart toilet approaches have been proposed to obtain health-related information from different configurations, e.g., devices snapped on the toilet bowl [54] or integrated in the toilet seat [16, 55]. Notably, Park et al. [16] introduced a defecation monitoring module that uses sensors and computer vision to acquire basic properties of human excreta from sensors integrated in a commercially available electronic bidet. The acquired images are fed offline to machine learning algorithms for analysis. However, cameras and illumination device in the toilet seat create an

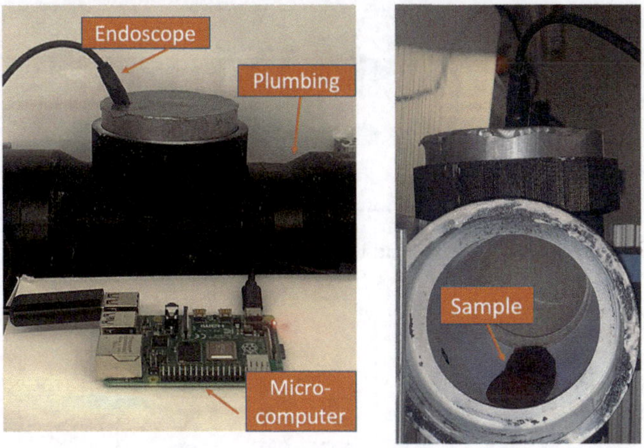

(a) (b)

Fig. 3 (**a**) The setup for stool image analysis. (**b**) Cross-section view of the plumbing with surrogate specimen [57]

uncomfortable environment for the user, as highlighted by the results of a user survey regarding the technology [16].

An alternative approach, which avoids the adoption barrier due to user discomfort, is a technology that integrates sensors in the toilet plumbing where they are not visible to users. A toilet manufacturer has reported such a configuration for urine analysis [56].

We have developed a novel approach to image feces in the plumbing at the outlet of the toilet after the user has flushed. Our design offers a unique opportunity for real-time inline sensing approach specific to feces without engendering user discomfort. Using fingerprint recognition on the toilet flush button, users within a residential setting can be individually tracked. The hardware setup used for stool image analysis is shown in Fig. 3a. A borescope featuring six illuminating LEDs is installed in a viewing port in the toilet plumbing and connected to a microcomputer (Raspberry Pi 3B+) for real-time image collection and analysis. Figure 3 is a cross-sectional view of the immobilization region in the plumbing. The computational capability of the system must be optimized to provide rapidly and inexpensively analysis results to the user from the cloud, ultimately through a phone app.

3.2 Edge-Cloud Interplay

Figure 4 provides an illustration of the overall smart toilet system. A commercially available fingerprint sensor placed on the toilet flush button is used to identify users. The image captured by the smart toilet is fed to the module of stool

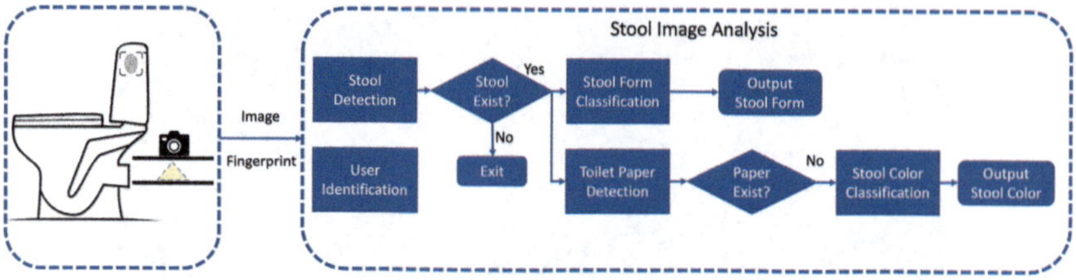

Fig. 4 Framework for the IoT-based smart toilet system [57]

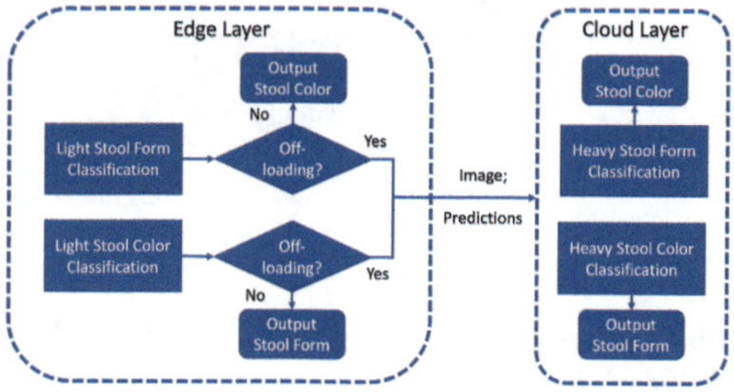

Fig. 5 Interplay between the edge and the cloud [57]

detection to discriminate defecation events from urination and blank toilet flushes. Images containing stool are then fed to two parallel modules, namely, stool form classification and toilet paper detection. When toilet paper is not detected, stool color analysis is performed.

As shown in Sects. 5.1 and 5.2, in order to achieve higher accuracy for stool form and stool color classification, multiple heavy deep learning models need to be utilized. However, to constrain the cost of the smart toilet, the adopted edge device has limited computational capability for real-time inferencing multiple deep learning models. Moreover, even with Nvidia Jetson TX1 ($600), previous research shows that offloading deep learning tasks can easily double the battery life of the device while still meeting the real-time requirements [58]. Therefore, we utilize the cloud layer where data can be offloaded to the cloud for analysis, but the time cost for data offloading from the edge to the cloud has to be considered. Therefore, efficient data offloading schemes are critical.

We propose an architecture with models both at the edge and in the cloud and offload only a subset of the data to the cloud. We then investigate the criteria for data offloading to ensure low transmission cost in terms of communication delay. This interplay between the edge and the cloud is shown in Fig. 5.

We design lightweight models at the edge and heavyweight models in the cloud. We handle the output from the edge models as follows: For a sample c, instead of outputting the predicted label y_{edge} only, we also output the assigned score of x_i being in each of the classes $\{p_1(x_i), \ldots, p_n(x_i)\}$. We scale the score to the range of [0, 1] and use the maximum score $p_{\text{edge}} = \max\{p_1(x_i), \ldots, p_n(x_i)\}$ as the final score for the sample to make the data transmission decision. Assuming that the decision threshold is c, samples with scores below the threshold are offloaded to the cloud, and the rest of the samples remaining at the edge are assigned to the class where they have the maximum output score.

For each sample offloaded to the cloud, we investigate two options for making predictions:

Option 1 We assign the predicted label y_{cloud} from the cloud model as the final output.

Option 2 We output the maximum score $p_{\text{cloud}} = \max\{p_1(x_i), \ldots, p_n(x_i)\}$ as the final score, as shown in Eq. (1).

$$\text{Out} = \begin{cases} y_{\text{cloud}}, & \text{if } p_{\text{cloud}} > p_{\text{edge}} \\ y_{\text{edge}}, & \text{otherwise} \end{cases} \tag{1}$$

Through this interplay between the edge and the cloud, we achieve better overall performance, as shown by the results in Sects. 5.1.2 and 5.2.3. The proposed solution requires brute-force exploration to select appropriate thresholds; this approach is feasible for small datasets. As part of future work, reinforcement learning-based schemes [59, 60] will be exploited to search for the optimal thresholds in order to make this solution applicable to larger datasets.

4 Stool Image Analysis

In this section, we describe the dataset used for analysis, as well as the machine learning techniques used for detection and classification.

4.1 Stool Image Dataset Preparation

No publicly available stool image dataset exists; thus, we developed our own image dataset. Of note, the stool image dataset used in [31] for their StoolNet model only contained 110 images (each rotated and used four times), and, furthermore, those images are not publicly available.

4.1.1 Data Collection

Our work leverages a dataset of 3,629 stool images spanning all 7 BSFS types obtained from two sources: the web and anonymous image uploads from the general public. A total of 2,720 online unique images were obtained through search engines such as Google and platforms such as Reddit with keywords such as "feces in toilet" and "bristol." Additionally, to collect images representing a wide range of bowel movements, we developed an institutional review board (IRB)-approved protocol

(Duke University IRB 2020–0569) to request the general public to take images of stool in their toilet and upload anonymously to a secure repository site. We created a website, then advertised on social media, and leveraged the physician collaborators' professional networks. Potential personally identifying information unintentionally contained in image metadata, such as GPS coordinates, were removed from the images uploaded to the depository prior to further analysis. We obtained 909 public-uploaded images. A total of 256 online images and 98 public-uploaded images were rejected because the image quality did not allow accurate annotation; therefore, the study used a total of 3,275 images.

4.1.2 Stool Form Annotation

A total of 552 images were annotated by two gastroenterologists, and an additional 2,723 were rated by one of them. We used the online platform Labelbox [61] and assigned to each image a BSFS score from 1 through 7, importantly, indicating the presence/absence of blood. Despite being a clinical standard, the BSFS score does not capture the full variety of stool forms and does not account for the presence of stools of more than one BSFS category in the same image. From a clinical point of view, the important information is whether the bowel movement is normal (types 3, 4, and 5) or abnormal, which is further classified as constipated (types 1 and 2) and diarrhea (types 6 and 7). The correlation between the labels assigned by the two gastroenterologists was measured using Cohen's Kappa statistic [62], which ranges from 1 (perfect classification) to -1 (extreme misclassification). The Cohen's Kappa statistic metric was calculated to be $k = 0.44$ for BSFS and $k = 0.54$ for the consolidated categories. These values show a satisfactory agreement between the two gastroenterologists. Due to heterogeneity of feces, large inter-rater variability is not surprising, and our $k = 0.54$ is similar to $k = 0.58$ reported by Hachuel et al. [22].

4.1.3 Stool Color Annotation

Color annotation was conducted to flag abnormal color of stool with the understanding that the abnormal color may be caused by either food or other substance intakes instead of a pathology. Color annotation was conducted in two stages: In the first stage, the clinicians were asked to flag images of abnormal color, and 167 out of 3,275 were flagged by one of them based on color. We then sought to increase the resolution in abnormal color classification to colors of physiological relevance as listed in Fig. 2. After a preliminary evaluation, images containing toilet paper were removed to overcome the algorithm challenge in classifying color gray, and images containing blue color from toilet bowl cleaners were also removed for a total of 2,732 images for color analysis. For the second stage of color annotation, the image set was reduced to images containing at least 10% of a non-brown color (based on method in Sect. 4.3) for a total of 575 images; for the remaining images,

Table 1 Class distribution in the dataset used for stool color classification

Total	Normal	Abnormal		
		Red	Black	Pale
2732	2629	47	38	18

we use the color annotation "normal" from the first stage. The set of 575 images were annotated by the gastroenterologists. To standardize the annotation criteria, we present a newly developed stool color card in four categories, one for regular color and three for abnormal color to flag as red, pale, and black, as discussed in Sect. 5.2. The correlation between the two gastroenterologist for color annotation was $k = 0.55$. The distribution of the annotation results is shown in Table 1 for a total of 103 images of abnormal color (3.8% of the total).

4.2 Stool Form Classification

The first part of our classification problem is to determine the bowel movement type based on the stool form.

4.2.1 Hierarchical Architecture

Visual separability between different BSFS categories is uneven, for example, it is difficult to distinguish type 3 from type 4, while it is easy to tell a type 1 from type 3. As introduced in Sect. 2.1.1, types 3 and 4 are defined as the same consolidated category "normal." To leverage the hierarchical structure of stool form categories, we deploy a treelike hierarchical architecture [35, 63], as shown in Fig. 6.

The main components in our architecture include the base CNN, a coarse-grained classifier, and three fine-grained classifiers. For an input image, the base CNN extracts low-level features. The extracted features are fed to the coarse-

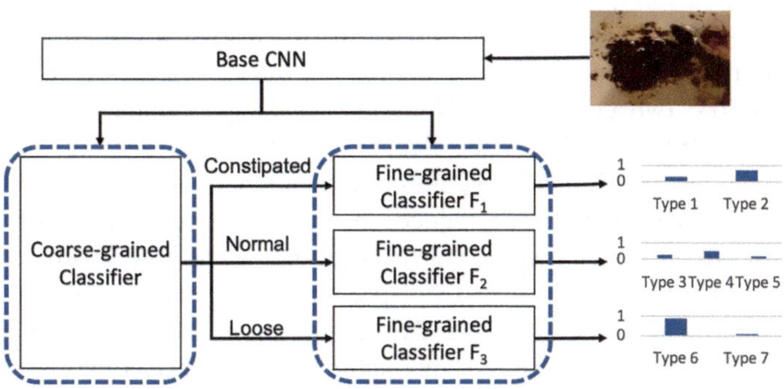

Fig. 6 Hierarchical architecture for stool form classification [57]

Table 2 Configurations used in the fine-grained classifiers

Layer	Type	Depth	Activation	Stride	Padding
1	Convolution	320	ReLU	3	1
2	Fully connected	2 or 3	N/A	N/A	N/A

grained classifier which produces a consolidated prediction over three categories ("loose," "normal," and "constipation"). The consolidated prediction enables the corresponding fine-grained classifier to reuse the extracted features and make the prediction over the seven BSFS types, for example, the consolidated prediction "constipated" triggers the fine-grained classifier F_1 to make the prediction over BSFS type 1 and type 2.

For both coarse-grained classifier and fine-grained classifiers, we evaluated several options for the number of hidden layers and the number of neurons per layer. We found that a simple two-layer CNN, as described in Table 2, is effective for the prediction of the stool form types. The depth of layer 2 is three for the coarse-grained classifier and the fine-grained classifier F_2 and two for the fine-grained classifier F_1 and classifier F_3.

4.2.2 Base CNN Design

In recent years, state-of-the-art CNN designs have advanced from 8 layers (AlexNet) to 152 layers (ResNet) [32, 64]. The progression of deeper networks has dramatically increased the computational requirement for CNN model deployment. To address this problem, we first consider utilizing the cloud layer and offloading the image data to the cloud for analysis. We investigate two popular CNN designs, namely, VGG16 [33] and ResNet50 [64] as the base CNN. VGG is a classic CNN with a simple architecture composed of Conv, ReLU, and Pooling. It is heavily used due to its generalization ability on a variety of tasks [65]. ResNets introduced novel features including shortcut connections and batch normalization, which can effectively smooth out the loss optimization process and thereby facilitate model training [66].

However, cloud computing suffers from high latency and energy consumption during data transmission from the endpoint IoT devices to the cloud. Therefore, we also explore two compact CNN designs, namely, MobileNetV2 [67] and ShuffleNetV2 [68], which can be deployed at the edge devices with limited computational power. MobileNetV2 introduces a module of inverted residual with linear bottleneck and significantly reduces the memory footprint required for inference. ShuffleNet [69] employs two new operations, point-wise group convolution and channel shuffle, and achieves 13× speedup over AlexNet while maintaining comparable accuracy. ShuffleNetV2 further increases the model efficiency by introducing a channel split operator that splits feature channels into branches and concatenates them after convolution.

The base CNN is used for feature extraction. We utilize these CNN designs as the base CNN by removing their last classification layers. Specifically, we remove the last two layers (one dropout layer and one fully connected layer) for MobileNetV2 and remove the last one layer (fully connected layer) for VGG16, ResNet50, and ShuffleNetV2.

4.2.3 Loss Function

The softmax cross-entropy loss has been commonly used for CNN training [70, 71]. For an input sample x with class label y, assuming that the predicted output from the model for all classes is $z = [z_1, z_2, \ldots, z_C]$, where C is the total number of classes. The softmax cross-entropy (CE) loss for this sample is defined as

$$\mathrm{CE}(\mathrm{z}, y) = -\log\left(\frac{\exp(z_y)}{\sum_{j=1}^{C}\exp(z_j)}\right). \tag{2}$$

Considering the problem of class imbalance in the training data, the network is trained with a class-balanced softmax cross-entropy loss [43], which reweights loss inversely with the effective number of samples per class. The class-balanced softmax cross-entropy (CB) loss for this sample is defined as

$$\mathrm{CB}(\mathrm{z}, y) = -\frac{1-\beta}{1-\beta^{n_y}}\log\left(\frac{\exp(z_y)}{\sum_{j=1}^{C}\exp(z_j)}\right) \tag{3}$$

where n_y is the number of training samples in class y and $\frac{1-\beta}{1-\beta^{n_y}}$ is the weighting term for the loss function, with hyperparameter $\beta \in [0, 1)$.

4.3 *Stool Color Classification*

Figure 7 outlines the steps of the method adopted in stool color classification. The original image was fed to a white balance script, where the image illumination is adjusted. Then, a pretrained deep learning model was used to segment the stool for further color analysis. We used perceptual color quantization and color projection to analyze all the colors relevant to stools. We then adopted a feature-based classification method, in which multiple machine learning techniques were investigated for the classification algorithm.

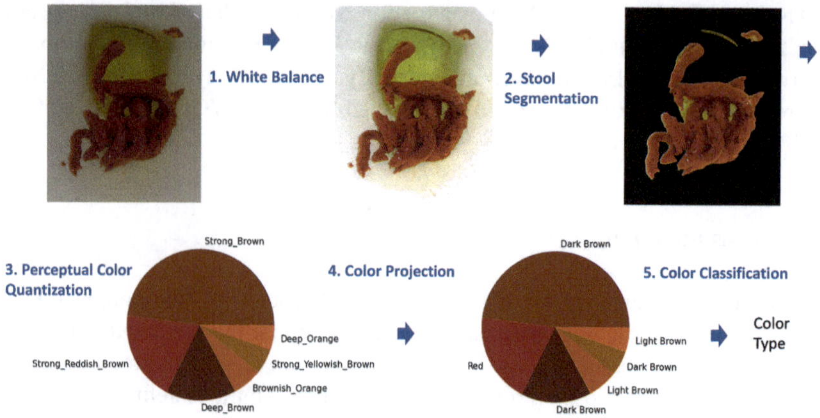

Fig. 7 Process sequence adopted in stool color analysis [57]

4.3.1 White Balance

Due to variable lighting conditions in the collected images, we utilized a white balance algorithm to adjust the color fidelity for stool representations under a uniform light source. The white balance algorithm corrects the colors of the image by scaling red, green, and blue channels independently. Taking advantage of the fact that toilet ceramic background has the color of white, we replace the largest 0.05% of the pixels in the normalized image with the value of one in each channel and stretch the remaining pixels by multiplying with a scaling factor $\alpha_c = 1/P_{0.05}(C)$, where $P_i(C)$ is the i_{th} percentile of pixel values in channel C. Images from toilets that are not white or include blue toilet bowl cleaner were removed from the dataset to implement this step.

4.3.2 Stool Segmentation

For the task of stool segmentation, we employed the pretrained salient object detection model U^2-Net [72], a recently proposed method that has been shown to be effective for segmentation problems. U^2-Net has a two-level nested U-structure. The nested U-structure built upon ReSidual U-blocks can well capture local and global features from both shallow and deep layers. U^2-Net was trained on DUTS-TR dataset [73], which is the largest and most frequently used training dataset for salient object detection.

4.3.3 Perceptual Color Quantization

We use perceptual color quantization [74], which employs the CIE LAB color space [75] and the standard ISCC-NBS Color Names Dictionary [76]. The CIE LAB color space was designed to approximate human vision. The ISCC-NBS system defines 267 standard color partitions. Each color partition has a standard color name and a centroid color. The color names are designed as basic colors with one or more adjectives, such as “Vivid Red” and “Light Grayish Yellowish Brown.” For an input image I, each pixel is assigned the closest ISCC-NBS centroid color based on the euclidean distance in CIE LAB space.

4.3.4 Color Projection

In preliminary studies, we attempted to classify the 267 colors of the ISCC system to color coordinate for stool colors reported in the literature [26, 29, 30] for a total of 18 colors, however we found that a smaller number of critical colors associated with physiological values, as in Fig. 2, is both more manageable and relevant. Therefore, we projected 267 colors to 8 colors based on the conversion chart shown in Table 3.

4.3.5 Color Classification

Out of the eight colors in Table 3, green and blue occur only in a handful of images in our dataset. The color gray and light brown are found to be not specific to stool because of the image background. Thus, for stool color classification, we use the frequency of yellow, black, dark brown, and red colors as the features. We investigate three machine learning techniques, namely, decision tree (DT), k-nearest neighbors (KNN), and logistic regression (LR). An advantage of using DT, KNN, and LR is that the analysis results are easy to interpret and explain, which facilitates interactions with the medical team.

Table 3 The criteria used for color projection

Color	Description
Gray	Gray, grayish (not “dark grayish”), light (not “light and brown”), pale, white – for all hues
Yellow	Greenish yellow, yellow (not dark/deep/strong yellow)
Light brown	Orange, deep/dark/strong yellow, light and brown, reddish brown
Dark brown	Dark/deep/strong/moderate brown, dark grayish, reddish brown
Green	Green, olive green
Red	Red, pink, reddish (not reddish brown), purple, violet (if any of these is light or pale, go to gray), strong reddish red
Black	Black, blackish
Blue	All others

A DT is a treelike model that consists of two types of nodes, leaf (terminal) nodes and decision (internal) nodes [77]. Leaf nodes refer to the nodes that do not branch and contain prediction information and, each leaf node holds a class label. Decision nodes refer to the nodes that can branch to multiple child nodes or leaf nodes. Class labels are denoted by the elements of the set $Y = \{Y_0, Y_1\}$, where Y_0 indicate images of healthy stool and Y_1 indicate images containing blood content. These class labels are the leaf nodes in the decision tree. Our selected color features $C = \{C_0, C_1, \ldots C_i\}$ are encoded in the decision nodes.

KNN is a distance-based technique [78]. Assuming that the training dataset is C_{train} and the test dataset is C_{test}, then, for each instance C^j_{test} in C_{test}, we calculate its distance to all instances C^i_{train} in C_{train}. The list of distances obtained in this manner is sorted in ascending order, and the most common label among the first k elements in the sorted distance list is assigned to instance C^j_{test}.

LR is a widely used statistical algorithm to find the best fitting model to describe the relationship between a set of independent variables and response variable that follows a binomial distribution [79]. The main difference between linear regression and LR is that in LR the dependent variable is categorical and has binary value.

4.4 Toilet State Classification

As a critical step toward practical implementation of the system, we need to address two additional distinct needs: (1) defecation event detection (discriminating images of empty toilet from images containing stool) and (2) toilet paper detection (check if there is a toilet paper in the image). Due to the clear separability between images of empty toilet and stool images, we utilize the color classification method described in Sect. 4.3 for defecation event detection based on the brown color and added 14 images of empty toilets to the dataset.

When toilet paper is detected, the confidence of the prediction results will be low, as part of the stool might be covered by the toilet paper. Moreover, the color analysis is not performed, because our method has been developed thus far on images excluding toilet paper. Here, we utilize a lightweight CNN model, such as MobileNetV2, for toilet paper detection.

5 Experiments and Results

Experiments were conducted to evaluate the effectiveness of the proposed approach for classifying stool form and color. We preprocessed the stool images by cropping them to remove noise, e.g., due to the toilet seat. The stool image dataset is imbalanced in terms of stool forms, stool colors, and toilet states. Therefore, for each task, the data with the minority classes (i.e., stool images labeled with the red color) are randomly over-sampled so a balanced dataset is obtained.

5.1 Results on Stool Form Classification

The balanced accuracy (BA) metric was utilized to evaluate different model architectures trained with various loss functions. The BA metric is defined as the average recall obtained on each class [80], as shown in Eq. (4),

$$BA = \frac{1}{n}\sum_{i=1}^{n} \frac{tp_i}{tp_i + fn_i} \times 100\% \tag{4}$$

where n is the number of classes and tp_i and fn_i are the number of true positive and false negative predictions for class i, respectively.

5.1.1 Hierarchical Architecture

The training process for the hierarchical CNN includes three steps: We first initialize the base CNN with pretraining on ImageNet [81]. After initialization, we train the coarse-grained classifier and the base CNN together over three consolidated categories. In the last step, the base CNN is kept fixed, and fine-grained classifiers are trained over seven BSFS scales. We utilized 552 images annotated by 2 gastroenterologists for testing and 2,723 images rated by one of them for training. We resized the images to 224 × 224 pixels.

We used PyTorch [82] to implement and train the CNN using stochastic gradient descent with momentum. Experiments were executed on a Linux platform integrated with an 11-GB memory GPU (Nvidia GeForce RTX 2080 Ti). The training was performed with mini-batches of size 14. In each batch, the numbers of samples for each class were restricted to be the same. We used the best BA during the training process as the quality metric.

We consider two compact CNN designs described in Sect. 4.2 and two traditional CNN designs, namely, VGG16 [33] and ResNet50 [64] as the base CNN in the hierarchical architecture. Furthermore, we train the models with flat and hierarchical architectures and with both cross-entropy loss and class-balanced cross-entropy loss for comparisons. We evaluate the performance of coarse-grained classification over three consolidated categories (constipated, normal, and loose) and fine-grained classification over the BSFS scale with seven values. The results are shown in Table 4 and summarized as follows:

(a) For all models except ShuffleNetV2, CB loss improves the performance in terms of the BA metric. One possible explanation of the performance decrease in

ShuffleNetV2 with CB loss is that its more complex architecture and small size lead sensitivity to the reweighting strategy in CB loss.

(b) Hierarchical architectures improve the performance of both coarse-grained and fine-grained classification in terms of the BA metric for ResNet50 as well as for MobileNetV2. The hierarchical architecture brings slight increase in required memory and $FLOPs$ for inferencing, because it has three more classifiers than the flat architecture.

(c) Hierarchical architectures with MobileNetV2 as the base CNN achieve the best performance (81.7% BA in coarse-grained classification and 54.6% in fine-grained classification). Moreover, MobileNetV2 only requires 0.35 $GFLOPs$ for inferencing and the memory requirement is only 15.6 MB.

5.1.2 Edge-Cloud Interplay

As shown in Table 4, MobileNetV2 outperforms ShuffleNetV2, and ResNet50 outperforms VGG16 in terms of the metrics of BA and computational costs ($FLOPs$ and required memory). To evaluate the effectiveness of the edge-cloud framework, we conduct experiments with MobileNetV2 and ResNet50 as the base CNN in the hierarchical architecture for edge model and cloud model, respectively. As different CNN designs can extract different sets of features from the input image, we adopted Option 2 as described in Sect. 3.2.

In order to investigate the trade-off between the overall accuracy and the communication cost during data transmission from the edge to the cloud, we sampled a series of decision thresholds to obtain the corresponding partitioned samples and the overall performance. To compute the communication latency during data off-loading to the cloud, we first obtained the Wi-Fi upload bandwidth (2.7 Mbps) and date rate (337.5 KB/s) using TestMy.net [83]. For each decision threshold, we calculated the total size of the data offloaded to the cloud in KB by multiplying the number of samples with the average size of the test image (809 KB). We then divided the total cloud sample size by the upload internet speed to obtain the communication latency. The results are shown in Fig. 8 and summarized as follows:

(a) When the threshold is below 0.6, no sample is offloaded to the cloud, and the edge-cloud model is equivalent to using edge model only. As the decision threshold increases, more samples are sent to the cloud, resulting in an increase in the balanced accuracy.

(b) When the threshold is 1, all samples are offloaded to the cloud, and the edge-cloud model achieves the best performance 57.6% BA in fine-grained classification. However, a disadvantage is the relatively high latency (1,323 seconds).

(c) With the threshold of 0.7, the edge-cloud model achieves the best performance in coarse-grained classification (84.4% BA) and comparable performance in

Table 4 Balanced accuracy of flat and hierarchical architectures with various base CNN designs on the stool image dataset

Models	Architecture	Coarse-grained	Fine-grained				
		CB softmax	Softmax	CB Softmax	Softmax	$FLOPs$	Memory required
VGG16	Hierarchical	73.50%	65.78%	50.03%	37.58%	15.39 G	61.5 MB
	Flat	75.17%	65.83%	43.41%	39.70%	15.38 G	59.5 MB
ResNet50	Hierarchical	80.15%	74.66%	51.32%	44.75%	4.16 G	104.8 MB
	Flat	71.92%	62.95%	41.33%	39.36%	4.13 G	96.9 MB
MobileNetV2	Hierarchical	81.66%	69.61%	54.58%	42.10%	0.35 G	15.6 MB
	Flat	74.55%	61.45%	44.25%	41.01%	0.33 G	10.7 MB
ShuffleNetV2	Hierarchical	33.33%	74.12%	22.86%	43.59%	0.17 G	10.4 MB
	Flat	47.60%	64.22%	26.53%	36.59%	0.16 G	6.5 MB

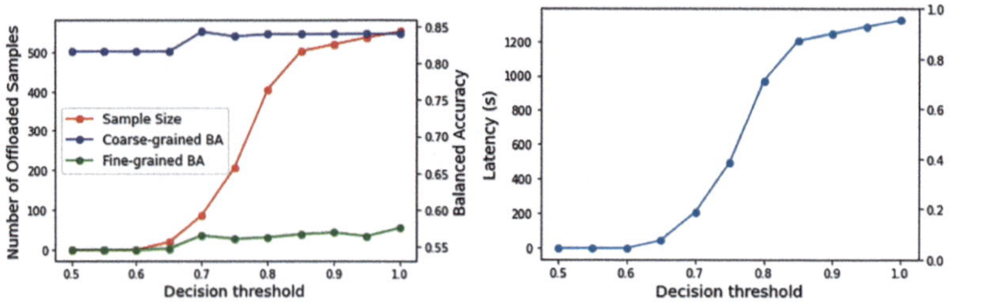

(a) (b)

Fig. 8 Results on edge-cloud interplay for stool form classification [57]. (**a**) Accuracy and sample partitioning obtained at different thresholds. (**b**) Communication latency at different thresholds

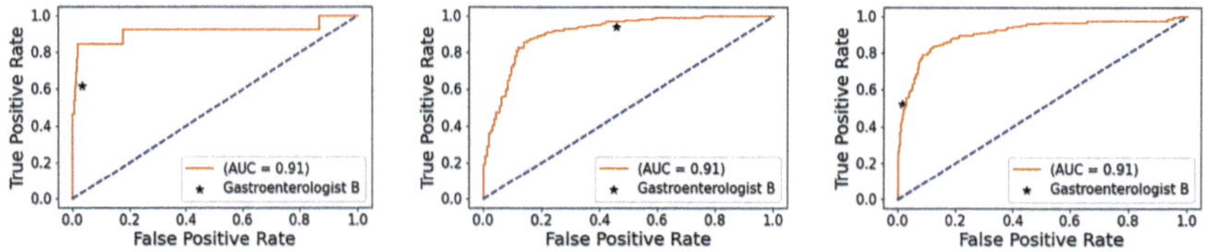

(a) (b) (c)

Fig. 9 ROC analysis for the stool form classification results of our edge-cloud model [57]. Using the labels assigned by the first gastroenterologist (A) as the ground truth, all AUCs are greater than 0.90 and comparable with the second gastroenterologist (B). (**a**) Constipated (**b**) Normal (**c**) Loose

fine-grained classification (56.5% BA), with an 84.2% reduction in latency compared with using cloud model only.

5.1.3 Performance Analysis

We first analyze the performance of our edge-cloud model for the task of coarse-grained classification over three consolidated categories. Figure 9 shows the receiver operating characteristic (ROC) analysis. Here, ROCs were reduced to a dichotomous classification based on the one versus rest approach, treating the corresponding class as positive and all of the other classes as negative. Table 5 shows the analysis based on the metrics of sensitivity and specificity. We note that the edge-cloud model achieves satisfactory performance in terms of coarse-grained classification.

We further analyze the performance of our model for the task of fine-grained classification over the BSFS scale with seven values. Figure 10 shows the confusion matrices, which describe the classification conformance among our CNN model and the two gastroenterologists regarding the BSFS. The agreement between our CNN model and gastroenterologist A is measured by Cohen's Kappa statistical metric and has the value of $k = 0.41$, which is close to the agreement between the two

Table 5 Sensitivity and specificity of our hierarchical architecture with our edge-cloud model. C, N, and L stand for constipated, normal, and loose, respectively

	Coarse-grained			Fine-grained						
	C	N	L	1	2	3	4	5	6	7
Sensitivity	0.85	0.84	0.85	0.33	0.80	0.65	0.49	0.31	0.72	0.64
Specificity	0.98	0.87	0.87	1.00	0.97	0.80	0.91	0.91	0.86	0.97

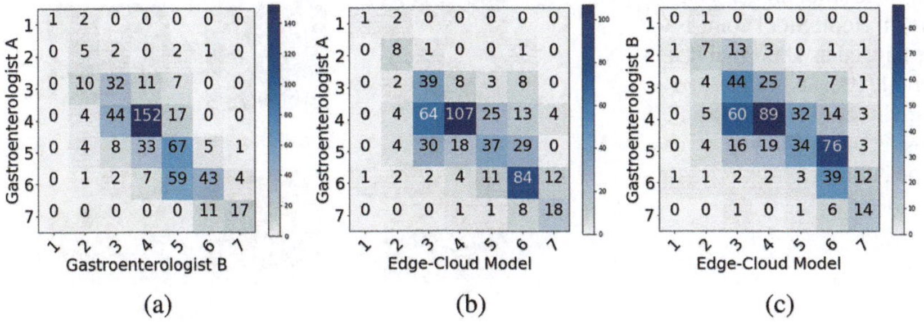

Fig. 10 Confusion matrices for comparison between the stool form classifications made by (**a**) the two gastroenterologists, (**b**) gastroenterologist A and our edge-cloud model, (**c**) gastroenterologist B and our edge-cloud model [57]. The values on the axes indicate the BSFS

gastroenterologists ($k = 0.44$, as discussed in Sect. 4.1). The agreement between the CNN and gastroenterologist B has $k = 0.26$, and this can be explained by the fact that in 2,723 training images, 1,928 were annotated by gastroenterologist A, and only 248 were annotated by gastroenterologist B. The analysis based on the metrics of sensitivity and specificity is shown in Table 5. For BSFS 1, the metric is not meaningful because we have only three examples for BSFS 1. The reason behind the poor performance for BSFS 4 and 5 is that these two values are often misclassified with each other, as shown in Fig. 10.

5.2 *Results on Stool Color Classification*

5.2.1 Stool Color Card

The image dataset of this study contains a variety of stool colors and color mixtures. While stool color cards exist for some specific applications, there is no established clinical scale for color as the Bristol Stool Scale is for stool form. After a first stage of color annotation flagging abnormal colors on the dataset, we found it necessary to provide the clinicians performing the color annotation a stool color card and specific criteria. The color card of Fig. 11 was constructed with the agreement of the two gastroenterologists of this study based on images from this dataset and associated criteria. Several stool colors that may appear unusual for the untrained eye were selected to be in the normal category.

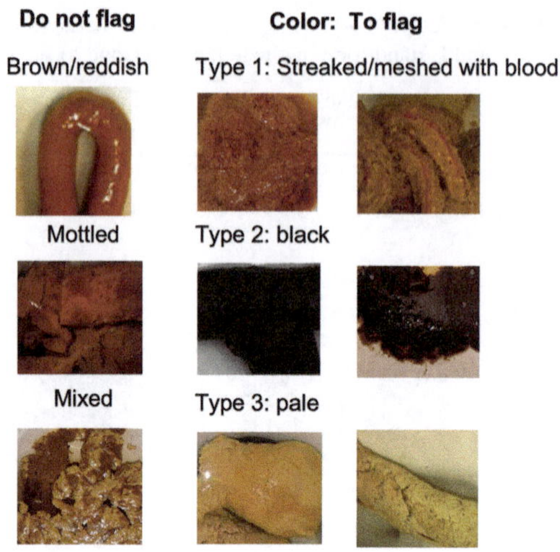

Fig. 11 Stool color card developed for annotation and consisting of stool photographs [57]. Apart from brown color, stool color is not to be flagged if stool is (i) reddish but no blood can be distinguished, (ii) mottled, or (iii) pale stool mixed with brown stool. Stool color is to be flagged as abnormal if it is type 1, streaked or meshed with blood (hematochezia); type 2, pale yellow/gray; or type 3, black

Because of the difficulty of judging the reddish color of stool, whether it is physiological or associated with bleeding, the criteria was formulated to flag red color only if the image contains streaks or clots of blood and not if the reddish color is uniform. Normal stool can also be mottled with patches of light and dark brown. While pale stool may be pathological, if pale feces are mixed with brown feces, the image will not be flagged.

The color annotation flagged as abnormal color images with type 1, streaked or meshed with blood (hematochezia); type 2, pale yellow/gray; or type 3, black. The green color is almost always normal unless it is associated with diarrhea, so while it was annotated as its own class, we have combined those images in the "normal" category.

5.2.2 Classification Performance

We randomly sample 80% (2,185 images) of the data for training and the rest 20% (547 images) for testing. We consider three machine learning models as the classifier for stool color classification and train them with and without stool segmentation for comparisons. The results are shown in Table 6 and summarized as follows:

(a) For all classifiers, stool segmentation improves the performance in terms of the BA metric. This can be explained by the fact that segmentation module can remove most of the non-stool area from the image, which reduces the irrelevant information for the classifiers.
(b) The LR classifier with stool segmentation provides 81.88% BA. LR consistently outperforms KNN and DT with and without stool segmentation. One possible explanation is that KNN and DT are more sensitive to class imbalance.

Table 6 Balanced accuracy of three classifiers for stool color classification

	Without segmentation	With segmentation
KNN	52.05%	58.58%
DT	73.74%	74.86%
LR	74.09%	81.00%

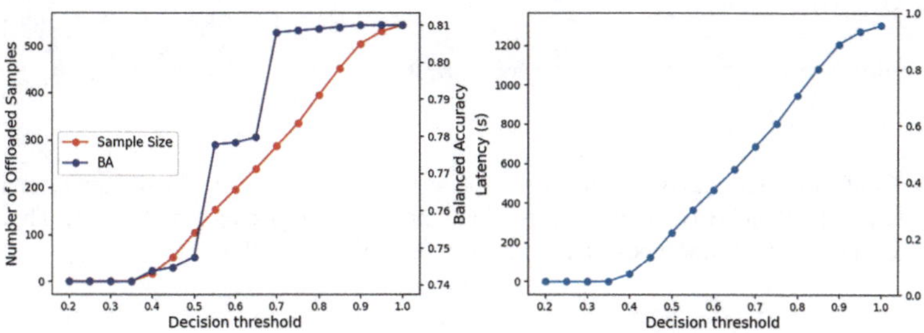

(a) (b)

Fig. 12 Results on edge-cloud interplay for stool color classification [57]. (**a**) Accuracy and sample partitioning obtained at different thresholds. (**b**) Communication latency at different thresholds

5.2.3 Edge-Cloud Interplay

To evaluate the effectiveness of the edge-cloud framework for stool color classification, we utilize LR without and with the stool segmentation module for the edge model and the cloud model, respectively. As stool segmentation significantly improves performance in terms of the BA metric, we adopt Option 1 as described in Sect. 3.2. The results are shown in Fig. 12 and summarized as follows:

(a) When the threshold is above 0.35, as the decision threshold increases, more samples are sent to the cloud, resulting in an increase in the balanced accuracy.
(b) When the threshold is 1, all 547 samples in the testset are offloaded to the cloud; we achieve the best performance (81.0% BA). However, with the threshold of 0.7, the edge-cloud model achieves 80.8% BA, requiring only 286 samples in the cloud, and provides a 47.3% reduction in communication latency compared with using the cloud model only.

Figure 13 shows the confusion matrices, which shows that the proposed edge-cloud model provides comparable performance compared with off-loading all samples to the cloud model. Figure 13 also shows that the model not only accurately identifies abnormal colors but also flags many "normal" images.

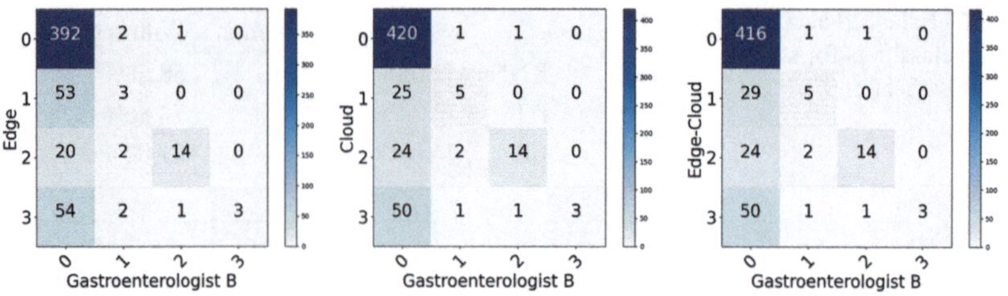

(a) (b) (c)

Fig. 13 Confusion matrices for comparison between the stool color classifications made by gastroenterologist B and (**a**) edge model, (**b**) cloud model, (**c**) edge-cloud model [57]. The values on the axes indicate the color types (0, normal; 1, red; 2, black; 3, pale)

5.3 Results on Toilet State Classification

5.3.1 Defecation Event Detection

We consider three machine learning models as the classifier for stool color classification and train them with and without stool segmentation for comparisons. Due to the limited number of the images of empty toilets, we used an m-fold ($m = 5$) cross-validation method to evaluate the performance of the three classifier. An m-fold cross-validation method randomly partitions separate experimental dataset into m groups. In each round of an experiment, one group is regarded as the test dataset, while all the other groups are used for training. The results are shown in Table 7 and summarized as follows:

(a) Unlike stool color classification, stool segmentation doesn't bring the performance improvement in terms of the BA metric. This can be explained by the fact that the non-stool area can contribute useful color features for defecation detection.
(b) All three classifiers provide promising results in terms of the BA metric. Notably, the LR classifier without stool segmentation provide 98.5% BA.

Table 7 Balanced accuracy of the three classifiers for defecation event detection

	Without segmentation	With segmentation
KNN	94.91%	88.17%
DT	91.65%	88.00%
LR	98.54%	93.35%

Table 8 Balanced accuracy and resource requirements of the four CNN-based models for toilet paper detection

Models	BA	FLOPs	Memory required
VGG16	94.71%	15.38 G	59.5 MB
ResNet50	91.03%	4.13 G	96.9 MB
MobileNetV2	95.63%	0.33 G	10.7 MB
ShuffleNetV2	50.00%	0.16 G	6.5 MB

5.3.2 Toilet Paper Detection

In the annotated 3,275 images, 470 images included toilet paper. We randomly sample 80% of the data for training and the rest 20% for testing. We consider four CNN-based models as the classifier for toilet paper detection. Table 8 shows that MobileNetV2 achieves the best performance (95.6% in BA). Moreover, MobileNetV2 can be deployed at the edge, as it only requires 0.35 $GFLOPs$ for inferencing, and the memory requirement is only 15.6 MB.

6 Limitations and Discussion

The development of an ML-AI program to automatically classify stool images for form (Bristol scale) and color requires a large number of annotated photos of stool in a toilet. There was no publicly available database available to us; therefore, we developed our own dataset with over 3,000 images. A limitation of this approach is that the photos had no clinical data associated with them, and while they spanned the full spectrum of the Bristol scale, the representation of associated GI conditions or symptoms was unknown. Additionally, while the use of the Bristol scale helps standardize stool evaluation, there remains some variability in assessment even among GI specialists. However, the agreement between the two gastroenterologists in this study was satisfactory for form and, after development of a color card, for color. Finally, in the development of a model detecting abnormal stool colors of pathological relevance, we removed images containing toilet paper which may weigh color assessment toward abnormal gray color; we anticipate that future refinement of the model is needed ahead of practical implementation of the system.

The definition of abnormal stool form from the BSFS was obtained from population studies, and abnormal color associated with pathophysiology has been defined based on patient populations [23]. However, a large amount of inter-individual variability exists for very common gastrointestinal disorder such as IBS. Personalization based on biomarkers has been shown to improve IBS diagnosis [84] and interventions [85]. We envision that with the future deployment of the smart toilet hardware prototype for use by human subjects, we will be able to collect time-series data from individual subjects and periodically retrain the model in an incremental manner at the cloud. We expect that stool image data collection from the controlled environment will result in more consistent lighting and even background that will enhance the model accuracy. Moreover, operation of the smart toilet is expected to generate personal health data akin to wearable sensors measuring heart functions or glucose level. As part of future work, our IoT framework will also implement data security and privacy protection similar to platform developed for wearables [86].

A smart toilet with machine learning image analysis capability to determine stool frequency, form, and color will provide important diagnostic data that can help identify specific food intolerance (e.g., foods that exacerbate chronic diarrhea) and effects of medication (e.g., medications taken for diarrhea or constipation) and can trigger timely evaluation (e.g., IBD flare with bloody diarrhea). Additionally, ongoing development of the technology includes stool specimen sampling for biochemical marker analysis that will provide highly specific disease data. We envision that the smart toilet time-series data collected from individuals will be integrated with machine learning predictive models and provide a valuable diagnostic and surveillance tool for GI, infectious disease, and other specialties.

7 Conclusion

Stool appearance is a valuable GI diagnostic indicator, and in this research, we have presented an edge-assisted IoT-based solution for processing data captured by a smart toilet to provide clinically and physiologically relevant information on two characteristics of stool, form, and color for intestinal health assessment. In this chapter, we present the design of hierarchical CNN architectures for training a stool form classifier and a pipeline that employs perceptual color quantization coupled with machine learning techniques to optimize the color feature space for the classification of stool color. Specifically, we showed that the interplay between a hierarchical CNN based on MobileNetV2 at the edge and ResNet50 in the cloud (both trained with class-balanced softmax entropy loss) can achieve a balanced accuracy of 84.4% in coarse-grained classification and 56.5% in fine-grained classification. Additionally, we demonstrated an 84.2% reduction in communication latency compared with a cloud-based model only. We obtained a balanced accuracy of 80.8% using perceptual color quantization coupled with stool segmentation and using logistic regression as the classifier a 47.3% reduction in communication latency.

Acknowledgments This research was funded in part by the Duke Bass Connection Program. We thank Ms. Mara Shurgot for helping with the recruitment of participants through social media. We thank Ms. Shiyi Jiang for suggestions on the IoT-architecture design.

Data Availability Statement The data of this study are available to other researchers upon request to sonia.grego@duke.edu. Fulfillment of the request will be subject to limitations due to ethical restrictions.

Competing Interests SG is an inventor on a patent application related to the excreta sampling toilet and co-founder of Coprata Inc., a company that develops excreta sampling toilet technology. JRR is an advisor for Coprata. Other authors do not have competing interest.

References

1. J. Zhou et al.'s "Stool Image Analysis for Digital Health Monitoring By Smart Toilets" in *IEEE Internet of Things Journal*, vol. 10, no. 5, pp. 3720–3734, 1 March 2023, https://doi.org/10.1109/JIOT.2022.3158886
2. A. D. Sperber, D. Dumitrascu, S. Fukudo, C. Gerson, U. C. Ghoshal, K. A. Gwee, A. P. S. Hungin, J.-Y. Kang, C. Minhu, M. Schmulson, A. Bolotin, M. Friger, T. Freud, and W. Whitehead, "The global prevalence of IBS in adults remains elusive due to the heterogeneity of studies: a Rome Foundation working team literature review," *Gut*, vol. 66, no. 6, pp. 1075–1082, 2017.
3. A. Peery, S. Crockett, C. Murphy, J. Lund, E. Dellon, J. Williams, E. Jensen, N. Shaheen, A. Barritt, S. Lieber, B. Kochar, E. Barnes, Y. Fan, V. Pate, J. Galanko, T. Baron, and R. Sandler, "Burden and cost of gastrointestinal, liver, and pancreatic diseases in the United States: Update 2018," *Gastroenterology*, vol. 156, no. 1, pp. 254–272.e11, 2019.
4. G. Zuckerman, D. Trellis, T. Sherman, and R. Clouse, "An objective measure of stool color for differentiating upper from lower gastrointestinal bleeding," *Digestive Diseases and Sciences*, vol. 40, pp. 1614–1621, 2005.
5. E. Halmos, J. Biesiekierski, E. Newnham, R. Burgell, J. Muir, and P. Gibson, "Inaccuracy of patient-reported descriptions of and satisfaction with bowel actions in irritable bowel syndrome," *Neurogastroenterology & Motility*, vol. 30, no. 2, 2018.
6. H. Majid, P. Emery, and K. Whelan, "Definitions, attitudes, and management practices in relation to diarrhea during enteral nutrition: a survey of patients, nurses, and dietitians," *Nutrition in Clinical Practice*, vol. 27, pp. 252–260, 2012.
7. Y. A. Qadri, A. Nauman, Y. B. Zikria, A. V. Vasilakos, and S. W. Kim, "The future of healthcare internet of things: A survey of emerging technologies," *IEEE Communications Surveys Tutorials*, vol. 22, no. 2, pp. 1121–1167, 2020.
8. F. Alshehri and G. Muhammad, "A comprehensive survey of the internet of things (IoT) and AI-based smart healthcare," *IEEE Access*, vol. 9, pp. 3660–3678, 2021.
9. L. Greco, G. Percannella, P. Ritrovato, F. Tortorella, and M. Vento, "Trends in IoT based solutions for health care: Moving AI to the edge," *Pattern Recognition Letters*, vol. 135, pp. 346–353, 2020.
10. M. Mozafari, F. Firouzi, and B. Farahani, "Towards IoT-enabled multimodal mental stress monitoring," in *2020 International Conference on Omni-layer Intelligent Systems (COINS)*, 2020.
11. "Internet of things in healthcare market size report, 2019–2025." [Online]. Available: http://www.grandviewresearch.com/industry-analysis/internet-of-things-iot-healthcare-market.
12. S. Jiang, F. Firouzi, K. Chakrabarty, and E. B. Elbogen, "A resilient and hierarchical IoT-based solution for stress monitoring in everyday settings," *IEEE Internet of Things Journal*, pp. 1–1, 2021.
13. J. Zhou, N. DeCapite, J. McNabb, J. R. Ruiz, D. A. Fisher, S. Grego, and K. Chakrabarty, "Stool image analysis for precision health monitoring by smart toilets," in *Proceedings of the 6th Machine Learning for Healthcare Conference*, 2021.
14. Z. I. Attia, D. M. Harmon, E. R. Behr, and P. A. Friedman, "Application of artificial intelligence to the electrocardiogram," *European Heart Journal*, vol. 42, no. 46, pp. 4717–4730, 09 2021.
15. R. L. Draelos, D. Dov, M. A. Mazurowski, J. Y. Lo, R. Henao, G. D. Rubin, and L. Carin, "Machine-learning-based multiple abnormality prediction with large-scale chest computed tomography volumes," *Medical Image Analysis*, vol. 67, p. 101857, 2021.

16. S. Park, D. Won, B. Lee, D. Escobedo, A. Esteva, A. Aalipour, T. Ge, J. Kim, S. Suh, E. Choi, A. Lozano, C. Yao, S. Bodapati, F. Achterberg, J. Kim, H. Park, Y. Choi, W. Kim, J. Yu, A. Bhatt, J. Lee, R. Spitler, S. Wang, and S. Gambhir, "A mountable toilet system for personalized health monitoring via the analysis of excreta," *Nature Biomedical Engineering*, vol. 4, p. 624–635, 2020.
17. C. Wald, "Diagnostics: A flow of information," *Nature*, vol. 551, 2017.
18. J. Bae and H. Lee, "User health information analysis with a urine and feces separable smart toilet system," *IEEE Access*, vol. 6, pp. 78 751–78 765, 2018.
19. P. Ghosh, D. Bhattacharjee, and M. Nasipuri, "Intelligent toilet system for non-invasive estimation of blood-sugar level from urine," *IRBM*, vol. 41, no. 2, pp. 94–105, 2020.
20. M. Ra, M. S. Muhammad, C. Lim, S. Han, C. Jung, and W. Kim, "Smartphone-based point-of-care urinalysis under variable illumination," *IEEE Journal of Translational Engineering in Health and Medicine*, vol. 6, pp. 1–11, 2018.
21. H. Kim and D. Allen, "Using digital filters to obtain accurate trended urine glucose levels from toilet-deployable near-infrared spectrometers," *Journal of Analytical & Bioanalytical Techniques*, vol. 7, p. 338, 2016.
22. D. Hachuel, A. Jha, D. Estrin, A. Martinez, K. Staller, and C. Velez, "Augmenting gastrointestinal health: A deep learning approach to human stool recognition and characterization in macroscopic images," *arXiv preprint arXiv:1903.10578*, 2019.
23. S. Lewis and K. Heaton, "Stool form scale as a useful guide to intestinal transit time," *Scandinavian Journal of Gastroenterology*, vol. 32, no. 9, pp. 920–924, 1997.
24. A. Markland, O. Palsson, P. Goode, K. Burgio, J. Busby-Whitehead, and W. Whitehead, "Association of low dietary intake of fiber and liquids with constipation: Evidence from the national health and nutrition examination survey," *The American Journal of Gastroenterology*, vol. 108, no. 5, p. 796, 2013.
25. J. Blake, M. Raker and K. Whelan, "Validity and reliability of the Bristol Stool Form Scale in healthy adults and patients with diarrhoea-predominant irritable bowel syndrome," *Aliment Pharmacol Ther.*, vol. 44, pp. 693–703, 2016.
26. H. Ohno, H. Murakami, K. Tanisawa, K. Konishi, and M. Miyachi, "Validity of an observational assessment tool for multifaceted evaluation of faecal condition," *Scientific Reports*, vol. 9, no. 1, pp. 1–9, 2019.
27. S.-M. Chen, M.-H. Chang, J.-C. Du, C.-C. Lin, A.-C. Chen, H.-C. Lee, B.-H. Lau, Y.-J. Yang, T.-C. Wu, C.-H. Chu, M.-W. Lai, and H.-L. Chen, "Screening for biliary atresia by infant stool color card in Taiwan," *Pediatrics*, vol. 117, no. 4, pp. 1147–1154, 2006.
28. Y.-H. Gu, K. Yokoyama, K. Mizuta, T. Tsuchioka, T. Kudo, H. Sasaki, M. Nio, J. Julian, T. Ohkubo, and A. Matsui, "Stool color card screening for early detection of biliary atresia and long-term native liver survival: A 19-year cohort study in Japan," *The Journal of Pediatrics*, vol. 166, no. 4, pp. 897–902.e1, 2015.
29. A. Franciscovich, D. Vaidya, J. Doyle, J. Bolinger, M. Capdevila, M. Rice, L. Hancock, T. Mahr, and D. B. Mogul, "PoopMD, a mobile health application, accurately identifies infant acholic stools," *PLOS ONE*, vol. 10, no. 7, pp. 1–9, 07 2015. [Online]. Available: https://doi.org/10.1371/journal.pone.0132270
30. K. Fine, A. Nelson, T. Ellington, and A. Mossburg, "Comparison of the color of fecal blood with the anatomical location of gastrointestinal bleeding lesions: Potential misdiagnosis using only flexible sigmoidoscopy for bright red blood per rectum," *American Journal of Gastroenterology*, vol. 94, no. 11, pp. 3202–3210, 1999.
31. Z. Yang, L. Leng, and B. Kim, "StoolNet for color classification of stool medical images," *Electronics*, vol. 8, p. 1464, 2019.
32. A. Krizhevsky, I. Sutskever, and G. E. Hinton, "Imagenet classification with deep convolutional neural networks," in *Advances in Neural Information Processing Systems*, 2012.
33. K. Simonyan and A. Zisserman, "Very deep convolutional networks for large-scale image recognition," in *International Conference on Learning Representations*, 2015.

34. V. Murthy, V. Singh, T. Chen, R. Manmatha, and D. Comaniciu, "Deep decision network for multi-class image classification," in *IEEE Conference on Computer Vision and Pattern Recognition*, 2016.
35. Z. Yan, H. Zhang, R. Piramuthu, V. Jagadeesh, D. DeCoste, W. Di, and Y. Yu, "HD-CNN: Hierarchical deep convolutional neural network for large scale visual recognition," in *IEEE Conference on Computer Vision and Pattern Recognition*, 2015, pp. 2740–2748.
36. M. Oquab, L. Bottou, I. Laptev, and J. Sivic, "Learning and transferring mid-level image representations using convolutional neural networks," in *IEEE Conference on Computer Vision and Pattern Recognition*, 2014, pp. 1717–1724.
37. N. Chawla, K. Bowyer, L. Hall, and W. Kegelmeyer, "SMOTE: Synthetic minority over-sampling technique," *Journal of Artificial Intelligence Research*, vol. 16, no. 1, p. 321–357, Jun. 2002.
38. A. Ali-Gombe, E. Elyan, and C. Jayne, "Multiple fake classes GAN for data augmentation in face image dataset," in *2019 International Joint Conference on Neural Networks (IJCNN)*, 2019, pp. 1–8.
39. A. B. Qasim, I. Ezhov, S. Shit, O. Schoppe, J. C. Paetzold, A. Sekuboyina, F. Kofler, J. Lipkova, H. Li, and B. Menze, "Red-GAN: Attacking class imbalance via conditioned generation. Yet another medical imaging perspective," in *Proceedings of the Third Conference on Medical Imaging with Deep Learning*, 06–08 Jul 2020, pp. 655–668.
40. E. Nazari and P. Branco, "On oversampling via generative adversarial networks under different data difficulty factors," in *Proceedings of the Third International Workshop on Learning with Imbalanced Domains: Theory and Applications*, vol. 154, 17 Sep 2021, pp. 76–89.
41. M. Zhao, Y. Cong, and L. Carin, "On leveraging pretrained GANs for limited-data generation," in *ICML*, 2020.
42. Y. Wang, D. Ramanan, and M. Hebert, "Learning to model the tail," in *Advances in Neural Information Processing Systems*, 2017.
43. Y. Cui, M. Jia, T. Lin, Y. Song, and S. Belongie, "Class-balanced loss based on effective number of samples," in *IEEE Conference on Computer Vision and Pattern Recognition*, 2019.
44. B. Zadrozny, J. Langford, and N. Abe, "Cost-sensitive learning by cost-proportionate example weighting," in *IEEE International Conference on Data Mining*, 2003.
45. P. Pace, G. Aloi, R. Gravina, G. Caliciuri, G. Fortino, and A. Liotta, "An edge-based architecture to support efficient applications for healthcare industry 4.0," *IEEE Transactions on Industrial Informatics*, vol. 15, no. 1, pp. 481–489, 2019.
46. B. Farahani, F. Firouzi, V. Chang, M. Badaroglu, N. Constant, and K. Mankodiya, "Towards fog-driven IoT eHealth: Promises and challenges of IoT in medicine and healthcare," *Future Generation Computer Systems*, vol. 78, pp. 659–676, 2018.
47. V. Bianchi, M. Bassoli, G. Lombardo, P. Fornacciari, M. Mordonini, and I. De Munari, "IoT wearable sensor and deep learning: An integrated approach for personalized human activity recognition in a smart home environment," *IEEE Internet of Things Journal*, vol. 6, no. 5, pp. 8553–8562, 2019.
48. M. A. Rahman and M. S. Hossain, "An internet-of-medical-things-enabled edge computing framework for tackling COVID-19," *IEEE Internet of Things Journal*, vol. 8, no. 21, pp. 15 847–15 854, 2021.
49. A. Chowdhery, P. Warden, J. Shlens, A. Howard, and R. Rhodes, "Visual wake words dataset," *arXiv preprint:1906.05721*, 2019.
50. J. Tang, D. Sun, S. Liu, and J.-L. Gaudiot, "Enabling deep learning on IoT devices," *Computer*, vol. 50, no. 10, pp. 92–96, 2017.
51. F. Firouzi, B. Farahani, and A. Marinšek, "The convergence and interplay of edge, fog, and cloud in the ai-driven internet of things (IoT)," *Information Systems*, p. 101840, 2021.
52. M. Min, X. Wan, L. Xiao, Y. Chen, M. Xia, D. Wu, and H. Dai, "Learning-based privacy-aware offloading for healthcare IoT with energy harvesting," *IEEE Internet of Things Journal*, vol. 6, pp. 4307–4316, 2019.
53. Z. Ning, P. Dong, X. Wang, X. Hu, L. Guo, B. Hu, Y. Guo, T. Qiu, and R. Y. K. Kwok, "Mobile edge computing enabled 5G health monitoring for internet of medical things: A decentralized

game theoretic approach," *IEEE Journal on Selected Areas in Communications*, vol. 39, pp. 463–478, 2021.
54. D. Hall, D. Crismon, J. Larsen, K. Campbell, J. Reynolds, and J. Blake, "US20200225121A1: Toilet equipped to provide fecal analysis," 2020.
55. N. Conn, K. Schwarz, and D. Borkholder, "In-home cardiovascular monitoring system for heart failure: comparative study," *JMIR mHealth and uHealth*, vol. 7, no. 1, p. e12419, 2019.
56. W. Tsang, Y. Liong, L. Raman, Z. Wai, D. Consigliere, and E. Chiong, "Validation of the TOTO Flowsky uroflowmetry device," *European Urology Supplements*, vol. 16, no. 3, pp. e1959–e1960, 2017.
57. J. Zhou, J. McNabb, N. DeCapite, J. R. Ruiz, D. A. Fisher, S. Grego, and K. Chakrabarty, "Stool image analysis for digital health monitoring by smart toilets," *IEEE Internet of Things Journal*, vol. 10, no. 5, pp. 3720–3734, 2023.
58. J. Tang, Y. Ren, and S. Liu, "Real-time robot localization, vision, and speech recognition on nvidia jetson tx1," 2017.
59. Y. He, J. Lin, Z. Liu, H. Wang, L. Li, and S. Han, "AMC: Automl for model compression and acceleration on mobile devices," in *Proceedings of the European Conference on Computer Vision*, 2018.
60. H. Cai, J. Lin, Y. Lin, Z. Liu, K. Wang, T. Wang, L. Zhu, and S. Han, "Automl for architecting efficient and specialized neural networks," *IEEE Micro*, vol. 40, no. 1, pp. 75–82, 2020.
61. Labelbox, "https://labelbox.com," 2019.
62. M. Banerjee, M. Capozzoli, L. McSweeney, and D. Sinha, "Beyond kappa: A review of interrater agreement measures," *Canadian Journal of Statistics*, vol. 27, no. 1, pp. 3–23, 1999.
63. Y. Seo and K. Shin, "Hierarchical convolutional neural networks for fashion image classification," *Expert Systems with Applications*, vol. 116, pp. 328–339, 2019.
64. K. He, X. Zhang, S. Ren, and J. Sun, "Deep residual learning for image recognition," in *IEEE Conference on Computer Vision and Pattern Recognition*, 2016.
65. X. Ding, X. Zhang, N. Ma, J. Han, G. Ding, and J. Sun, "Repvgg: Making vgg-style convnets great again," in *Proceedings of the IEEE/CVF Conference on Computer Vision and Pattern Recognition*, 2021.
66. Y. L. Hong, J. Li, T. Y. Lang, and X. Zhu, "Dynamics of an axially moving unidirectional plate partially immersed in fluid under two frequency parametric excitation," *International Journal of Non-Linear Mechanics*, vol. 99, pp. 31–39, 2018.
67. M. Sandler, A. Howard, M. Zhu, A. Zhmoginov, and L. Chen, "Mobilenetv2: Inverted residuals and linear bottlenecks," in *IEEE Conference on Computer Vision and Pattern Recognition*, 2018.
68. N. Ma, X. Zhang, H. Zheng, and J. Sun, "Shufflenet v2: Practical guidelines for efficient CNN architecture design," in *Proceedings of the European Conference on Computer Vision*, 2018.
69. X. Zhang, X. Zhou, M. Lin, and J. Sun, "Shufflenet: An extremely efficient convolutional neural network for mobile devices," in *IEEE Conference on Computer Vision and Pattern Recognition*, 2018.
70. I. Goodfellow, Y. Bengio, and A. Courville, *Deep Learning*. MIT Press, 2016.
71. K. Murphy, *Machine Learning: A Probabilistic Perspective*. MIT Press, 2012.
72. X. Qin, Z. Zhang, C. Huang, M. Dehghan, O. Zaiane, and M. Jagersand, "U2-Net: Going deeper with nested U-structure for salient object detection," *Pattern Recognition*, vol. 106, p. 107404, 2020.
73. L. Wang, H. Lu, Y. Wang, M. Feng, D. Wang, B. Yin, and X. Ruan, "Learning to detect salient objects with image-level supervision," in *Proceedings of the IEEE Conference on Computer Vision and Pattern Recognition (CVPR)*, July 2017.
74. D. Crandall and J. Luo, "Robust color object detection using spatial-color joint probability functions," in *IEEE Conference on Computer Vision and Pattern Recognition*, 2004.
75. E. Giorgianni and T. Madden, *Digital Color Management: Encoding Solutions*. Reading, MA: Addison-Wesley, 1997.
76. K. Kelly and D. Judd, *Color Universal Language and Dictionary of Names*. National Bureau of Standards, 1976.

77. R. Quinlan, "Induction of decision trees," *Machine Learning*, vol. 1, no. 1, pp. 81–106, Mar 1986.
78. N. Altman, "An introduction to kernel and nearest-neighbor nonparametric regression," *The American Statistician*, vol. 46, no. 3, pp. 175–185, 1992.
79. S. H. Walker and D. B. Duncan, "Estimation of the probability of an event as a function of several independent variables," *Biometrika*, vol. 54, pp. 167–179, 07 1967.
80. K. Brodersen, C. Ong, K. Stephan, and J. Buhmann, "The balanced accuracy and its posterior distribution," in *International Conference on Pattern Recognition*, 2010.
81. J. Deng, W. Dong, R. Socher, L. Li, K. Li, and F. Li, "Imagenet: A large-scale hierarchical image database," in *IEEE Conference on Computer Vision and Pattern Recognition*, 2009.
82. A. Paszke, S. Gross, F. Massa, A. Lerer, J. Bradbury, G. Chanan, T. Killeen, Z. Lin, N. Gimelshein, L. Antiga, A. Desmaison, A. Kopf, E. Yang, Z. DeVito, M. Raison, A. Tejani, S. Chilamkurthy, B. Steiner, L. Fang, J. Bai, and S. Chintala, "Pytorch: An imperative style, high-performance deep learning library," in *Advances in Neural Information Processing Systems*, 2019.
83. TestMy.net Internet Speed Test. Accessed: 2020-12. [Online]. Available: https://testmy.net
84. C. J. Han, K. Pike, M. E. Jarrett, and M. M. Heitkemper, "Symptom-based latent classes of persons with irritable bowel syndrome," *Research in Nursing & Health*, vol. 42, no. 5, pp. 382–391, 2019.
85. M. Simrén and J. Tack, "New treatments and therapeutic targets for IBS and other functional bowel disorders," *Nat Rev Gastroenterol Hepatol*, vol. 15, p. 589–605, 2018.
86. A. Bahmani, A. Alavi, T. Buergel, and et al, "A scalable, secure, and interoperable platform for deep data-driven health management," *Nature Communications*, vol. 12, 2021.

Accelerating AI Applications with Silicon Photonics: Status and Perspective

Amin Shafiee, Febin Sunny, Ebadollah Taheri, Mahdi Nikdast, and Sudeep Pasricha

> *The limits of the possible can only be defined by going beyond them into the impossible.*
>
> *—Arthur C. Clarke*

Abstract Deep learning has achieved remarkable success in areas like computer vision, natural language processing, and pattern recognition, driven by advancements in training techniques, deeper neural networks, and hardware improvements. Application-specific integrated circuit (ASIC) accelerators have gained attention for their superior performance and energy efficiency compared to CPUs and graphics processing unit (GPUs). However, these accelerators face challenges due to limitations in complementary metal oxide semiconductor (CMOS) scaling and metallic interconnects, which hinder computational efficiency and energy performance. Silicon photonics offers a promising CMOS-compatible solution by leveraging light for communication and computation, potentially addressing these bottlenecks. This chapter explores the potential of silicon photonics to revolutionize deep learning accelerators, discussing its capabilities and limitations through a bottom-up perspective on design abstractions.

Keywords Silicon photonics · Deep learning · Neuromorphic computing

1 Introduction

Deep learning, a subfield of AI, has revolutionized problem-solving in areas like computer vision, natural language processing, and big data analysis. Once abandoned in the 1990s, it gained traction after Hinton et al.'s 2006 work achieved

---

A. Shafiee (✉) · F. Sunny · E. Taheri · M. Nikdast · S. Pasricha
Department of Electrical and Computer Engineering, Colorado State University, Fort Collins, CO, USA
e-mail: Amin.Shafiee@colostate.edu

B. Farahani et al. (eds.), *Smart and Connected Health: Applications and Real-World Use Cases*, https://doi.org/10.1007/978-3-032-09253-3_6

>98% precision in handwritten digit recognition [1]. Today, deep learning powers technologies like search engines, recommendation systems, speech recognition, and IoT applications while enabling innovations such as self-driving cars [2], robotics [3], fake news detection [4], pandemic prediction [5], anomaly detection [6], and language translation [7]. The magic behind deep learning owes much to our brain's architecture. As far back as 1943, the neurophysiologist Warren McCulloch and mathematician Walter Pitts presented a simplified model of how biological neurons work together in animal brains to perform complex computations [8]. This was the first artificial neural network (ANN) architecture, and it inspired a race to build intelligent machines that could rival and eventually surpass the capabilities of the human brain. The introduction of the perceptron in 1957 by Frank Rosenblatt was another landmark, showing how the simple ANN could be trained to solve classification problems [9]. However, the limited capabilities of hardware to run even moderately complex ANNs led researchers to abandon the study of ANNs in the late 1960s. Even though new architectures and better training techniques emerged in the 1980s and early 1990s, progress was limited due to several factors, most crucially the lack of powerful machines to train and run these models. Fortunately, over the past decade, ever-improving capabilities of complementary metal oxide semiconductor (CMOS) fabrication technology have enabled extremely powerful Tera Floating-point Operations Per Second (TFLOP)-class graphics processing unit (GPU) and CPU processing chips with billions of transistors in small form factors that have made it possible to train and use deep ANN (i.e., multilayer perceptron (MLP)) architectures in a timely and cost-effective manner. Coupled with the availability of large datasets in the IoT and Big Data era, theoretical advances in training algorithms, and the emergence of new deep ANN architectures such as convolutional neural networks (CNNs), deep learning has now established its dominance over other machine learning models for many problems of interest in the domains of computer vision, natural language processing, and general pattern recognition.

With researchers creating deeper and more complex MLP and CNN architectures to push deep learning performance levels to new heights, the underlying hardware platform must consistently deliver better performance levels while also satisfying strict power dissipation limits. This endeavor to achieve higher performance-per-watt has driven hardware architects to design application-specific integrated circuit (ASIC) accelerators for deep learning that have much higher performance-per-watt than conventional general-purpose CPUs and GPUs. IBM's 4096 core TrueNorth chip that was released in 2014 was one of the earliest high-profile ASIC deep learning accelerators [10]. Since then, many other accelerators have become available, including Intel's Loihi [11] and Google's Tensor Processing Units (TPU) [12]. Several academic efforts have also led to the design of new types of ASIC and field-programmable gate array (FPGA)-based deep learning accelerators [13–17]. Even conventional GPUs and CPUs have evolved to speed up deep learning model execution, e.g., Nvidia GPUs now include tensor cores [18], and CPUs support increasingly advanced vector instructions [19], both of which are designed to accelerate common matrix and vector operations in deep learning processing.

Beyond digital domain solutions, accelerators have also been proposed that work in the analog domain [20–22] or the analog-digital mixed signal domain [23–25].

Electronic accelerator architectures are increasingly constrained by fundamental limits in the post-Moore's law era, where processing improvements have slowed significantly [26]. A major challenge is the energy and bandwidth bottleneck caused by data movement on metallic wires [27]. This issue stems from the von Neumann architecture's inherent separation of memory and processing elements, which demands extensive data transfer. Inspiration can be drawn from the human brain's efficient structure, processing capabilities, and low power consumption. This interdisciplinary approach, known as neuromorphic engineering or neuromorphic computing, aims to mimic the brain using advanced neuron models and neural systems. Recently, neuromorphic computing has come to encompass biologically inspired computing techniques and artificial neural networks implemented in non-von Neumann architectures.

Photonic interconnects present a promising solution to data movement challenges, with photonic links already replacing metallic ones for high-speed information transmission across various computing levels and now being considered for chip-scale integration [28]. Silicon photonics, enabling cost-effective integration of optical components using CMOS manufacturing, has been a key driver for chip-scale photonic interconnects [29]. Notably, computations crucial to deep learning, such as matrix-vector multiplications, can be executed entirely in the optical domain [30]. This development positions silicon photonics as a potential foundation for deep learning accelerators that leverage optical communication and computation. These accelerators could deliver unprecedented energy efficiency and parallelism, achieving energy footprint efficiency up to 1000× greater than current electronic accelerators for multiply-and-accumulate (MAC) operations [31]. Additionally, photonic MACs can operate at bandwidths nearing photodetection rates in the hundreds of GHz, vastly outperforming electronic systems, which are limited to clock rates of a few GHz [32].

This chapter presents a comprehensive overview of silicon photonics for accelerating deep learning, with a bottom-up approach spanning device-level innovations to system-level architectures [33–35] such as Broadcast and Weight (B&W) [31, 36–39]. It explores CMOS-compatible materials (e.g., Ge, SiN) and devices like Mach-Zehnder interferometers (MZIs) and microring resonators (MRs) as building blocks for photonic neurons. Proposed architectures include noncoherent B&W [40] and coherent artificial linear neurons (COLN) [36], enabling scalable photonic neural networks while balancing energy, performance, and reliability tradeoffs. This classification highlights silicon photonics' potential and challenges in deep learning acceleration.

2 An Overview of Deep Learning

Deep learning is a subset of machine learning (ML), which itself is a subset of the broader field of AI. Deep learning aims to emulate the deep architecture of a human brain, which has billions of interconnected neurons acting as computational units. Human brains also work hierarchically, starting from simpler concepts and then combining them to learn more abstract ideas. This mode of learning is reflected in deep learning models, which break down input data into features and then recombine them to perform the task at hand (e.g., detection, classification). Once relevant features have been learned by a deep learning model in the training phase, the model can be applied to tasks of a similar nature, with no human intervention.

As mentioned earlier, deep learning has gained a lot of attention in recent years. But the concept is not new. The idea to make machines as intelligent as humans is the very basis of the analytical engine conceived by Charles Babbage in 1837. The field of AI and the research into making machines capable of thinking like humans started as far back as the mid-twentieth century, with the computational model for neural networks and neuron operation developed by Warren McCulloch and Walter Pitts in 1943 [8]. The perceptron algorithm was invented by psychologist Frank Rosenblatt in his seminal 1957 paper [9]. Leveraging this algorithm, Rosenblatt created the first single-layer perceptron (see Fig. 1a) that is an electronic computational device adhering to the biological principles behind how the human brain functions.

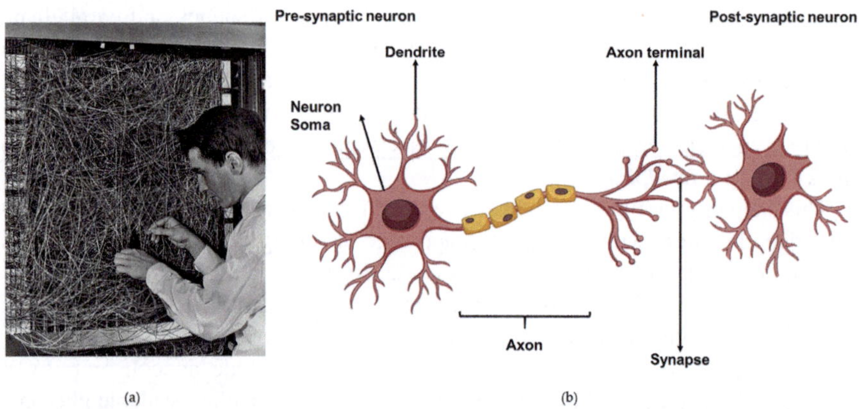

Fig. 1 (**a**) Frank Rosenblatt with his Mark-1 single-layer perceptron; (**b**) a depiction of the neuron and the synaptic connection to another neuron. This is a simplistic depiction of a neuron, showing only the most basic components

2.1 Artificial Neural Networks

The emergence of the computational model for representing neural activities paved the way for artificial neural networks (ANNs). When compared to spiking neural networks (SNNs) [41], ANNs are markedly abstract in their approach to implementing brain functions. The weights, which represent synaptic plasticity, are simple scalars. The neurons utilized are also much simpler and tasked with the accumulation of input-weight products followed by passing the resulting output through a nonlinear function.

ANNs emulate brain activity by simulating a collection of interconnected neurons, arranged in layers. The simplest representation will have three layers: an input layer, an output layer, and a hidden layer in between these two layers (Fig. 2a). The input layer accepts data from outside the ANN; the hidden layer is where the computation happens; and the output layer is where we can get the results from the neural network. The activation functions also play an important part in simulating intelligence. Mathematically, without appropriate activation functions, the neuron model is a simple linear model, which multiplies and accumulates input-weight products. To introduce nonlinearity into the network and make it possible for the model to approximate more complex functions, we need to use appropriate nonlinear activation functions, such as *sigmoid*, *rectified linear unit (ReLu)*, and *tanh* to list a few. Utilizing these functions, ANNs are able to learn very complex nonlinear relationships between input features.

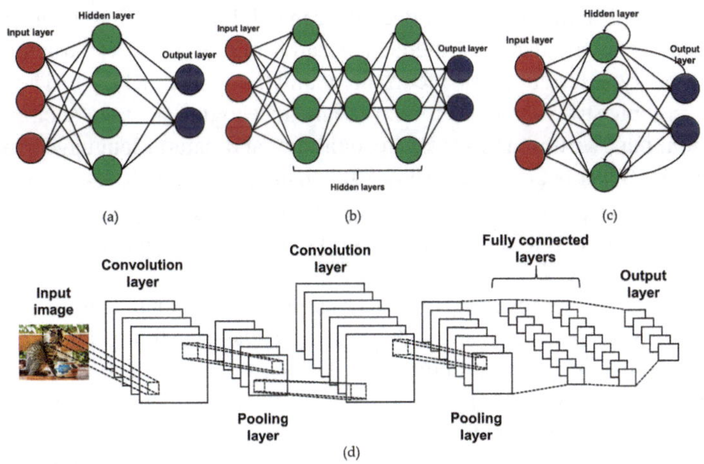

Fig. 2 (**a**) Layered architecture of an ANN; this figure shows a shallow ANN with one hidden layer; (**b**) as the number of hidden layers in the ANN increases, we achieve DNNs. Note that the layers of an ANN are populated by neurons, with associated weights and biases; (**c**) a fully connected RNN, which shows the feedback connections in its hidden layer and simulates memory or saved states in this architecture; (**d**) representation of various layers and operations in a CNN

An important distinction to be made here is between ANNs and traditional ML algorithms such as support vector machines (SVMs), K-nearest neighbors (KNN), and random forests (RF). What makes ANNs distinct is their ability to handle large quantities of data, with minimal human intervention. Traditional ML algorithms usually require a human expert to provide the necessary rule set on which they operate. Often, assistance for feature extraction from the data is also needed, e.g., for kernel selection in SVMs.

Deep neural networks (DNNs) are ANNs with multiple hidden layers (see Fig. 2b), which can utilize the complex interconnectivity among the neurons to compute and efficiently represent very complex nonlinear relationships after being trained. During the training phase, input activations traverse a forward path from the input to the hidden layers and finally to the output layer. The error (often called the loss) between the DNN output and the expected output is backpropagated through the model to update the neuron weights and biases, in a manner that reduces the loss. This process is iteratively repeated until the model output (e.g., image class prediction) is as close as possible to the expected outputs, i.e., the loss is minimized. After training, the model can make predictions given an input, in what is referred to as the inference phase. The training phase for DNNs is a time- and resource-intensive process, compared to the inference phase. The notable learning architectures which utilize DNNs include multilayer perceptrons (MLPs), recurrent neural networks (RNNs), deep Boltzmann machines (DBMs), stacked auto-encoders (SAEs), and convolutional neural networks (CNNs).

MLPs only include feedforward fully connected (FC) layers as shown in Fig. 2b, where each neuron in a layer is connected to each neuron in the preceding and following layers. Some model architectures can exhibit temporal dynamic behavior and possess an internal state or memory because of their network structure. These can be broadly referred to as recurrent neural networks (RNNs). The inherent memory in their structure makes them ideal for recognition tasks such as pattern recognition, handwriting and speech recognition, and natural language processing. Research on RNNs began with David Rumehalt in 1986 [42]. As of today, many RNN variants are popular, including long short-term memory (LSTM) networks, gated recurrent units (GRUs), continuous time RNNs (CTRNNs), etc. A simplified RNN is shown in Fig. 2c.

CNNs target the processing of 2D or higher-dimensional features instead of the 1D ones in MLPs. They are widely used for classification problems in image and video processing. The structure of a CNN is depicted in Fig. 2d. A typical CNN contains three types of layers: convolutional (Conv), pooling (Pool), and fully connected (FC). In Conv and Pool layers, there are multiple channels (called feature maps) that extract different local features from the input data. These layers combine the lower-level features from multiple channels of the previous layer into higher-level features that are passed to the next layer, till the final classification layer where an output prediction is generated. Conv layers have much fewer parameters than FC layers, but involve a high computational footprint due to the many convolution operations that are required between filter weights and the input activations across all their channels. Pooling layers generate output activations based only on the local receptive field in the corresponding input feature map (e.g., a single "pooled" output from a group of 2x2 inputs). The two widely used variants of pooling layers are max and average pooling, and they produce the maximum or average value of each receptive field, respectively. Lastly, FC layers follow Conv and Pool layers and act to work as a classifier with the extracted features, similar to how these layers are used in MLPs.

DNNs are beginning to be widely used in real-world applications such as autonomous driving, robotics, and IoT processing. The resource intensity needed for training DNN models was met by the emergence of GPUs that are used for significantly reducing the training time of DNNs, due to the greater data and thread-level parallelism supported in GPUs than CPUs. Much like for SNNs, there is growing interest in designing energy-efficient ASIC accelerators for DNNs. Such DNN accelerators, e.g., the neural processing unit (NPU) [47], are designed to accelerate the inference phase, although a few accelerators are also aimed at improving training performance. An example of a DNN accelerator that has been very successful for accelerating both training and inference with DNNs is the TPU [12] from Google. The TPU has dedicated matrix-multiplication units and distributed memory management that makes it ideal for handling the heavy lifting needed to train DNN models and also for inference tasks. TPUs are deployed widely in Google's data centers. Newer GPU architectures have also adopted similar tensor cores for DNN acceleration [44]. Researchers have also suggested utilizing nonvolatile memory technology and processing-in-memory (PIM) for DNN accelerators. PRIME [45] and ISAAC [20] are examples of such accelerators that utilize resistive random-access memories (ReRAMs) and PIM to accelerate DNN execution.

In all of these different ANN models, the neurons are interconnected using weighted links to form neural networks. In ANNs, neurons are grouped by layer, and all the neurons in a layer get activated together. The neuron inputs are summed and weighted and have nonlinearity applied to them to generate outputs that become the input to the neurons in subsequent layers. This is different from the techniques utilized by SNNs to emulate biological interactions and complex neuronal behaviors. SNNs implement more complex behaviors, which are captured using more detailed neuron models, such as the Hodgkin-Huxley Model [43], leaky integrate-and-fire (LIF) neuron models [41], and the Izhikevich spiking neuron model [46]. In addition, the interaction between neurons occurs in the form of 1-bit spike communication, as opposed to multiple-bit weight and activation representation adopted in ANNs. The threshold and resultant spike generation is count dependent in SNNs, i.e., the number of input spikes a neuron receives determines whether it will fire or not. This also increases the importance of spiking time-dependent plasticity (STDP) implementation in SNNs, which acts as local memories between neurons, while ANNs represent the synaptic plasticity as simple scalar weights. The clock-less asynchronous communication method between neurons adapted in SNNs may also make their implementations more complex than that of the synchronous per-layer communication approach in ANNs. This consideration is reflected in research efforts as well, where there has been a greater focus on developing ANN implementations than SNN implementations in hardware. We will now provide an overview of silicon photonics technology (Sect. 3), followed by in-depth discussions on various deep learning architectures built using photonic neurons (Sect. 4).

3 An Overview of Silicon Photonics

Optical communication has been widely employed in communication networks wherever low-cost and high-bandwidth communication at low power consumption and over large distances is required, e.g., in long-haul telecommunication networks. In recent years, silicon photonics has enabled CMOS-compatible integrated photonics and gained widespread adoption in commercial offerings for low-cost optical interconnects in data centers. Optical interconnects are now being aggressively considered at much smaller scales to connect multiple processing chips at the board level and even to connect cores within a single computing chip. As the name suggests, silicon photonics employs light, which is guided through the silicon (Si) medium on a CMOS chip, for communication. In a silicon-on-insulator (SOI) fabrication platform, the high refractive index contrast between the waveguide core (silicon) and the waveguide's cladding and substrate (e.g., silicon dioxide) results in guided optical signal propagation through total internal reflection. A single waveguide can be used to carry multiple wavelengths of light simultaneously, each capable of carrying data at high speed and high frequency, and without any interference. This is possible using a technique called wavelength-division multiplexing (WDM). The number of wavelengths in a waveguide is referred to as the WDM degree of the waveguide. The WDM degree can be increased to 64 or beyond, at which point the multiplexing is often referred to as dense wavelength-division multiplexing (DWDM). For chip-scale communication with silicon photonics, digital data from electronic components (e.g., processor, memory) can be encoded into an optical signal using electronic to optical (E/O) conversion with devices such as microring resonator (MR) modulators, subsequently transmitted over a waveguide with multiple carrier wavelengths and then detected at a receiver, where optical-to-electronic (O/E) conversion is performed with devices such as photodetectors (PDs).

There has been growing interest in using silicon photonics for more than just communication. In particular, silicon photonic devices can also be used to perform computation in the optical domain. Together, such light speed communication and computation can significantly accelerate the execution of deep learning workloads. While silicon photonic devices face several challenges for robust computation and communication at the chip-scale (e.g., they are sensitive to thermal and fabrication-process variations [48–51]), they also offer several advantages (e.g., high speed, high bandwidth, and low power) to support inter-neuron communications and implement different neural functions required in photonic neural networks. Such neural functions and their implementations are discussed in the next section. In this section, we review some of the fundamental silicon photonic devices that are employed to implement photonics-based ANNs and SNNs.

3.1 Lasers

A laser is a key requirement in optical circuits and neural networks, serving as the light source to support optical communication and computation. Lasers can be either off-chip or on-chip. Although off-chip lasers offer better light emission efficiency, they necessitate the use of couplers to couple the off-chip optical signal to the chip, where such couplers impose high optical power losses. On the other hand, on-chip lasers provide better integration density and lower optical loss, as there is no need to couple light from an off-chip source. However, on-chip lasers suffer from low emission efficiency and instability against thermal variations [55].

Lasers are used in photonic neural networks to implement different neural functions and requirements in such systems. In directly modulated lasers (see Fig. 3a [56]), the laser itself modulates the data onto an optical signal, while in another arrangement, as shown in Fig. 3b, the laser output can be modulated by a modulator, which is responsible to modulate the data onto the optical signal. Indeed, employing a laser in conjunction with modulators is common in optical interconnection networks [57, 58]. In photonic neural networks, this laser configuration can be used to design a scalable neural network [52, 54], where an off-chip laser source in combination with modulators can support multiple on-chip inter-neuron communications. In addition to the modulation, lasers can also be used to implement neural activation functions [59] because lasers have shown potential to mimic neural activation functions [60–62] where an optical stimulus in the input

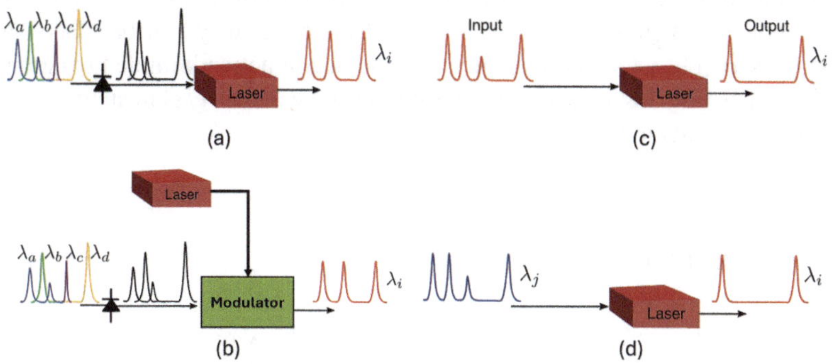

Fig. 3 Lasers used in photonic neural networks [52, 53]. (**a**) Directly modulated laser; (**b**) laser connected to a modulator [54, 55]; (**c**) coherent laser where the wavelength in the input and output is the same; (**d**) incoherent laser with different wavelengths at its input and output. Note that the signal waveform in black is an electrical signal and the other colored signal waveforms are optical ones. Different colors represent different wavelengths

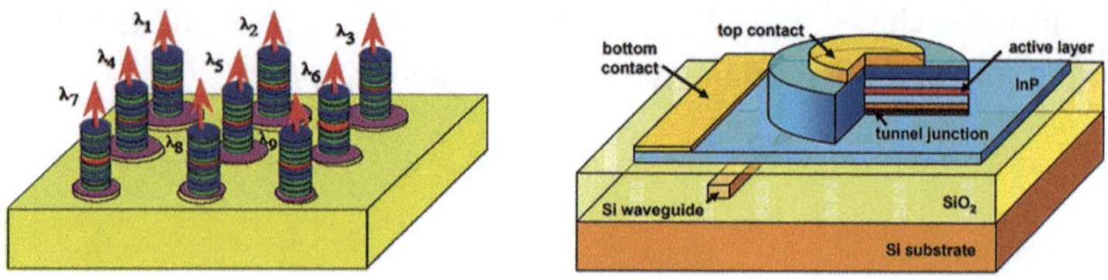

(a) (b)

Fig. 4 (**a**) A VCSEL array [63]; (**b**) a microdisk laser [64]

of the laser can result in an optical output based on an activation function (see Fig. 3c and d). Thus lasers are used in all photonic neural networks to not only support inter-neuron communication but also, in some cases, implement different neural functions in the optical domain.

A laser can be implemented in different ways. A vertical cavity surface emitting laser (VCSEL) is a semiconductor laser diode with laser beam emission perpendicular to the chip surface, as shown in Fig. 4a. Such a feature allows for several VCSELs to be placed in an array to power a large number of optical neurons and hence design scalable neural networks [60, 61, 65]. In addition to the scalability advantage, VCSEL-based neural networks can be realized using both off-chip and on-chip VCSELs [66]. Moreover, VCSELs have shown excitability behaviors of neurons [67], in which the laser emits light when the combination of inputs reaches a threshold. As a result, VCSELs offer scalability, efficiency, and several functionalities required for photonic neural network designs. Microdisk lasers, shown in Fig. 4b, are another type of lasers in which a ring resonator is formed by successive total internal reflections inside a circularly shaped waveguide [69]. Compared to VCSELs, microdisk lasers are more area efficient (a laser apparatus radius—radius of microcavity—of a few microns) and deliver a lower threshold current and maximum on-chip optical power [70]. Moreover, microdisk lasers offer a low optical loss [68], and similar to VCSEL arrays, they can be placed in an array of lasers [69] to enable scalable photonic neural network implementations. In addition, microdisk lasers have shown excitability dynamics [71] to support spiking neuron implementations.

3.2 *Waveguides*

A silicon photonic waveguide is analogous to a metallic wire, enabling optical signal transmission and routing in photonic neural networks. As shown in Fig. 5, waveguides can be classified into ridge and strip waveguides. Ridge waveguides are often employed in active devices and networks as they allow for electrical connections to be made to the waveguide (e.g., through PN junctions) where the characteristics of the optical signal can be actively controlled and altered using electro-optic or

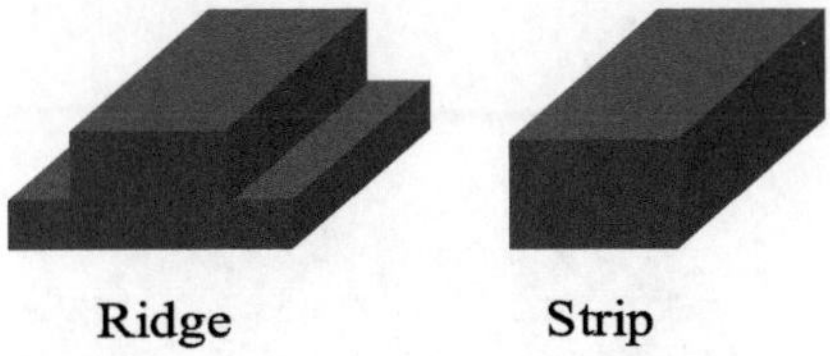

Fig. 5 Silicon photonic waveguides [28]: a ridge waveguide and a strip waveguide

thermo-optic effects in silicon [28]. On the other hand, strip waveguides are usually employed in passive devices and networks to passively route optical signals [28]. As discussed earlier, a single waveguide can support simultaneous transmission of multiple optical wavelengths with no interference (using WDM). This allows for ultrahigh bandwidth communication, which is of great interest in neural-network designs to support demanding inter-neuron communication.

When an optical signal traverses a waveguide, it experiences some optical loss (i.e., the propagation loss, often characterized in dB/cm) imposed due to, for example, some imperfections in the waveguide structure (e.g., waveguide sidewall roughness). Minimizing such optical loss in silicon photonic waveguides is essential as it limits the scalability of photonic neural networks and substantially degrades the power and energy efficiency in such networks. There have been a lot of efforts to minimize the propagation loss in silicon photonic waveguides, and SOI waveguides with propagation losses as low as 0.026 dB/cm have been proposed [72]. In general, this propagation loss in waveguides depends on precise geometry adjustment in these devices, and hence any shape distortion in a waveguide (e.g., angular sidewalls) reduces its transmission efficiency (i.e., increases the propagation loss). In addition to propagation loss, waveguide bends create optical bending loss where an optical signal will be attenuated due to the mode mismatch and radiation loss in waveguide bends. This bending loss is proportional to the radius of the waveguide bend.

3.3 Couplers

Silicon photonic couplers, as shown in Fig. 6, are used to couple an optical signal from an optical fiber (e.g., connected to an off-chip laser source) to an on-chip waveguide due to the significant mismatch between the cross-section of optical fibers (tens of microns) and that of silicon photonic waveguides (hundreds of nanometers). Such mismatch usually imposes some optical loss (i.e., coupling loss), which is considered a significant portion of optical loss in optical networks employing off-chip lasers. Two major coupling solutions are surface grating coupling and edge coupling. Surface-grating couplers, shown in Fig. 6a, are advantageous in terms of a simpler and low-cost fabrication process but at the cost of low coupling efficiency, while edge couplers, shown in Fig. 6b, provide better coupling efficiency but require a more complex fabrication and packaging process [28]. In

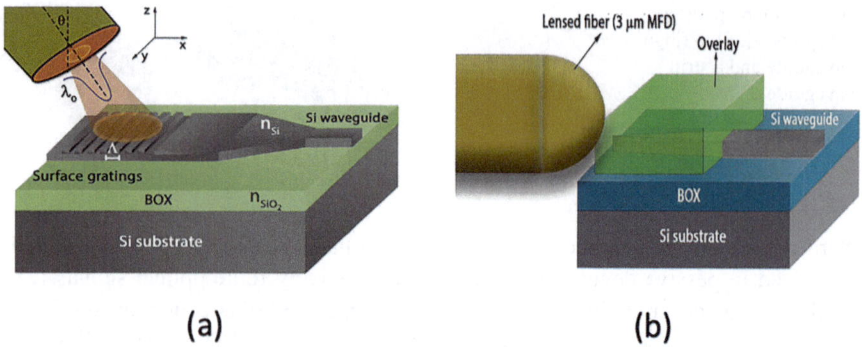

Fig. 6 Couplers [73]: (**a**) a surface-grating coupler; (**b**) an edge coupler

edge couplers, as shown in Fig. 6a, a tapered waveguide is used to couple light from the fiber to the chip. Surface-grating couplers couple the input light from a fiber to a waveguide using diffractive gratings, where a periodic structure splits and diffracts light and eventually couples the light into the waveguide. Diffractive coupling is a common means of optical coupling between VCSELs because of its simple implementation [74] and is also useful to implement photonic reservoir computing [75].

3.4 Modulators, Filters, and Switches

Microring resonators (MRs) are widely employed to design modulators, switches, and optical filters in optical interconnection networks [76, 77]. In addition to such applications in interconnection networks, they are promising devices to implement artificial neural synapses [59, 78, 79] and excitation function of neurons [80, 81], which are further discussed in Sect. 4.

MRs, as shown in Fig. 7a, are made with a ring-shaped waveguide in proximity with an input waveguide and a drop waveguide (also known as add-drop filter). When the drop waveguide is missing (e.g., in modulators and some filters), the MR is an all-pass filter (see Fig. 7b). An MR can be in two different states: on- or off-resonance, based on which the optical signal can be switched to different ports. As shown in Fig. 7a, when the MR is in the off-state, the input signal is routed to the through port, because the ring is not in resonance with the input optical signal. On the other hand, when the MR is in the on-state, the ring couples the input optical signal and drops it on the drop port. The resonant wavelength of an MR can be tuned to realize various functionalities needed to design optical modulators, switches, and filters. Here, tuning refers to sweeping the resonant wavelength of an MR by leveraging electro-optic or thermo-optic effects of silicon that can alter the optical signal characteristics, and hence the resonant wavelength in the case of an

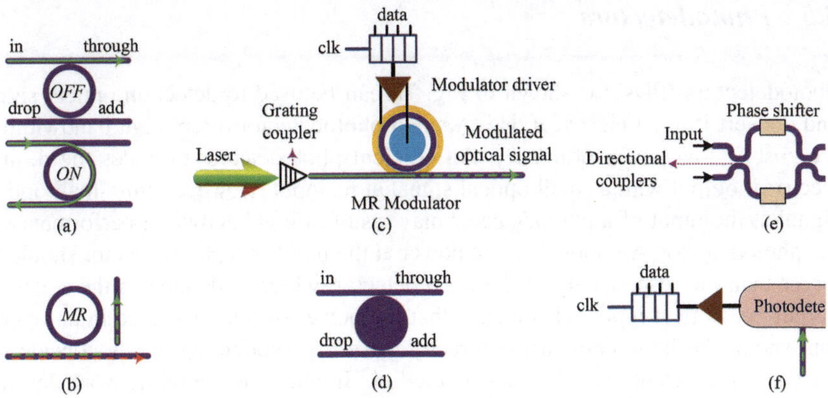

Fig. 7 Silicon photonic switching devices [28]. (**a**) MR add-drop filter/switch; (**b**) MR all-pass filter; (**c**) MR modulator; (**d**) Microdisk resonator; (**e**) MZI; (**f**) photodetector

MR. Compared to tuning mechanisms based on electro-optic effects, those based on thermo-optic effects are slower (a few microseconds versus tens of nanoseconds in tuning techniques based on electro-optic effects) but are more power efficient. Figure 7c shows an example of an MR-based modulator that is responsible for modulating electronic data onto an optical signal. The modulator can modulate electronic data onto a specific optical wavelength, and this modulated optical signal can be filtered with a wavelength-selective MR-based filter at the receiver (see Fig. 7b) and then detected and converted to electronic data through a photodetector.

Compared to an MR, a microdisk resonator (see Fig. 7d), which employs a disk instead of the ring structure, offers better optical confinement to provide a smaller disk size and potentially lower power consumption [82]. MZIs, as shown in Fig. 7e, are made of two waveguides with directional couplers and phase shifters. The phase shifters implemented using electro-optic or thermo-optic tuning change the optical phase in one or both arms of the MZI, introducing constructive or destructive interferences at the output to switch an optical signal between the output ports. Similar to MRs, MZIs have been applied to the design of optical modulators, switches, and filters. In comparison to MZIs, MRs have a smaller footprint and lower power consumption. On the other hand, MZIs provide high bandwidth and better tolerance to thermal variations.

In summary, MRs, microdisks, and MZIs are widely employed to design modulators, switches, and filters. In photonic neural networks, a set of optical filters, in each of which the optical transmission can be adjusted, can be grouped into a weight bank to support weighting of activation signals as part of a photonic neuron [40].

3.5 Photodetectors

Photodetectors (PDs), as shown in Fig. 7f, can be used to detect an optical signal and convert it to an electrical one. A small photodetector offers high bandwidth at the cost of low power efficiency. An efficient photodetector provides the desired electrical output with a small optical signal at its input. However, this small optical signal at the input of a photodetector may result in low bandwidth performance in the photodetector. An optical signal power at the input of a photodetector should be larger than the responsivity of the photodetector, which is defined as the electrical output per optical input. This means that the power of a laser source in an optical link should be large enough to correctly drive a photodetector while considering the sum of different optical losses on the link. In photonic neural network designs, photodetectors not only convert optical signals to electrical ones but also combine (i.e., sum the magnitudes of) several optical signals over different wavelengths [40, 54, 79], which is a useful function in designing a silicon photonic neuron.

3.6 Devices Based on Phase-Change Materials

Devices that utilize phase-change materials (PCMs) for tuning are of great interest in silicon photonic circuits to design modulators [83–91], MZI-based switches [92], and low-loss phase shifters [117]. The main principle in these PCM-based photonic devices is to employ a PCM (e.g., GST: $Ge_2Sb_2Te_5$ used in [83]) to efficiently induce high refractive index changes for efficient phase tuning. Unlike electro- and thermo-optic tuning, PCM-based tuning is nonvolatile: it only requires power for transition between the amorphous and crystalline states in the PCM [93]. This can allow for low-overhead tuning of silicon photonic devices, e.g., from on-resonance to off-resonance in a PCM-based MR. In some photonic neural networks, PCM-based devices [94–96] are proposed as part of the design of neurons. For example, in the synapse design proposed in [94], several PCMs are placed on a waveguide to control optical transmission in the waveguide and implement the function of a synapse. Moreover, PCMs are useful to implement summation [95] and weighting functions [94, 95, 97].

3.7 Other Devices

A semiconductor optical amplifier (SOA) is a device in which a semiconductor is used to add a gain to an optical signal without electro-optical or opto-electrical conversions. SOAs are mainly used to compensate for optical losses in optical communication systems. In photonic neural networks, SOAs can be employed to implement learning functions [98, 99]. However, SOAs suffer from poor coupling efficiency to optical fibers and are sensitive to polarization because of their planar shape [100]. Vertical cavity semiconductor optical amplifiers (VCSOAs) provide a better coupling efficiency and a lower sensitivity to polarization, and they can also be integrated into 2D arrays [81]. Moreover, based on the proposed learning function implementation in [101], VCSOAs can offer low-power consumption to implement learning functions in neural networks.

Lastly, silicon photonic arrayed-waveguide gratings (AWG) are commonly used as optical (de)multiplexers in WDM systems. These devices are capable of multiplexing many wavelengths into a single optical fiber, thereby considerably increasing the transmission capacity of optical networks. AWGs have been used to implement matrix multiplication [103] and CNNs [104].

4 Silicon Photonic Neural Network Architectures

At the architecture level, prior work focuses on implementing different types of neural network models (discussed in Sect. 2): ANNs (MLPs, CNNs, RNNs) and SNNs. The overarching innovation of a particular work is governed by the basic optical devices used in the architecture and the fundamental principles, such as optical resonance and optical interference, which govern those devices. These principles ultimately have the biggest impact on the performance-, power-, and reliability-centric design decisions and also the inherent limitations of the architecture built using them. Thus, devices and the driving principles behind them drive the innovations required to realize the architectures with silicon photonics technology. Hence, our classification in this section will be based on the primary photonic principles used to construct the neural network architectures.

4.1 Optical Resonance-Based Network Architectures

Optical resonance-based neural network implementations usually rely on the wavelength specificity of MRs or microdisks, which leads to the utilization of WDM-based implementations where multiple wavelengths are utilized in a waveguide. Most of these architectures utilize or build on the B&W protocol, illustrated in Fig. 8, for setting and updating the weights as it was demonstrated in [159] to have isomorphism to continuous time recurrent neural networks (CTRNNs). The feedback loops, which are characteristic of RNNs, can be emulated by MRs when they reach optical bistability. Under favorable conditions pertaining to the resonant material and incident transmission intensity, the output transmission of the resonator can enter a hysteresis cycle, with two stable transmission levels. This is referred to as optical bistability of resonators. The work in [107] also suggested using Mach-Zehnder modulators (MZMs) to generate the *sigmoid* activation function. This specific work provided the proof-of-concept that B&W-based MR architectures can

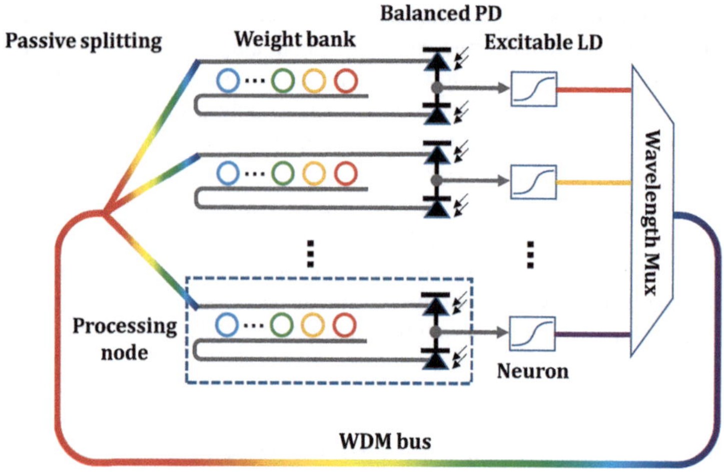

Fig. 8 The Broadcast and Weight (B&W) protocol as illustrated in [37]

be used to implement neural networks and that they can yield better performance over traditional CPU-based CTRNNs. For benchmarking, they considered a Lorenz attractor [108] simulation application and reported a 294× acceleration with their photonic architecture compared to CPU-based simulations.

The B&W protocol is a multiwavelength analog networking protocol in which multiple all-photonic neuron outputs are multiplexed and distributed to all-neuron inputs. These architectures tend to make use of the parallelism that is inherent in photonic architectures, employing multiple wavelengths to transfer data in parallel using WDM. Different wavelengths in a waveguide represent the input signals to the neuron. Weights are reconfigured by tuning the MRs, so that the characteristics of a specific wavelength are modified. MR weight banks comprise tunable MRs, which can be tuned to drain energy from their resonant wavelength so that intensity of the wavelengths reflect the weights or the kernel values. The change in intensity is read using photodetectors (PDs) and summed to obtain the output values from the weight bank. The obvious advantage of this approach is the utilization of the well-studied and mature MR technology to implement photonic neural networks, which makes the hardware implementation and integration easier. However, an issue this protocol can face is the number of MRs needed to implement it for real-world applications, and depending on the feature map and the kernel size for CNNs, this can become exorbitant. The research utilizing this protocol tries to work around this issue.

The authors in [109] utilize the B&W protocol with an MZM to implement the *sigmoid* nonlinearity as in [106] to propose a CNN accelerator, dubbed photonic CNN accelerator (PCNNA). PCNNA implements one CNN layer and reuses that layer sequentially, with varying kernel sizes to implement the whole CNN. The input feature map and the required kernels are loaded from an off-chip DRAM. The results from the execution of individual layers are fed back into the memory. This is a sequential execution of kernels using an optical core, which runs at a higher clock frequency than its electrical components. The optical core mentioned here is comprised of the weighting MR banks and digital to analog converters (DACs) that feed data into the MR banks and the laser diodes (LDs). The authors argue that because CNNs use kernels with the same dimensions per layer, they share the same receptive fields, and hence convolution computations for different kernels can be performed in parallel. They demonstrated the effectiveness of this work by implementing *AlexNet*, which is a deep CNN architecture with eight layers, five convolution layers, and three fully connected layers. The authors showed how their filter-based approach to implementing *AlexNet* had substantially fewer number of MRs than an approach which does not consider any optimizations for implementing *AlexNet* (they claimed a reduction in the number of MRs from one-billion to 100,000 range).

Another architecture which utilizes photonic weight banks for implementing CNNs is described in [110]. The authors have described an architecture which implements the entirety of the CNN layers using connected convolution units which are comprised of weight banks, where the tuned MRs assume the kernel values by using phase tuning to manipulate the energy in their resonant wavelengths. The architecture was tested using the *MNIST* dataset [111] and was shown to have better execution time than GPU-based classification, with the AMD Vega FE, AMD M125, NVIDIA Tesla P100, and NVIDIA GTX 1080 Ti GPUs. However, they do not consider any optimization methods on the model to reduce the MR count required to represent it, and they report a very high 100 W power utilization for a 1024 MR modulator array in their proposed architecture.

A CNN accelerator implemented using MRs and memristors is described in [102]. In this work, weights are fed into the MR-weight bank through memristors, which in turn get their weight values from off-chip memories via static random access memory (SRAM) buffers. The architecture includes individual layers needed for CNN implementation. The convolution layer is comprised of the memristor-based photonic weight bank. The activation layer (*ReLU* layer) is built using SOAs. The work also uses an all-optical analog comparator, proposed in [112], to implement the maxpooling layer. These three layers form a single feature-extraction layer. Two feature-extraction layers are interconnected using an interface layer, which demodulates the output from the previous maxpool layer, generates the corresponding electronic voltage values, and then feeds them into the memristors of the next feature extraction layer. This work focused on the recognition of handwritten digits, using the *MNIST* dataset and shows better execution time against the FPGA-based Caffeine accelerator [113] and the memristor-based ISAAC accelerator [20] on various benchmarks. The architecture was further extended by the authors in [156].

A variation of the B&W protocol for MLP implementations was explored in [37] where the authors described the "hitless weight-and-aggregate" architecture. This method to accumulate weight values from the weight banks was devised to overcome the possible corruption of weight values from thermal variations and thermal crosstalk. The proposed "hitless weight-and-aggregate" architecture for

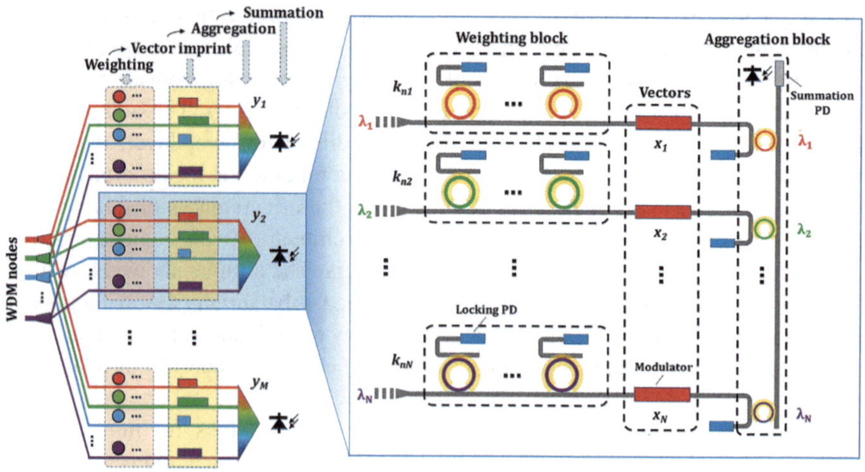

Fig. 9 The hitless weight and aggregate architecture for optical matrix-vector multipliers (OMMs), from [37]. The architecture aims to avoid the thermal crosstalk-based weight corruption that can occur in the B&W protocol-based architectures

MR-weight banks separates the wavelengths and weighs them in a parallel manner instead of the cascaded approach proposed in [107]. The process of updating the input matrix is simplified to counteract the delay induced by updating the weights in the weight bank. This is done by encoding the kernel directly into the input matrix of the optical matrix multiply (OMM) unit, which is illustrated in Fig. 9. The control of the hitless MR bank architecture is given to an FPGA, which uses PDs and analog to digital converter (ADCs) to obtain the summed signals from the OMM. The OMMs were implemented using MRs, but the vector storage was implemented using MZIs. Given that the main issue with B&W is the large MR count, it is unclear if the modified architecture can address that issue, as the work does not elaborate on the MR count, even with the input matrix minimization approach discussed in the chapter.

An MR-based neural network accelerator, called CrossLight, designed to be resilient to on-chip fabrication-process variations (FPV) and thermal variations, was proposed in [114]. The proposed CrossLight accelerator utilized FPV-resilient MR designs coupled with a thermal eigen decomposition-based tuning approach and intelligent MR placement to combat on-chip variations. These device and circuit-level considerations help increase the weight resolution achievable to 16 bits for a single MR. Also, in this work, the weight matrices are decomposed to component vectors, which effectively convert matrix multiplication to a collection of vector dot product operations. The hardware design considered splitter losses and waveguide propagation losses during the design of vector-granularity aware, compact vector dot product (VDP) units. The VDPs operate based on the B&W protocol, and the partial sums generated are added photonically before passing to the electronic control unit. The Crosslight architecture was shown to have 9.5× lower energy per bit and 15.9×

higher performance (frames per second) per watt on average than the DEAP-CNN [110] and Holylight [117] photonic accelerators.

An MR-based CNN accelerator that used a WDM optical patching scheme (WDM-OPS) as the data feeding structure for increased energy efficiency was proposed in [115]. The photonic MAC units, which follow the B&W protocol, already use WDM so WDM-OPS is a good fit for feeding data to these structures. The proposed architecture used optical delay lines to buffer the data as it was being fed into the MAC units, which is claimed to lead to superior operation speed and energy efficiency compared to electronic counterparts. The architecture was evaluated using the MNIST-handwritten digit classification task. The evaluation was done at 5 GHz MR modulation rate and claimed a 100 TMAC/s computation speed and an energy efficiency of 4.06 pJ/MAC. The work also claimed to achieve 0.16 pJ/MAC without weighting power consumption (power consumed by phase shifters to imprint weights onto MRs), which outperforms electronic accelerator energy efficiencies (~1 pJ/MAC).

The work in [116] used MR-based matrix multipliers and the Winograd filtering algorithm to accelerate CNN inference. The core of the proposed architecture is a photonic element-wise multiplication unit, which accepts CNN input feature maps and filters that have undergone Winograd transformation, and performs Winograd multiplications. The work also notably modifies Google Tensorflow to account for power consumption and noise from physical devices on chip. The accelerator performance was analyzed using 3×3 filters of the VGG16 network. The work reports 45 GOP/s and an energy efficiency of 10,000 GOP/s/W for the accelerator.

As discussed so far, MRs are prominently used to implement the B&W protocol, but an architecture which considered microdisks over MRs for its lower area and power consumption is described in [117]. This work explores designing and implementing photonic matrix-vector multipliers (MVMs), adders, and shifters, which are the fundamental computing components for CNN inference, using microdisks (Fig. 10). The MVM (Fig. 10a) uses the transmissivity of the microdisk array to represent the elements of one matrix and the input power into the microdisk array from the LD array to represent the other matrix. The output power from the microdisk array is the product of the two matrices. The electro-optic full adder (Fig. 10b) utilizes CMOS logic gates to calculate the propagate bit (P_n) and the generate bit (G_n) needed for the full adder, and microdisks to calculate the sum and carry. The P_n and G_n values are used to modulate the microdisks. By modulating the light intensity, one of the optical combiners at the output implements an XOR gate, while the other generates an OR gate, thereby implementing all the necessary operations for the sum and carry operations. The authors also propose a binary shifter using microdisks as shown in Fig. 10c. The shifting operation is performed by configuring the on/off states of the microdisk crossing switches. The authors also describe simplified CNN models called power-of-two quantized CNN (P2Q-CNN) models to avoid reliance on ADCs and boost CNN inference accuracy with negligible drop in accuracy (below 1%). This alternative architecture uses a photonic adder and shifter combination instead of the MVM. For testing the architecture, the authors used benchmarks based on *MNIST* and *ImageNet* datasets. They compared

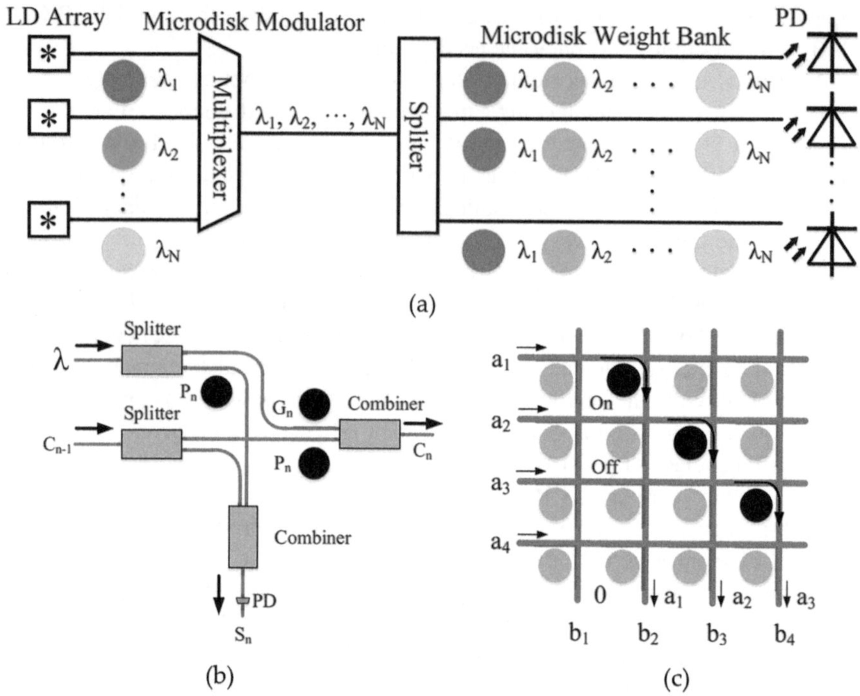

Fig. 10 (**a**) Microdisk-based on-chip matrix-vector multiplier (MVM); (**b**) electro-optic full adder; (**c**) photonic bit shifter, from [117]

this architecture to a ReRAM-based PIM accelerator ISAAC [20] and showed 13× better performance-per-watt.

Another architecture that leveraged microdisks along with photonic memory to obtain better speed of operation and energy efficiency was described in [118]. The work, named LightBulb, utilized photonic racetrack memory [119] to operate at 50 GHz, thus obtaining high inference throughput. To obtain better energy efficiency, the architecture used binarized CNNs. The architecture was tested using MNIST classification and ImageNet classification using MobileNet, ShuffleNet, and ResNet-18 CNN architectures. The work was compared against the previous work from the same authors, HolyLight [117], and showed a 16.9× better frames-per-second (FPS) over it. The increase in performance was attributed to the photonic memory, whereas [117] relied on electronic memory which limited its speed of operation to 5 GHz. LightBulb also claimed ~17.5× better FPS/Watt over [117].

A recent work [120] combined MRs and MZIs to design the basic optical multiply and sccumulate (OMAC) unit, which is used in an accelerator for CNNs, called PIXEL. The work describes two versions of the accelerator: a hybrid version that multiplies optically and accumulates electrically, and a fully optical version that multiplies and accumulates optically. The hybrid version uses MRs to implement an AND function with the MRs controlled using a synapse-lane, with shift-accumulation being done electrically. The bitwise AND operation along with shift-accumulate is used as an alternative to the opto-electrical MAC operation. The all-optical shift and ADD design uses MZIs to perform low-latency, low-power shift-accumulate operation optically and by cascading MZIs together. Synchronization of the signals from AND output is achieved with the help of propagation delay in MZI arms. Large dimensional MZIs and 6 mm waveguide arms between these MZIs are used to induce propagation delay in optical signals. The output signal from these interconnected MZIs effectively bit-shifts the input. The proposed architecture has register files for filter weight storage in each OMAC. The OMACs are arranged in a grid, and neuron outputs are passed through photonic interconnects in both x- and y-dimensions. The synapses are preloaded into the OMAC, and the proposed design assumes timed firing of the neurons to implement the MAC functionality. The hybrid and all-optical approaches were compared against an all-electrical architecture via simulations for the ResNet, GoogleNet, and ZFNet models. The all-optical approach shows better energy efficiency than the all-electrical approach and is comparable to the hybrid approach in this regard. The hybrid approach, which relies only on MRs, occupies significantly smaller area than the all-optical version of the architecture, which uses both MRs and MZIs.

The B&W protocol has also been used for SNNs as discussed in [90]. The suggested architecture utilizes laser neurons in conjunction with two different MR weight banks to interact with the WDM signals. The weight banks are used to represent excitatory and inhibitory weights. The weights are accumulated using a balanced PD pair before being used to excite a laser neuron. One of the PDs in the pair accepts the signal power from the excitatory weight bank and the other accepts power from the inhibitory weight bank. A short wire is incorporated to perform a subtraction operation, thereby considering the values from the inhibitory bank as negative values. The combination of the weight banks, the PDs, and the LD acting as the firing mechanism simulates a basic spiking neuron and is called a processing network node (PNN) in this work. The WDM signals are transmitted between these PNNs using a broadcast loop (BL). Multiple broadcast loops can be connected together in hierarchical manner via interfacing PNNs. Interfacing PNNs are PNNs which are tasked with the purpose of accepting output values from one BL and passing it to another, essentially acting as an optical router for the signals. The authors of [120] explored this strategy to allow for spectrum reuse and improve parallel processing. The work does not provide an experiment section to demonstrate the capabilities of the proposed architecture. Rather this work explored the feasibility of the B&W protocol-based spiking neural networks. The key observation in the work involves how utilizing the hierarchical broadcast loop architecture would allow for better spatial layout freedom than other conventional hardware neuromorphic systems.

There are other instances where MRs are used to implement SNNs other than through the B&W protocol. For example, in [121], MRs are employed to implement STDP on a chip. The authors of [121] incorporated $Ge_2Sb_5Te_5$ (GST), a popular and well-studied PCM material [122–125] on top of the ring waveguide in the ring resonator. This allowed the control of light propagation through the ports by merely changing the state of the GST. In this case, the PCM and its different phases act as the memory in the synapse. The authors of [121] also discussed potential integration of the integrate-and-fire neuron using MRs and GST in an SNN framework consisting of bipolar weights (weights with positive and negative values). The positive and negative weighted sums are computed using two separate dot-product engines and input to two different MRs. The bidirectional integrating action of the two ports of the MR is leveraged to calculate the effective membrane potential under the action of the bipolar weighted sums. Output spikes are generated when the effective membrane potential of the neuron crosses a threshold. Upon receiving the dot product stimulus, the neurons integrate their membrane potential at that time-step. The work tested this architecture using the *MNIST* dataset and assumed that the dot product engine has perfect operation. The work highlights the viability of PCM-based STDP in neuromorphic architectures and showcased this by demonstrating faster read/write operations and low energy consumption for the photonic architecture when compared to an electronic counterpart. In their simulations, a testing accuracy of 98.3% was achieved using this architecture. A related work [126] discussed how MRs can be used to incorporate spike delays into a photonic SNN.

A few works have also proposed MR-based photonic reservoir computing (RC) architectures, such as the 5x5 MR reservoir for high-bit-rate digital pattern classification in [127]. In this work, the reservoir was formed by randomly interconnected MRs. The simulated architecture was able to achieve a classification error of only 0.5% while offering bit rates up to 160 Gbps for eight-bit-length digital words for bit-pattern recognition on a custom dataset. The authors in [128] explored a 4x4 swirl topology-based reservoir design which utilizes MRs. The work also demonstrated basic Boolean operations. In such architectures, the nodes are comprised of nonlinear elements (MRs) and are part of the recurrence of the network, which is a departure from the original swirl topology introduced in [134]. This architecture has been widely used in photonic RC research. The swirl in the data paths allows for sufficient mixing of the input signals/weight matrix. Traditionally, reservoir architectures set their nodes at near instability for proper operation of the reservoir to ensure that they have sufficient memory of past inputs and respond well to new inputs. The MRs in [128] are set to this operating point after detailed analysis of MR stability in operation and resonance at various input power values, as well as temperature-induced optical detuning from resonance.

A recent work explored how to use MRs as nonlinear neurons in a delayed feedback reservoir [129]. In this work, active MRs are used as nonlinear reservoir nodes and a low-loss photonic waveguide acts as the delay feedback loop. The performance of this MR-based delayed feedback reservoir computing (DFRC) accelerator was evaluated using two time series prediction tasks (NARMA10 [130] and Santa-Fe [145]) and a third nonlinear channel equalization task [131]. Normalized root mean square error (NRMSE) and symbol error rate (SER) were used to measure the computational capabilities, and the architecture was compared against an all-optical MZI-based DRFC accelerator [132] and an electronic DRFC accelerator [133]. Evaluations with benchmark tasks show that this MR-based

DFRC accelerator achieves 35% and 98.7% lower NRMSE, for NARMA10 and Santa-Fe time series prediction tasks, respectively. It also achieves up to 58.8% less average SER, and up to 98× faster training time compared to the MZI DFRC accelerator. However, the analyses showed the electrical accelerator displaying superior performance in all aspects except the training time, for which the MR-based accelerator showed 93× acceleration.

In summary, noncoherent neuron microarchitectures that use MRs are one of the most prolifically used components to implement photonic neural network architectures. These noncoherent architectures that use MRs span SNNs and ANNs (MLPs, CNNs). The majority of the SNN, MLP, and CNN architectures utilize the B&W protocol for propagating weights through the network. Some efforts have identified several issues with this protocol, such as heterodyne crosstalk corrupting the weight values and the increasingly large number of MRs needed for the implementation of larger networks, especially when larger WDM degrees are utilized [37]. Other works have suggested reusing implemented layers as in [109] and also methods to reduce energy consumption and increase speed of operation by reducing the involvement of electronic components [117]. Moreover, there have been suggestions to use microdisks instead of MRs to further increase the integration density in chips as in [117]. MRs have also been used for weight propagation in SNNs using the B&W protocol as in [90] and for implementing STDP in photonic SNNs [121, 135].

4.2 *Optical Interferometry-Based Network Architectures*

Optical interferometry-based neural network implementations usually rely on the manipulation of the electrical field phase and amplitude of a single optical wavelength. These are coherent architectures and utilize the coherent neuron microarchitectures. Coherent architectures rely extensively on MZIs and have been widely employed in MLP implementations. MZIs have less commonly been used to play the part of intensity modulators in some SNN implementations. The large number of nodes needed in the reservoir and the large area requirement of MZIs make them not very popular in RC implementations. MZI-based coherent architectures utilize universal linear meshes of MZIs to implement the required matrix multiplications needed by neural networks. The weights are controlled by controlling the phase and amplitude of optical signals, which is done by implanting attenuators and phase shifters on the MZI arms. This was demonstrated in [136] where 2×2 beam splitters and phase shifters in the form of an MZI were programmed to enable independent control of amplitude and phase of light for a set of optical channels.

The work in [137] fabricates and demonstrates an MZI-based 4x4 optical matrix multiplier. Here, the architecture is constructed based on the premise that an ideal NxN multiport reconfigurable MZI-based interferometer represents a special unitary (SU) group of degree N, SU(N), which is comprised of n MZIs with N optical channels forming a unitary transformation matrix. In [137], the structure is made

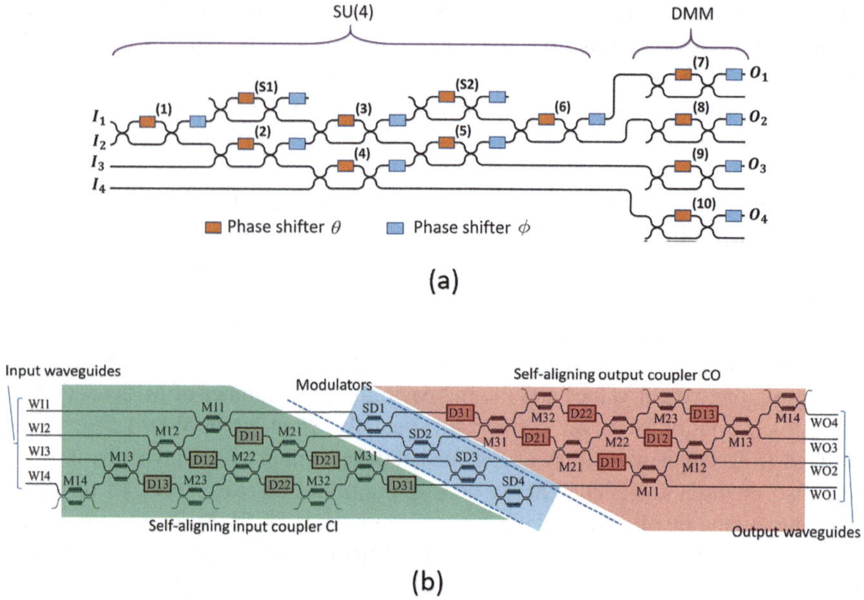

Fig. 11 (**a**) The 4 × 4 MZI-based reconfigurable linear optical processor from [137]; SU is special unitary group and DMM is the diagonal matrix multiplication unit; (**b**) the universal linear mesh for MZI-based singular value decomposition (SVD) as described in [138]. The matrices involved in SVD implemented: V in green, diagonal matrix Σ in blue, and U^T in red

of an SU($N = 4$) section followed by a diagonal matrix multiplication (DMM) section (Fig. 11a). The DMM can be extended depending on the application and can form a complete SVD through cascading. This 4x4 optical matrix multiplier was used for implementing a single-layered neural network. The performance of the neural network was evaluated by tasking it to classify 50 data samples of a synthetic linearly separable multivariate Gaussian dataset, for which it was able to achieve a 72% accuracy. The work in [138] describes another matrix SVD-based mesh implementation, which can implement arbitrary non-unitary matrices using MZIs. The singular value decomposition (SVD)-based methodology is used to perform decomposition of matrices to unitary matrices, and these simplified matrices are implemented on-chip. More specifically, SVD is the process by which a matrix can be decomposed into three matrices, two unitary matrices V and U, and a diagonal matrix comprised of nonzero singular values Σ. The SVD process can be implemented in an MZI mesh by using a diagonal matrix, which implements the amplitude and phase, while the universal unitary matrices follow the designs as proposed by [136, 139]. The final architecture of this approach is shown in Fig. 11b.

The authors in [32] proposed an architecture that utilizes an SVD-based approach for implementing the necessary matrix calculations (Fig. 12), where vectors were encoded in the intensity and phase of light and then fed into each layer of the network. SVD is used to decompose the matrices to be multiplied into unitary

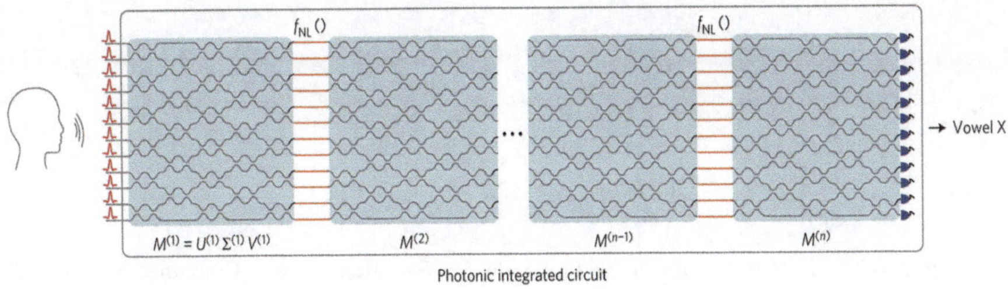

Fig. 12 The architecture from [32] that utilizes SVD to implement the matrix multiplications for vowel recognition using a photonic integrated circuit (PIC)

matrices, which can be encoded into the MZI mesh. The SVD operation and encoding signals for the MZIs were generated using a digital computer. Once the matrices are encoded into the MZI mesh, matrix multiplication between them can be performed by allowing the optical signal to pass passively through the mesh. The key advantages of SVD are the reduced complexity of operation and reduction in dimensionality, which help with reduced cost of operation of the DNN model at hand. Each layer of this proposed model is comprised of an optical interference unit (OIU) and an optical nonlinearity unit (ONU). In this work, ONU functionality was implemented with digital electronics, while the OIU was implemented in a photonic integrated circuit, which performed optical matrix multiplications using the SVD approach as described in [138]. The work in [32] discussed how to use such an architecture for vowel recognition. They also utilized forward propagation with finite difference method instead of backpropagation to train the architecture. The architecture was able to achieve 76.7% accuracy in classifying vowels, which is lower than the 91.7% achieved by the same architecture implemented on a conventional 64-bit digital computer. The authors attribute the lower accuracy to the limited computational resolution (24 bits as opposed to the 64 bits of the conventional computer) of the optical neural network.

MZIs have been noted to have a much larger footprint (which can be up to a few millimeters [120] or tens to hundreds of micrometers) than their counterparts (e.g., MRs), and as [102] noted, this large footprint in combination with the accumulation of phase errors throughout an MZI-based mesh can limit the scalability of neural networks built with MZIs. There has been research, such as in [140], focusing on reducing the overall area consumption by these architectures, whether it be by utilizing methods to prune the weight matrices represented by MZI meshes or by utilizing other devices in tandem.

The work in [140] demonstrated a fast Fourier transform (FFT)-based methodology to reduce area and energy footprint of MZI meshes used to implement MLPs. This is achieved by sparsifying the network through reducing the overall number of weights used, thereby compressing the neural network. The proposed architecture is based on structured neural networks with circulant matrix representation. Structured neural networks are a class of neural networks that are specially designed for computational complexity reduction, whose weight matrices are regularized using

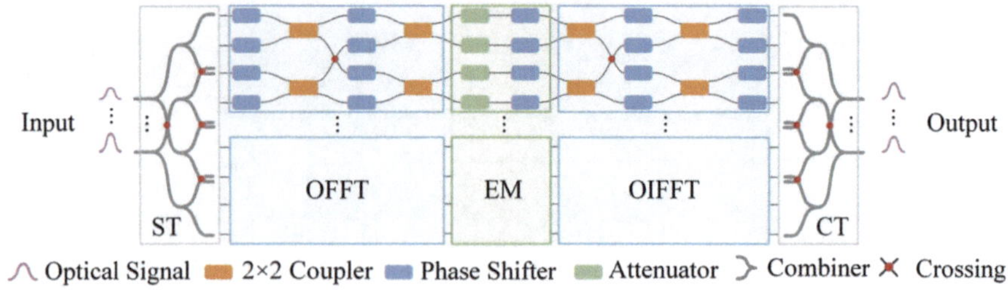

Fig. 13 FFT-IFFT-based photonic structured neural network architecture described in [140]. The FFT-based analyses of the model aim to simplify the model and thereby reduce energy and area consumption of the MZI meshes

the composition of structured sub-matrices [141, 142]. Structured neural networks utilize circulant weight matrices, which can be efficiently calculated using FFT and inverse FFT (IFFT). The weight matrices are further pruned using group Lasso regularization [144], and these operations can be implemented in MZI meshes using cascaded attenuator/amplifiers and phase shifters (Fig. 13). The authors of [140] adapted this methodology because of the difficulty in pruning SVD-based architectures. The architecture was tested using the *MNIST* dataset against SVD-based architectures to show how effective their method is in reducing the overall area consumption of MZI meshes. The results indicate the architecture was able to achieve close to 98.5% testing accuracy while substantially reducing the overall area consumption.

FFT-based 4-f architectures (Fig. 14) are at the heart of optical accelerators from the start-up Optalysis. They have used lens-based diffraction to convert optical signals to the Fourier domain, and the multiplications in the Fourier domain are performed using MZIs. They have proposed and built a high-resolution (two million pixels), low-speed (kHz range of operation) optical CNN accelerator [105, 106]. The prototype, dubbed FT:X 2000, can interface with a digital computing host using the PCIe interface and is proposed to be used for cryptography and DNN acceleration applications.

The authors of [106] proposed a coherent MZI-based binary neural network implementation with weights restricted to $+1$ or $-\ 1$. The activation function is a symbolic decision function which binarizes any real number mapped to it to $+1$ or $-\ 1$. The weights of binarization are encoded onto the MZI by shifting voltages on the internal and external phase shifters on MZI arms. The real and imaginary parts of the two-way polarized in-phase and quadrature (IQ) component-modulated optical signal are used for training the model in simulation. The input to the model is the real and imaginary part of the signal, while the output is the prediction of the input position by the neural network. This work tested the architecture, comprised of seven hidden layers, for classifying to nearest neighbors the constellation formed by real and imaginary parts of a 100 GHz DP-QPSK signal. Close to 100% accuracy in

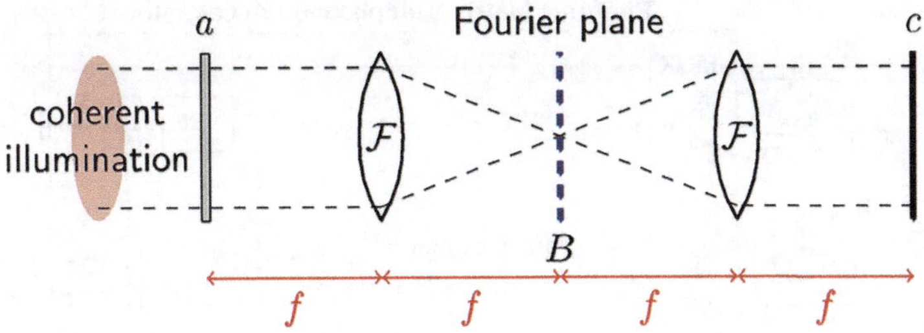

Fig. 14 4-f architecture for Fourier transformation and optical convolution operation [143]. Fourier transform of the signal is obtained at *B*, and multiplication to some other signal, also Fourier transformed, can be done here. The IFFT by the second lens provides the convolution output at "*c*"

classification was achieved for high signal-to-noise ratio (SNR) input signals, while an accuracy close to 90% was achieved for low SNR signals.

As mentioned earlier, MZI-based architectures are typically coherent and utilize only a single wavelength. But MZIs have been used to implement WDM-based or noncoherent architectures as well. For instance, the work in [103] demonstrated a photonic matrix multiplication accelerator using MZIs and arrayed waveguide grating (AWG)-multimode interferences (MMIs). A single unit of AWG-MMI coupler balanced detector can successfully perform matrix multiplication by using WDM and coherent homodyne detection scheme. MZIs are utilized as intensity modulators, which feed into the multiplier (Fig. 15). The work in [105] demonstrated an all-optical WDM RNN utilizing an SOA-MZI as an activation unit incorporated into the feedback delay loop. In order to emulate a fully functional gated-recurrent unit (GRU), the authors integrated a gating mechanism (the SOA-MZI) to allow for agile reconfiguration of forget functions within a GRU. The SOA are embedded into the arms of the MZI and act as cross gain modulation wavelength converters. An RNN was constructed using this unit and tested using a four-input WDM. The utility of the RNN was tested by running a finance forecasting benchmark application using the FI-2010 dataset. The gated optical RNN was able to achieve a higher F1 score (41.85%) than the optical and regular RNNs.

In summary, coherent neuron microarchitectures that make use of interferometer devices such as MZIs have been widely used in photonic neural network architectures because of their ability to effectively represent matrices for neural network operations, but at the cost of a larger area overhead than MRs and susceptibility to phase-noise corruption. The basic principle of MZIs for neural network operations relies on phase and amplitude tuning of a passing optical wavelength, which can be easily achieved by integrating phase and amplitude tuners in MZI arms. MZIs are typically arranged in a mesh configuration in the works that use them, with SVD also being used to efficiently represent matrices. A reduction in MZI area footprint and phase-noise corruption is attempted via regularization approaches (e.g., [32, 106]) or by utilizing niche neural network models to reduce the overall MZI count ([140]). Usually, architectures which make use of MZIs use coherent

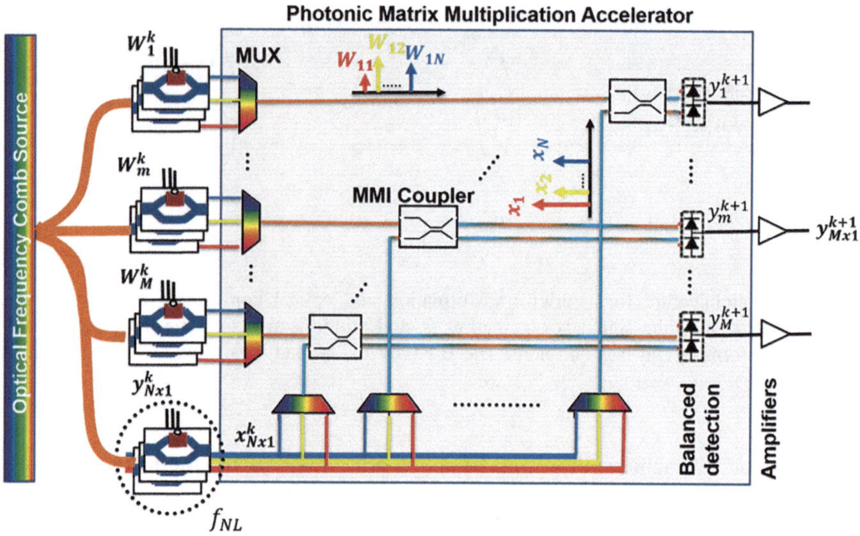

Fig. 15 Cascadable analog feed-forward artificial neural network structure with photonic matrix-vector multiplier circuit and Mach-Zehnder modulator nonlinearity, as depicted in [103]

principles to function, but MZIs have also been used in noncoherent approaches which make use of WDM, e.g., the RNN implementation in [103] which utilizes SOAs and MZIs in combination. They are not found to be used in many RC-based architectures, probably because of the large area requirement an MZI-based implementation would require to realize the large number of nonlinear nodes in an RC implementation.

5 Challenges and Opportunities

State-of-the-art silicon photonic devices have shown great promise to implement artificial neurons. Deep learning architectures built with photonic neurons support high parallelism in transmitting and processing weights by utilizing WDM, fast execution time, and low energy expenditure. However, there are several outstanding challenges to efficiently implement different neuron functions with silicon photonic devices, as well as to achieve high reliability, scalability, and cascadability in architecture implementations. Here we summarize challenges and opportunities for future research needed to overcome these challenges.

- **Coherency Challenges:** Mach-Zehnder interferometers (MZIs) are key to photonic neural networks for matrix operations but face challenges like large area requirements and phase noise from thermal and fabrication variations, impacting accuracy. Recent work [146, 147] addresses these issues during training, but further research is needed to improve area efficiency and noise resilience.

- **Noncoherency Challenges:** The broadcast-and-weight (B&W) protocol is commonly used in photonic neural networks but faces challenges like heterodyne crosstalk and the large number of microring resonators (MRs) required for larger networks, especially with DWDM. Recent efforts propose solutions, such as parallel MR arrangements to reduce thermal crosstalk and improve weight resolution [37], layer reuse [109], and reducing electronic component involvement to lower energy use and increase speed [117]. Alternatives include using microdisks for higher integration density [120]. However, as noted in [117], noncoherent architectures may suffer from low throughput due to slower electronic components. Further research is needed to address these limitations.
- **Variations and Reliability:** Silicon photonic devices, such as MRs and MZIs, are vulnerable to variations during design and runtime, including fabrication process [148], thermal crosstalk [58], and device aging [149]. These issues can degrade reliability and performance in photonic neural networks by causing crosstalk noise, phase shifts, resonance drifts, and mismatches in photodetection currents. For example, MR resonant wavelengths can shift by 4.79 nm due to fabrication variations [149] and by 0.1 nm/K from thermal changes [150]. Additionally, optical losses in these devices reduce energy efficiency and scalability [151], and limited phase-encoding precision introduces uncertainty in weight values. A recent study [152–155] found that these uncertainties can reduce inference accuracy by up to 70%. Further research is needed to improve the reliability of silicon photonic devices.

References

1. G. E. Hinton, S. Osindero, and Y.-W. Teh, "A fast learning algorithm for deep belief nets," in *Neural computation*, vol. 18, no. 7, pp. 1527–1554, 2006.
2. V. K. Kukkala, J. Tunnell, S. Pasricha, and T. Bradley, "Advanced driver-assistance systems: A path toward autonomous vehicles," in *IEEE Consumer Electronics Magazine*, vol. 7, no. 5, pp. 18–25, 2018.
3. S. Levine, P. Pastor, A. Krizhevsky, J. Ibarz, and D. Quillen, "Learning hand-eye coordination for robotic grasping with deep learning and large-scale data collection," in *The International Journal of Robotics Research*, vol. 37, no. 4-5, pp. 421–436, 2018.
4. F. Monti, F. Frasca, D. Eynard, D. Mannion, and M. M. Bronstein, "Fake news detection on social media using geometric deep learning," arXiv:1902.06673 [cs], Feb. 2019.
5. S. Lalmuanawma, J. Hussain, and L. Chhakchhuak, "Applications of machine learning and artificial intelligence for covid-19 (sars-cov-2) pandemic: A review," in *Chaos, Solitons & Fractals*, p. 110059, 2020.
6. K. Kukkala, S. V. Thiruloga, and S. Pasricha, "Indra: Intrusion detection using recurrent autoencoders in automotive embedded systems," arXiv preprint arXiv:2007.08795, 2020.
7. J. Gu, G. Neubig, K. Cho, and V. O. Li, "Learning to translate in real-time with neural machine translation," arXiv preprint arXiv:1610.00388, 2016.
8. W. S. McCulloch and W. Pitts, "A logical calculus of the ideas immanent in nervous activity," in *The bulletin of mathematical biophysics*, vol. 5, no. 4, pp. 115–133, 1943.
9. F. Rosenblatt, "The perceptron, a perceiving and recognizing automaton," in Report 85-460-1, Cornell Aeronautical Laboratory, 1957.

10. A. Merolla, J. V. Arthur, R. Alvarez-Icaza, A. S. Cassidy, J. Sawada, F. Akopyan, B. L. Jackson, N. Imam, C. Guo, Y. Nakamuraet al., "A million spiking-neuron integrated circuit with a scalable communication network and interface," in *Science*, vol. 345, no. 6197, pp. 668–673, 2014.
11. M. Davies, N. Srinivasa, T.-H. Lin, G. Chinya, Y. Cao, S. H. Choday, G. Dimou, P. Joshi, N. Imam, S. Jainet al., "Loihi: A neuromorphic manycore processor with on-chip learning," in *IEEE Micro*, vol. 38, no. 1, pp. 82–99, 2018.
12. P. Jouppi, C. Young, N. Patil, D. Patterson, G. Agrawal, R. Bajwa, S. Bates, S. Bhatia, N. Boden, A. Borcherset al., "In-datacenter performance analysis of a tensor processing unit," in *ACM/IEEE International Symposium on Computer Architecture (ISCA)*, 2017.
13. V. Gokhale, J. Jin, A. Dundar, B. Martini, and E. Culurciello, "A 240 G-ops/s Mobile Coprocessor for Deep Neural Networks," in *CVPR Workshop*, 2014.
14. Z. Du, R. Fasthuber, T. Chen, P. Ienne, L. Li, T. Luo, X. Feng, Y. Chen, and O. Temam, "ShiDianNao: Shifting Vision Processing Closer to the Sensor," in *ACM/IEEE International Symposium on Computer Architecture (ISCA)*, 2015.
15. C. Zhang, P. Li, G. Sun, Y. Guan, B. Xiao, and J. Cong, "Optimizing FPGA-based Accelerator Design for Deep Convolutional Neural Networks," in *FPGA*, 2015.
16. Y.-H. Chen, T.-J. Yang, J. Emer, and V. Sze, "Eyeriss v2: A flexible accelerator for emerging deep neural networks on mobile devices," in *IEEE Journal on Emerging and Selected Topics in Circuits and Systems (JETCAS)*, vol. 9, no. 2, pp. 292-308, June 2019.
17. A. Parashar, M. Rhu, A. Mukkara, A. Puglielli, R. Venkatesan, B. Khailany, J. Emer, S. W. Keckler, and W. J. Dally, "SCNN: An accelerator for compressed-sparse convolutional neural networks," in *ACM/IEEE International Symposium on Computer Architecture (ISCA)*, 2017.
18. S. Markidis, S. W. Der Chien, E. Laure, I. B. Peng, and J. S. Vetter, "Nvidia tensor core programmability, performance & precision," in *IEEE International Parallel and Distributed Processing Symposium Workshops (IPDPSW)*, 2018.
19. A. Sodani, R. Gramunt, J. Corbal, H.-S. Kim, K. Vinod, S. Chinthamani, S. Hutsell, R. Agarwal, and Y.-C. Liu, "Knights landing: Second-generation intel xeon phi product," in *IEEE Micro*, vol. 36, no. 2, pp. 34–46, 2016.
20. A. Shafiee, A. Nag, N. Muralimanohar, R. Balasubramonian, J. P.Strachan, M. Hu, R. S. Williams, and V. Srikumar, "Isaac: A convolutional neural network accelerator with in-situ analog arithmetic in crossbars," in *ACM SIGARCH Computer Architecture News*, vol. 44, no. 3, pp. 14–26, 2016.
21. L. Song, X. Qian, H. Li, and Y. Chen, "Pipelayer: A pipelined reram-based accelerator for deep learning," in *IEEE International Symposium on High Performance Computer Architecture (HPCA)*, pp. 541–552, 2017.
22. H. Tsai, S. Ambrogio, P. Narayanan, R. M. Shelby, and G. W. Burr, "Recent progress in analog memory-based accelerators for deep learning," in *Journal of Physics D: Applied Physics*, vol. 51, no. 28, p. 283001, 2018.
23. A. Amravati, S. B. Nasir, S. Thangadurai, I. Yoon, and A. Raychowdhury, "A 55nm time-domain mixed-signal neuromorphic accelerator with stochastic synapses and embedded reinforcement learning for autonomous micro-robots," in *IEEE International Solid - State Circuits Conference - (ISSCC)*, 2018.
24. H. Valavi, P. J. Ramadge, E. Nestler, and N. Verma, "A mixed-signal binarized convolutional-neural-network accelerator integrating dense weight storage and multiplication for reduced data movement," in *IEEE Symposium on VLSI Circuits*,2018.
25. A. Amaravati, S. B. Nasir, J. Ting, I. Yoon, and A. Raychowdhury, "A55-nm, 1.0–0.4 v, 1.25-pj/mac time-domain mixed-signal neuromorphic accelerator with stochastic synapses for reinforcement learning in autonomous mobile robots," *in IEEE Journal of Solid-State Circuits*, vol. 54,no. 1, pp. 75–87, 2018.
26. M. M. Waldrop, "The chips are down for Moore's law," in *Nature News*, vol.530, no. 7589, p. 144, 2016.
27. S. Pasricha, and N. Dutt. "On-Chip Communication Architectures," in *Morgan Kauffman, ISBN 978-0-12-373892-9*, Apr 2008.

28. S. Pasricha, M. Nikdast, "A Survey of Silicon Photonics for Energy Efficient Manycore Computing," in *IEEE Design and Test*, vol. 37, no. 4, pp. 60-81, Aug. 2020.
29. L. Chrostowski, H. Shoman, M. Hammood, H. Yun, J. Jhoja, E. Luan,S. Lin, A. Mistry, D. Witt, N. A. Jaegeret al., "Silicon photonic circuit design using rapid prototyping foundry process design kits," in *IEEE J. Sel. Top. Quantum Electron.*, vol. 25, no. 5, pp. 1–26, 2019.
30. D. A. Miller, "Silicon photonics: Meshing optics with applications," in *Nature Photonics*, vol. 11, no. 7, pp. 403–404, 2017.
31. A. R. Totović, G. Dabos, N. Passalis, A. Tefas, and N. Pleros, "Femtojoule per mac neuromorphic photonics: An energy and technology roadmap," in *IEEE J. Sel. Top. Quantum Electron.*, vol. 26, no. 5,pp. 1–15, 2020.
32. Y. Shen, N. C. Harris, S. Skirlo, M. Prabhu, T. Baehr-Jones, M. Hochberg, X. Sun, S. Zhao, H. Larochelle, D. Englundet al., "Deep learning with coherent nanophotonic circuits," in *Nature Photonics*, vol. 11, no. 7, p. 441, 2017.
33. G. Van der Sande, D. Brunner, and M. C. Soriano, "Advances in photonic reservoir computing," in *Nanophotonics*, vol. 6, no. 3, pp. 561–576, 2017.
34. A. Katumba, M. Freiberger, F. Laporte, A. Lugnan, S. Sackesyn, C. Ma,J. Dambre, and P. Bienstman, "Neuromorphic computing based on silicon photonics and reservoir computing," in *IEEE J. Sel. Top. Quantum Electron.*, vol. 24, no. 6, pp. 1–10, 2018.
35. G. Tanaka, T. Yamane, J. B. Héroux, R. Nakane, N. Kanazawa, S. Takeda, H. Numata, D. Nakano, and A. Hirose, "Recent advances in physical reservoir computing: A review," in *Neural Networks*, vol. 115, pp. 100–123, 2019.
36. A. N. Tait, M. A. Nahmias, Y. Tian, B. J. Shastri, and P. R. Prucnal, "Photonic neuromorphic signal processing and computing," in *Nanophotonic Information Physics*, 2014.
37. Q. Cheng, J. Kwon, M. Glick, M. Bahadori, L. P. Carloni, and K. Bergman, "Silicon photonics codesign for deep learning," in *Proceedings of the IEEE*, vol. 108, no. 8, pp. 1261-1282, Aug. 2020.
38. T. F. de Lima, H. Peng, A. N. Tait, M. A. Nahmias, H. B. Miller, B. J. Shastri, and P. R. Prucnal, "Machine learning with neuromorphic photonics," *IEEE IEEE Journal of Lightwave Technology*, vol. 37, no. 5, pp. 1515_1534, Mar. 1, 2019.
39. L. De Marinis, M. Cococcioni, P. Castoldi, and N. Andriolli, "Photonic neural networks: A survey," in *IEEE Access*, vol. 7, pp. 175 827–175 841,2019.
40. A. N. Tait, A. X. Wu, T. F. De Lima, E. Zhou, B. J. Shastri, M. A. Nahmias, and P. R. Prucnal, "Microring weight banks," *in IEEE J. Sel. Top. Quantum Electron.*, vol. 22, no. 6, pp. 312–325, 2016.
41. W. Maass. "Networks of spiking neurons: The third generation of neural network models," in *Neural Networks*, vol. 10, issue. 9, pp. 1659-1671, 1997.
42. F. A. Azevedo, L. R. Carvalho, L. T. Grinberg, J. M. Farfel, R. E. Ferretti, R. E. Leite, W. J. Filho, R. Lent, and S. Herculano-Houzel, "Equal numbers of neuronal and nonneuronal cells make the human brain an isometrically scaled-up primate brain," in *Journal of Comparative Neurology*, vol. 513, no. 5, pp. 532–541, 2009.
43. A.L Hodgkin, A.F. Huxley, "A quantitative description of membrane current and its application to conduction and excitation in nerve," in *The Journal of Physiology*, Aug 1952.
44. NVIDIA Corp., "NVIDIA A100 Tensor Core GPU Architecture," Whitepaper, 2020.
45. P. Chi, S. Li, C. Xu, T. Zhang, J. Zhao, Y. Liu, Y. Wang, and Y. Xie, "Prime: A novel processing-in-memory architecture for neural network computation in reram-based main memory," in *ACM SIGARCH Computer Architecture News*, vol. 44, no. 3, pp. 27–39, 2016.
46. E. M. Izhikevich, "Simple model of spiking neurons," in *IEEE Transactions on neural networks*, vol. 14, no. 6, pp. 1569–1572, 2003.
47. H. Esmaeilzadeh, A. Sampson, L. Ceze, and D. Burger, "Neural acceleration for general-purpose approximate programs," in *IEEE/ACM International Symposium on Microarchitecture*, 2012.
48. M. Nikdast, G. Nicolescu, J. Trajkovic, and O. Liboiron-Ladouceur, "Deeper: Enhancing performance and reliability in chip-scale optical interconnection networks," in *ACM Great Lakes Symposium on VLSI (GLSVLSI)*, 2018.

49. Z. Ghanaatian, A. Shafiee, M. Nikdast. Mastering Silicon Photonics Device Design for Scalable and Robust Optical Neural Networks. In Integrated Photonics Research, Silicon and Nanophotonics 2024 Jul 28 (pp. IW2B-1). Optica Publishing Group.
50. Z. Ghanaatian, A. Shafiee, M. Nikdast M. Variation-Aware Layout and Design Optimization of Silicon Photonic Mach–Zehnder Interferometers. In2023 IEEE Photonics Conference (IPC) 2023 Nov 12 (pp. 1-2). IEEE.
51. A. Mirza, A. Shafiee, S. Banerjee, K. Chakrabarty, S. Pasricha, M. Nikdast. Characterization and optimization of coherent MZI-based nanophotonic neural networks under fabrication non-uniformity. IEEE Transactions on Nanotechnology. 2022 Nov 23;21:763-71.
52. A. N. Tait et al., "Silicon photonic modulator neuron," in *Physical Review Applied*, vol. 11, no. 6, p. 064043, June 2019.
53. P. R. Prucnal, B. J. Shastri, T. F. de Lima, M. A. Nahmias, and A. N. Tait, "Recent progress in semiconductor excitable lasers for photonic spike processing," in *Adv. Opt. Photonics*, vol. 8, no. 2, pp. 228–299, Jun. 2016.
54. T. F. de Lima et al., "Noise analysis of photonic modulator neurons," in *IEEE J. Sel. Top. Quantum Electron.*, vol. 26, no. 1, pp. 1–9, Jan-Feb 2020.
55. D. Liang and J. E. Bowers, "Recent progress in lasers on silicon," in *Nature Photonics*, vol. 4, no. 8, Art. no. 8, Aug. 2010.
56. N. H. Zhu et al., "Directly modulated semiconductor lasers," in *IEEE J. Sel. Top. Quantum Electron.*, vol. 24, no. 1,pp. 1–19, Jan-Feb 2017.
57. D. Vantrease et al., "Corona: System implications of emerging nanophotonic technology," in *ACM SIGARCH Computer Architecture News*, vol. 36, no. 3, pp. 153–164, 2008.
58. A. Mirza, S. M. Avari, E. Taheri, S. Pasricha, and M. Nikdast, "Opportunities for Cross-Layer Design in High-Performance Computing Systems with Integrated Silicon Photonic Networks," in *IEEE/ACM Design, Automation Test in Europe Conference Exhibition (DATE)*, 2020.
59. A. N. Tait, M. A. Nahmias, B. J. Shastri, and P. R. Prucnal, "Broadcast and Weight: An Integrated Network For Scalable Photonic Spike Processing," in *IEEE Journal of Lightwave Technology*, vol. 32, no. 21, pp. 4029–4041, 2014.
60. S. Xiang, Y. Zhang, X. Guo, A. Wen, and Y. Hao, "Photonic generation of neuron-like dynamics using vcsels subject to double polarized optical injection," in *IEEE Journal of Lightwave Technology*, vol. 36, no. 19, pp. 4227–4234, 2018.
61. Z. W. Song, S. Y. Xiang, Z. X. Ren, S. H. Wang, A. J. Wen, and Y. Hao, "Photonic spiking neural network based on excitable vcsels-sa for sound azimuth detection," in *Optics Express*, vol. 28, no. 2, pp. 1561–1573, 2020; I. Aldaya, C. Gosset, C. Wang, G. Campuzano, F. Grillot, and G. Castanon, "Periodic and aperiodic pulse generation using optically injected dfb laser," in *Electronics Letters*, vol. 51, no. 3, pp. 280–282, 2015.
62. G. Sarantoglou, M. Skontranis, and C. Mesaritakis, "All optical integrate and fire neuromorphic node based on single section quantum dot laser," in *IEEE J. Sel. Top. Quantum Electron.*, vol. 26, no. 5,pp. 1–10, 2019.
63. F. Koyama, "Recent Advances of VCSEL Photonics," in *IEEE Journal of Lightwave Technology*, vol. 24, no. 12, pp. 4502–4513, 2006.
64. J. Van Campenhout, et al., "Electrically pumped InP-based microdisk lasers integrated with a nanophotonic silicon-on-insulator waveguide circuit," in *Optics express*, vol. 15, 2007.
65. J. Robertson, E. Wade, and A. Hurtado, "Electrically controlled neuron-like spiking regimes in vertical-cavity surface-emitting lasers at ultrafast rates," in *IEEE J. Sel. Top. Quantum Electron.*, vol. 25,no. 6, pp. 1–7, 2019.
66. J. Robertson, M. Hejda, J. Bueno, and A. Hurtado, "Ultrafast optical integration and pattern classification for neuromorphic photonics based on spiking vcsel neurons," in *Scientific reports*, vol. 10, no. 1, pp. 1–8, 2020.
67. A. Hurtado, K. Schires, I. Henning, and M. Adams, "Investigation of vertical cavity surface emitting laser dynamics for neuromorphic photonic systems," in *Applied Physics Letters*, vol. 100, no. 10, p. 103703, 2012.
68. A. Levi, "Microdisk lasers," in *Solid-state electronics*, vol. 37, no. 4-6, pp.1297–1302, 1994.

69. L. Mahler, A. Tredicucci, F. Beltram, C. Walther, J. Faist, B. Witzigmann, H. E. Beere, and D. A. Ritchie, "Vertically emitting microdisk lasers," in *Nature Photonics*, vol. 3, no. 1, pp. 46–49, 2009.
70. Yisu Yang, Gligor Djogo, Moez Haque, Peter R. Herman, and Joyce K. S. Poon, "Integration of an O-band VCSEL on silicon photonics with polarization maintenance and waveguide coupling," in *Optics Express*, 25, 2017.
71. K. Alexander, T. Van Vaerenbergh, M. Fiers, P. Mechet, J. Dambre, and P. Bienstman, "Excitability in optically injected microdisk lasers with phase controlled excitatory and inhibitory response," in *Optics Express*, vol. 21, no. 22, pp. 26182–26191, 2013.
72. M. Tran, D. Huang, T. Komljenovic, J. Peters, A. Malik, and J. Bowers, "Ultra-Low-Loss Silicon Waveguides for Heterogeneously Integrated Silicon/III-V Photonics," in *Applied Sciences*, vol. 8, no. 7, pp. 1139-, 2018.
73. S. Nambiar, S. Purnima, and S. K. Selvaraja, "Grating-assisted fiber to chip coupling for SOI photonic circuits," in *Applied Sciences*, vol. 8, 2018.
74. D. F. Siriani and K. D. Choquette, "Coherent Coupling of Vertical-Cavity Surface-Emitting Laser Arrays," in *Semiconductors and Semimetals*, vol. 86, pp. 226-264, 2012.
75. S. Maktoobi et al., "Diffractive Coupling for Photonic Networks: How Big Can We Go?," in *IEEE J. Sel. Top. Quantum Electron.*, vol. 26, no. 1, pp. 1–8, Jan. 2020.
76. X. Wu et al., "UNION: A Unified Inter/Intrachip Optical Network for Chip Multiprocessors," in *IEEE Trans. Very Large Scale Integr. VLSI Syst.*, vol. 22, no. 5, pp. 1082–1095, May 2014.
77. H. Shabani, A. Roohi, A. Reza, M. Reshadi, N. Bagherzadeh, and R. F. DeMara, "Loss-Aware Switch Design and Non-Blocking Detection Algorithm for Intra-Chip Scale Photonic Interconnection Networks," in *IEEE Trans. Comput.*, vol. 65, no. 6, pp. 1789–1801, Jun. 2016.
78. A. N. Tait, T. F. de Lima, M. A. Nahmias, B. J. Shastri, and P. R. Prucnal, "Microring Weight Banks for Neuromorphic Silicon Photonics," in *Conference on Lasers and Electro-Optics (CLEO)*, May 2018.
79. A. N. Tait, T. F. de Lima, M. A. Nahmias, B. J. Shastri, and P. R. Prucnal, "Multi-channel control for microring weight banks," in *Optics Express*, vol. 24, no. 8, pp. 8895–8906, 2016.
80. T. V. Vaerenbergh et al., "Cascadable excitability in microrings," in *Optics Express*, vol. 20, no. 18, pp. 20292–20308, Aug. 2012
81. J. Xiang, A. Torchy, X. Guo, and Y. Su, "All-Optical Spiking Neuron Based on Passive Microresonator," in *IEEE Journal of Lightwave Technology*, vol. 38, no. 15, pp. 4019-4029, Aug. 2020.
82. Z. Ying et al., "Comparison of microrings and microdisks for high-speed optical modulation in silicon photonics," in *Applied Physics Letters*, vol. 112, no. 11, p. 111108, Mar. 2018.
83. Z. Yu, J. Zheng, P. Xu, W. Zhang, and Y. Wu, "Ultracompact Electro-Optical Modulator-Based Ge2Sb2Te5 on Silicon," in *IEEE Photonics Technology Letters*, vol. 30, no. 3, pp. 250–253, Feb. 2018.
84. A. Shafiee, S. Pasricha, M. Nikdast. A survey on optical phase-change memory: The promise and challenges. IEEE Access. 2023 Jan 31;11:11781-803.
85. Shafiee A, Charbonnier B, Pasricha S, Nikdast M. Design space exploration for PCM-based photonic memory. In Proceedings of the Great Lakes Symposium on VLSI 2023 2023 Jun 5 (pp. 533-538).
86. Shafiee A, Banerjee S, Charbonnier B, Pasricha S, Nikdast M. Compact and low-loss PCM-based silicon photonic MZIs for photonic neural networks. In2023 IEEE Photonics Conference (IPC) 2023 Nov 12 (pp. 1-2). IEEE.
87. Sunny F, Shafiee A, Balasubramaniam A, Nikdast M, Pasricha S. OPIMA: Optical Processing-in-Memory for Convolutional Neural Network Acceleration. IEEE Transactions on Computer-Aided Design of Integrated Circuits and Systems. 2024 Nov 6;43(11):3888-99.
88. Sunny F, Shafiee A, Charbonnier B, Nikdast M, Pasricha S. COMET: A Cross-Layer Optimized Optical Phase-Change Main Memory Architecture. In 2024 Design, Automation & Test in Europe Conference & Exhibition (DATE) 2024 Mar 25 (pp. 1-6). IEEE.

89. Shafiee A, Charbonnier B, Yao J, Pasricha S, Nikdast M. Programmable phase change materials and silicon photonics co-integration for photonic memory applications: a systematic study. Journal of Optical Microsystems. 2024 Jul 1;4(3):031208-.
90. Shafiee A, Charbonnier B, Pasricha S, Nikdast M. A Multiphysics Simulation Approach for Photonic Devices Integrating Phase Change Materials. In 2023 International Conference on Numerical Simulation of Optoelectronic Devices (NUSOD) 2023 Sep 18 (pp. 119-120). IEEE.
91. Shafiee A, Chen L, Nikdast M, Yao J. Polarization-encodable photonic memory cells using next-generation 2D phase-change materials. Nano Research. 2024 Dec 16.
92. P. Xu, J. Zheng, J. Doylend, and A. Majumdar, "Non-Volatile Integrated-Silicon-Photonic Switches using Phase-Change Materials," in *Asia Communications and Photonics Conference (ACP)*, Nov. 2019.
93. N. Dhingra, J. Song, G. J. Saxena, E. K. Sharma, and B. M. A. Rahman, "Design of a Compact Low-Loss Phase Shifter Based on Optical Phase Change Material," in *IEEE Photonics Technology Letters*, vol. 31, no. 21, pp. 1757–1760, Nov. 2019.
94. Z. Cheng, C. Ríos, W. H. P. Pernice, C. D. Wright, and H. Bhaskaran, "On-chip photonic synapse," in *Sci. Adv.*, vol. 3, no. 9, p. e1700160, Sep. 2017.
95. C. D. Wright, Y. Liu, K. I. Kohary, M. M. Aziz, and R. J. Hicken, "Arithmetic and Biologically-Inspired Computing Using Phase-Change Materials," in *Advanced Materials*, vol. 23, no. 30, pp. 3408–3413, 2011.
96. J. Feldmann, N. Youngblood, C. D. Wright, H. Bhaskaran, and W. H. P. Pernice, "All-optical spiking neurosynaptic networks with self-learning capabilities," in *Nature*, vol. 569, no. 7755, Art. no. 7755, May 2019.
97. S. Kim et al., "NVM neuromorphic core with 64k-cell (256-by-256) phase change memory synaptic array with on-chip neuron circuits for continuous in-situ learning," in *IEEE International Electron Devices Meeting (IEDM)*, Dec. 2015.
98. M. P. Fok, Y. Tian, D. Rosenbluth, and P. R. Prucnal, "Pulse lead/lag timing detection for adaptive feedback and control based on optical spike-timing-dependent plasticity," in Optics Letters, vol. 38, no. 4, pp. 419–421, 2013.
99. R. Toole et al., "Photonic Implementation of Spike-Timing-Dependent Plasticity and Learning Algorithms of Biological Neural Systems," in *IEEE Journal of Lightwave Technology*, vol. 34, no. 2, pp. 470–476, Jan. 2016.
100. F. Marino and S. Balle, "Experimental study of a broad area vertical-cavity semiconductor optical amplifier," in *Opt. Commun.*, vol. 231, no. 1, pp. 325–330, Feb. 2004.
101. S. Xiang et al., "Numerical Implementation of Wavelength-Dependent Photonic Spike Timing Dependent Plasticity Based on VCSOA," in *IEEE J. Quantum Electron.*, vol. 54, no. 6, pp. 1–7, Dec. 2018.
102. D. Dang, J. Dass, R. Mahapatra, "ConvLight: A Convolutional Accelerator with Memristor integrated Photonic Computing," in *IEEE International Conference on High Performance Computing (HiPC)*, 2017.
103. M. B. On, H. Lu, H. Chen, R. Proietti and S. J. Ben Yoo, "Wavelength-Space Domain High-Throughput Artificial Neural Networks by Parallel Photoelectric Matrix Multiplier," in *Optical Fiber Communications Conference and Exhibition (OFC)*, 2020.
104. J. R. Ong, C. C. Ooi, T. Y. L. Ang, S. T. Lim and C. E. Png, "Photonic Convolutional Neural Networks Using Integrated Diffractive Optics," in *IEEE J. Sel. Top. Quantum Electron*, vol. 26, no. 5, pp. 1-8, Sept.-Oct. 2020.
105. G. M. Alexandris et al., "All-Optical WDM Recurrent Neural Networks With Gating," in *IEEE J. Sel. Top. Quantum Electron.*, vol. 26, no. 5, pp. 1-7, Sept.-Oct. 2020.
106. G. M. Alexandris et al., "Neuromorphic photonics with coherent linear neurons using dual-IQ modulation cells," in *IEEE Journal of Lightwave Technology*, vol. 38, no. 4, pp. 811–819, Feb. 2020.
107. A. N. Tait et al., "Neuromorphic photonic networks using silicon photonic weight banks," in *Scientific Reports*, vol. 7, no. 1, Aug. 2017.

108. E. N. Lorenz, "Deterministic nonperiodic flow," in *Journal of Atmospheric Sciences*, vol. 20, no.2, 1963.
109. A. Mehrabian, Y. Al-Kabani, V. J. Sorger, and T. El-Ghazawi, "PCNNA: A Photonic Convolutional Neural Network Accelerator," in *IEEE International System-on-Chip Conference (SOCC)*, 2018.
110. V. Bangari et al., "Digital Electronics and Analog Photonics for Convolutional Neural Networks (DEAP-CNNs)," in *IEEE J. Sel. Top. Quantum Electron.,* Volume: 26, Issue: 1, Jan.-Feb. 2020.
111. Y. LeCun, L. Bottou, Y. Bengio, and P. Haffner, "Gradient-based learning applied to document recognition," in *Proceedings of the IEEE*, vol. 86, no. 11, pp. 2278-2324, Nov. 1998.
112. P. Li et al, "All-optical Analog Comparator," in *Nature Communications*, 2016.
113. C. Zhang, Zhenman Fang, Peipei Zhou, Peichen Pan and Jason Cong, "Caffeine: Towards uniformed representation and acceleration for deep convolutional neural networks," in *IEEE International Conference On Computer-Aided Design*, 2016.
114. F. Sunny, A. Mirza, M. Nikdast, and S. Pasricha, "CrossLight: A Cross-Layer Optimized Silicon Photonic Neural Network Accelerator," in *IEEE/ACM Design Automation Conference (DAC)*, 2021.
115. S. Xu, J. Wang, and W. Zou, "Optical Convolutional Neural Network With WDM-Based Optical Patching and Microring Weighting Banks," in *IEEE Photonics Technology Letters*, vol. 33, no. 2, Jan 2021.
116. A. Mehrabin et al., "A Winograd-Based Integrated Photonics Accelerator for Convolutional Neural Networks," in *IEEE J. Sel. Top. Quantum Electron.*, vol. 26, no. 1, Jan/Feb 2020.
117. W. Liu, W. Liu, Y. Ye, Q. Lou, Y. Xie, L. Jiang, "HolyLight: A Nanophotonic Accelerator for Deep Learning in Data Centers," in *IEEE/ACM Design, Automation Test in Europe Conference Exhibition (DATE)*, 2019.
118. F. Zokae et al., "LightBulb: A Photonic-Nonvolatile-Memory-based Accelerator for Binarized Convolutional Neural Networks," in *IEEE/ACM Design, Automation Test in Europe Conference Exhibition (DATE)*, 2020.
119. M. L. Lalieu, R. Lavrijsen, and B. Koopmans, "Integrating all-optical switching with spintronics," in *Nature communications*, vol. 10, no.1, pp. 110, 2019.
120. K. Shiflett, D. Wright, A. Karanth, and A. Louri, "PIXEL: Photonic Neural Network Accelerator," in *IEEE International Symposium on High Performance Computer Architecture*, 2020.
121. I. Chakraborty, G. Saha, A. Sengupta and K. Roy, "Toward Fast Neural Computing using All-Photonic Phase Change Spiking Neurons," in *Nature*, August 2018.
122. M. Wuttig, N. Yamada, "Phase-change materials for rewriteable data storage," in *Nature Materials,* vol. 6, pp. 824–832, 2007.
123. S. R. Ovshinsky, Reversible electrical switching phenomena in disordered structures," in *Phys. Rev. Lett.*, vol. 21, pp. 1450–1453, 1968.
124. W. H. P. Pernice, H. Bhaskaran, "Photonic non-volatile memories using phase change materials," in *Applied Physics Letters*, vol. 101, no. 171101, 2012.
125. C. Rios, P. Hosseini, C. D. Wright, H. Bhaskaran, W. H. P. Pernice, "On-chip photonic memory elements employing phase-change materials," in *Advanced Materials*, vol. 26, pp. 1372–1377, 2014.
126. T. Van Vaerenbergh, M. Fiers, P. Bienstman and J. Dambre, "Towards integrated optical spiking neural networks: Delaying spikes on chip," in 2013 Sixth "Rio De La Plata" *Workshop on Laser Dynamics and Nonlinear Photonics*, 2013.
127. C. Mesaritakis, V. Papataxiarhis, and D. Syvridis, "Micro ring resonators as building blocks for an all-optical high-speed reservoir-computing bit-pattern-recognition system," in *J. Opt. Soc. Amer. B*, vol. 30, no. 11, pp. 3048–3055, Nov. 2013.
128. F. D. Coarer et al., "All-Optical Reservoir Computing on a Photonic Chip Using Silicon-Based Ring Resonators," in *IEEE J. Sel. Top. Quantum Electron.,* vol. 24, no. 6, Nov/Dec 2018.

129. S. S. Vatsavai, I. Thakkar, "Silicon Photonic Microring Based Chip-Scale Accelerator for Delayed Feedback Reservoir Computing," in *IEEE VLSID*, 2021.
130. H. Jaeger "Adaptive nonlinear system identification with echo state networks," in *Advance in Neural Information Processing Systems*, vol. 15, pp. 593—600, 2003.
131. H. Jaeger and H. Haas, "Harnessing nonlinearity: predicting chaotic systems and saving energy in wireless communication," in *Science*, vol. 304, pp. 78-80, 2004.
132. F. Duport, A. Smerieri, A. Akrout, M. Haelterman, and S. Massar, "Fully analogue photonic reservoir computer," in *Scientific Reports*, vol. 6, no. 22381, 2016.
133. L. Appeltant et al., "Information processing using a single dynamical node as complex system," in *Nature Communications*, 2, 468, 2011.
134. K. Vandoorne et al., "Experimental demonstration of reservoir computing on a silicon photonics chip," in *Nature*, March 2014.
135. C. Mesaritakis, M. Skontranis, G. Sarantoglou and A. Bogris, "Micro-Ring-Resonator Based Passive Photonic Spike-TimeDependent-Plasticity Scheme for Unsupervised Learning in Optical Neural Networks," in *Optical Fiber Communications Conference and Exhibition (OFC),* 2020.
136. M. Reck, A. Zeilinger, H. J. Bernstein, and P. Bertani, "Experimental realization of any discrete unitary operator," in *Phys. Rev. Lett.*, vol. 73, no. 1, pp. 58–61, Jul. 2002.
137. F. Shokraneh, S. Geoffroy-Gagnon, M. S. Nezami and O. Liboiron-Ladouceur, "A Single Layer Neural Network Implemented by a 4×4 MZI-Based Optical Processor," in *IEEE Photonics Journal*, vol. 11, no. 6, pp. 1-12, Dec. 2019.
138. D. A. B. Miller, "Self-configuring universal linear optical component [Invited]," in *Photon. Res.*, vol. 1, no. 1, Jun. 2013.
139. W. R. Clements, P. C. Humphreys, B. J. Metcalf, W. S. Kolthammer, and I. A. Walsmley, "Optimal design for universal multiport interferometers," in *Optica*, vol. 3, no. 12, pp. 1460–1465, Dec. 2016.
140. J. Gu et al., "Towards Area-Efficient Optical Neural Networks: An FFT-based Architecture," in *ACM ASPDAC*, 2020.
141. Z. Li, S. Wang, C. Ding et al., "Efficient recurrent neural networks using structured matrices in fpgas," in *ICLR Workshop*, 2018.
142. E. Cottle, F. Michel, J. Wilson, N. New, and I. Kundu, "Optical Convolutional Neural Networks—Combining Silicon Photonics and Fourier Optics for Computer Vision," in white paper, Optalysis.
143. "The Multiply and Fourier Transform Unit: A Micro-Scale Optical Processor," in white paper, Optalysis.
144. J. Friedman, T. Hastie, and R. Tibshirani, "A note on the group lasso and a sparse group lasso," in arXiv preprint arXiv:1001.0736, 2010.
145. A. S. Weigend and N. A. Gershenfeld, "Time series prediction: Forecasting the future and understanding the past," in 1993, [Online]. Available: http: //www-psych.stanford.edu/andreas/Time-Series/SantaFe.html.
146. Y. Zhu et al., "Countering Variations and Thermal Effects for Accurate Optical Neural Networks," in *IEEE International Conference on Computer-Aided Design*, 2020.
147. M.Y.-S. Fang et al., "Design of optical neural networks with component imprecisions," in *Optics Express* vol. 27, pp. 14009-14029, 2019.
148. M. Nikdast, G. Nicolescu, J. Trajkovic, and O. Liboiron-Ladouceur, "Modeling fabrication non-uniformity in chip-scale silicon photonic interconnects," in *IEEE/ACM Design, Automation & Test in Europe Conference & Exhibition (DATE)*, 2016.
149. J. S. Orcutt, A. Khilo, C. W. Holzwarth, M. A. Popovic, H. Li, J. Sun, T. Bonifield, R. Hollingsworth, F. X. Kartner, H. I. Smith, V. Stojanovic, and R. J. Ram "Nanophotonic integration in state-of-the-art CMOS foundries," in *Optics Express*, vol. 19, pp. 2335-2346, 2011.
150. A. Mahendra, C. Xiong, X. Zhang, B. J. Eggleton, and P. H. W. Leong, "Multiwavelength stabilization control of a thermo-optic system with adaptive reconfiguration," in *Applied Optics*, vol. 56, no. 4, pp. 1113-1118, 2017.

151. W. R. Clements, William R., P. C. Humphreys, B. J. Metcalf, W. S. Kolthammer, and I. A. Walmsley, "Optimal design for universal multiport interferometers," in *Optica*, vol. 3, no. 12, pp. 1460–1465, 2016.
152. Reference
153. Z. Ghanaatian, A. Mirza, A. Shafiee, S. Pasricha, M. Nikdast, "Bridging EDA and Silicon Photonics Design: Enabling Robust-by-Design Photonic Integrated Circuits". InProceedings of the 30th Asia and South Pacific Design Automation Conference 2025 Jan 20 (pp. 128-134).
154. S. Banerjee, M. Nikdast, and K. Chakrabarty, "Modeling Silicon-Photonic Neural Networks under Uncertainties," in *IEEE/ACM Design, Automation and Test in Europe (DATE) Conference and Exhibition*, 2021.
155. Shafiee A, Banerjee S, Chakrabarty K, Pasricha S, Nikdast M. Analysis of optical loss and crosstalk noise in mzi-based coherent photonic neural networks. Journal of Lightwave Technology. 2024 Mar 5.
156. D. Dang, S. V. R. Chittamuru, S. Pasricha, R. Mahapatra, D. Sahoo, "BPLight-CNN: A Photonics-based Backpropagation Accelerator for Deep Learning," in to appear, *ACM Journal on Emerging Technologies in Computing Systems (JETC)*, 2021.

Unraveling Freezing of Gait in Parkinson's Disease: Datasets, Algorithms, and Open-Source Solutions

Luigi Borzì, Luis Sigcha, and Florenc Demrozi

"An ounce of prevention is worth a pound of cure"
—Benjamin Franklin

Abstract Freezing of gait (FoG) is a complex and debilitating symptom of Parkinson's disease (PD), consisting of sudden motor block. FoG leads to an increased risk of falls and related injuries, as well as a reduction in autonomy and quality of life. Over the past two decades, a wide range of technological solutions has been proposed for the automatic detection of FoG, with heterogeneous experimental procedures, sensing technologies, algorithms, and results. Motion data can be collected using wearable sensors, environmental sensors, cameras, and radio sensors. Data can be analyzed using varying complex approaches, from simple threshold-based methods to more advanced machine and deep learning models. The encouraging results suggest that accurate and timely FoG detection can be achieved in various contexts. This chapter explores the broad and heterogeneous set of sensing systems and algorithms developed for effective FoG detection. Furthermore, public datasets and algorithms are summarized and comprehensively discussed. Finally, current limitations are identified, and possible solutions are provided. Overall, despite the research community's constant efforts, larger and more comprehensive datasets are needed to identify the best digital solution for the continuous detection and characterization of FoG. In addition, semi-supervised or unsupervised approaches are needed to handle unconstrained monitoring in daily life. Finally, sharing data and algorithms with the research community is highly

L. Borzì (✉)
Department of Control and Computer Engineering, Politecnico di Torino, Torino, Italy
e-mail: luigi.borzi@polito.it

L. Sigcha
ALGORITMI Research Centre, University of Minho, Guimarães, Portugal

F. Demrozi
Department of Electrical Engineering and Computer Science, University of Stavanger, Stavanger, Norway

B. Farahani et al. (eds.), *Smart and Connected Health: Applications and Real-World Use Cases*, https://doi.org/10.1007/978-3-032-09253-3_7

recommended. Future studies can use the information provided in this article to push research toward solutions aimed at efficient and effective FoG recognition methods that are ready to work continuously in real life, thus improving the quality of life of people with PD.

Keywords Internet of things (IoT) · Deep learning · Machine learning · Freezing of gait (FoG) · Parkinson's disease (PD) · Wearable sensors · Gait analysis · Vision-based health monitoring · Wi-Fi sensing · Personalized medicine · Daily life monitoring

1 Introduction

Parkinson's disease (PD) is a neurodegenerative disorder characterized by the loss of dopamine-producing neurons in the brain, particularly in the substantia nigra region [1]. This dopamine deficit leads to several motor symptoms, including tremors, rigidity, bradykinesia (slowness of movement), and postural instability. Tremors often manifest as rhythmic shaking, typically starting in one hand at rest and worsening with stress or fatigue. Rigidity increases resistance to movement, making simple activities difficult. Bradykinesia causes slowness of movement, making everyday activities such as walking or getting up from a chair difficult. Postural instability contributes to poor balance and an increased risk of falls [2].

Beyond motor symptoms, PD also presents with various nonmotor symptoms, significantly affecting patients' quality of life. These include cognitive changes like slowed thinking, memory difficulties, and impaired executive function. Mood disorders such as depression and anxiety are common, along with sleep disturbances such as insomnia and REM sleep behavior disorder. Autonomic dysfunction may lead to symptoms like constipation, urinary problems, and orthostatic hypotension [3].

Figure 1 provides an overview of motor and nonmotor symptoms and environmental conditions influencing such motor and nonmotor symptoms.

A particularly distressing phenomenon associated with PD is freezing of gait (FoG). FoG is a complex and debilitating phenomenon that affects 50–80% of people with PD (PwPD), often in the advanced stages of the disease [4]. It significantly impacts their quality of life and increases the risk of falls and related injuries [5]. FoG is characterized by sudden, transient episodes of gait disruption, in which individuals experience the feeling of "having the feet glued to the ground," despite having the intention to walk [6]. The duration of FoG episodes is variable, ranging from very short events (<1 s) to episodes lasting more than a minute. However, the distribution of the duration of FoG episodes is skewed toward short episodes, with at least half of the events lasting less than 2 s, most less than 10 s, and rarely lasting longer than 30 s [7]. FoG can manifest in different forms and situations and can be grouped into three categories based on the typical observed motion: (a) very small shuffling steps with minimal forward movement (shuffling

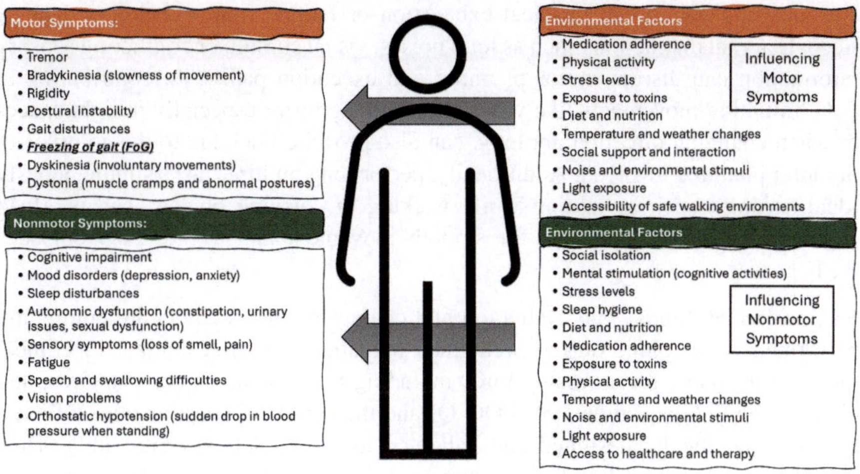

Fig. 1 Overview of motor and nonmotor symptoms in Parkinson's disease

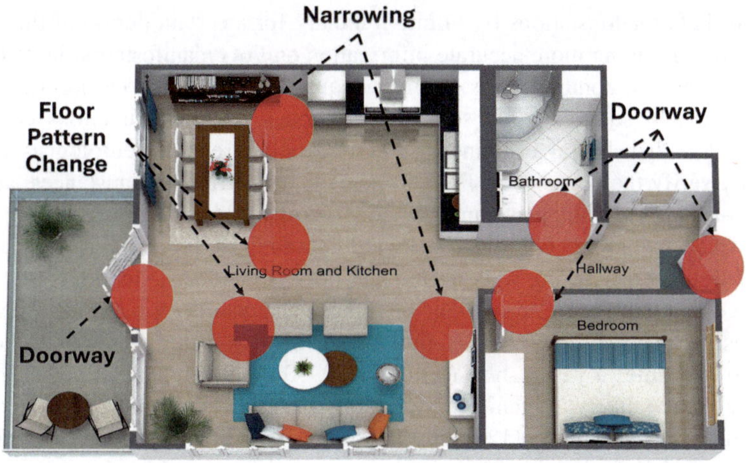

Fig. 2 Schematic of possible triggering factors related to the occurrence of freezing of gait in a daily life environment

steps), (b) leg trembling but no effective forward motion (trembling legs), and (c) complete akinesia, with no observable motion of the legs [7].

Triggering Situations FoG appearance (also known as triggering) is related to various factors, including complex environments such as crowded or cluttered spaces, narrow doorways, or sudden changes in floor patterns (Fig. 2). Emotional stress, anxiety, or the pressure to perform specific movements like turning or initiating walking can also provoke FoG episodes [8]. Fluctuations in medication effectiveness, particularly dopamine replacement therapies, may coincide with episodes of FoG, as can physical exhaustion or fatigue from exertion or lack of sleep. External distractions such as loud noises, visual stimuli, or conflicting sensory information can disrupt motor planning and execution pathways, contributing to FoG. Initiating movements like walking or making turns, especially in tight spaces or when changing direction abruptly, can also provoke FoG due to the complexity of motor planning involved. Additionally, performing multiple tasks simultaneously (dual-tasking), such as talking while walking or carrying objects, can overload cognitive resources necessary for smooth movement and trigger FoG episodes [9, 10].

Evaluation Methods The evaluation and characterization of FoG are commonly performed in the clinic during predefined outpatient visits. A number of clinical tools are available for assessing FoG, including rating scales, questionnaires, and diaries. The FoG questionnaire (FOG-Q) and the new FoG questionnaire (NFOG-Q) were specifically designed and validated to assess FoG severity [11]. They include questions regarding FoG frequency, duration, and impact of daily activities. However, other important aspects are not considered, such as triggering situations and the effect of therapy. Sometimes, patients and/or caregivers are asked to keep track of FoG manifestations by filling in a diary for a certain period of time. This allows for providing more accurate information on FoG manifestation in daily life. However, patient compliance is often inadequate, and the reported information is fragmented. In addition, only subjects with intact cognitive functions can keep diaries. Finally, clinical examinations carried out by expert neurologists aim to visually verify the presence of FoG, its severity, triggering circumstances, and the effect of therapy. Clinicians can ask patients to perform different gait and motor tasks, often including cognitive (backward counting) and motor (carrying a glass full of water) tasks during walking to elicit FoG. Sometimes, video recordings are captured and analyzed offline to annotate FoG characteristics carefully. However, the episodic nature of FoG makes it difficult to fully assess this symptom in sporadic (one to two times a year), short (typically 20–30 minutes) medical visits. Indeed, PwPDs are less likely to manifest FoG during the clinical examination than during daily activities at home [12], thus increasing the difficulty of correct clinical assessment of FoG characteristics [13].

Pharmacological Treatment Dopaminergic therapy, most commonly Levodopa, can improve FoG by reducing the number of episodes and their duration [1]. This is true for the most common form of FoG (dopaminergic-responsive FoG), where the phenomenon disappears during the ON phase due to the Levodopa treatment efficacy. However, about 30% of PwPD manifest dopaminergic-resistant FoG, where FoG is indifferent to the dopamine intake. Dopaminergic-induced FoG (provoked by the Levodopa assumption) is very uncommon. Finally, pseudo-on-state FoG is present in the ON state but ameliorates with increased dopaminergic treatment, while biphasic FoG appears in the transition phase between OFF and ON phases [14].

Non-pharmacological Treatment Physical rehabilitation is crucial in PD, and several approaches are applied to reduce FoG effect in research settings. These include action observation training, treadmill training, aquatic obstacle training, curved walking training, supervised slackline training, and home-based exercises [15]. In addition, visual (e.g., walking over parallel lines), auditory (i.e., regular, repeated audible sound), and somatosensory (e.g., shoulder tapping) cueing proved effective in ameliorating FoG by reducing the number and duration of episodes. While continuous cueing can produce habituation effects that reduce the efficacy of the stimulation, on-demand cueing promises to overcome this problem by providing automatic feedback triggered by FoG manifestation [16].

In such a direction, comprehensive and continuous monitoring of FoG is essential for fully understanding and characterizing such a complex and episodic phenomenon [12]. Current clinical assessment of FoG is subjective, performed sporadically, and often carried out through visual evaluations [17]. These limitations have pushed researchers to provide more objective and accurate digital solutions for recognizing and characterizing FoG automatically. A wide set of technologies have been proposed in the last two decades, promising to revolutionize the current clinical approaches [16, 18–20]. An overview of these digital tools is provided in the following sections.

2 Preliminaries on Sensing Technology and Algorithms for FoG Treatment

Understanding and effectively addressing FoG in PwPD necessitate a comprehensive exploration of sensing technologies and algorithmic solutions. Figure 3 provides an overview of the technologies most used in FoG detection; in addition, this section explores the technological underpinnings essential for the treatment and management of FoG.

From advanced wearable devices and sensor technologies to sophisticated algorithms and software tools, these preliminary insights provide a crucial framework for understanding how technological innovations are transforming the landscape of FoG treatment. Figure 4 provides an overview on the FoG solution taxonomy and organization of this section.

2.1 Sensing Technologies

The development of sensing technologies for FoG detection has evolved significantly over the past decades. Technological advances have paved the way for developing various sensing systems to detect and predict FoG, offering potential interventions to mitigate these episodes [21]. These technologies include cameras,

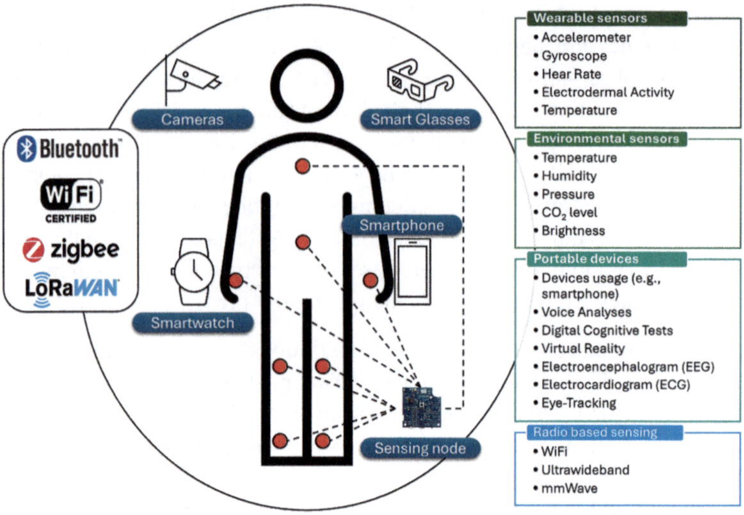

Fig. 3 Overview on the most used sensing and communication technology for motor symptom analysis in Parkinson's disease

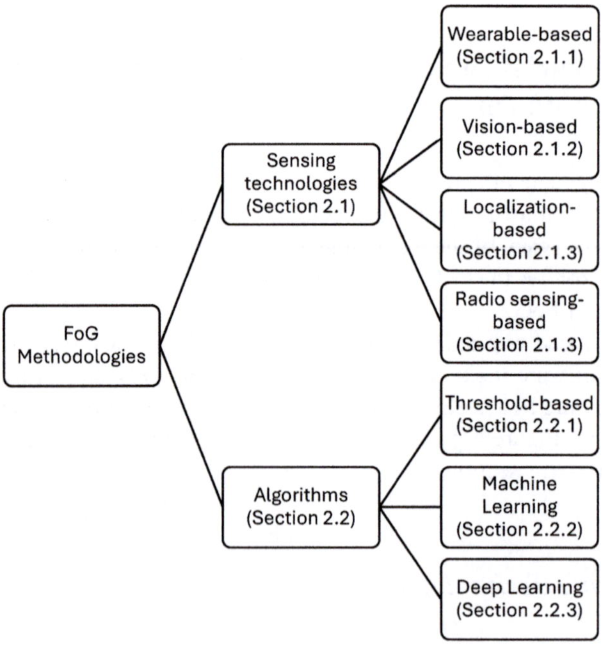

Fig. 4 Taxonomy of the analyzed FoG detection solutions

wearables, and localization/radar-based systems. Wearable devices enable remote monitoring using nonintrusive sensors. Vision-based systems currently employ advanced deep learning (DL) techniques for more accurate detection, while localization and radar-based systems offer noncontact monitoring solutions [22]. A detailed overview of these technologies is presented as follows.

Wearable Devices Wearable technology has been extensively used in FoG detection due to its unobtrusiveness and the ability to continuously monitor PwPD [23, 24]. Wearable devices commonly include sensors such as accelerometers, gyroscopes, or magnetometers [18]. Early initiatives in automatic FoG detection using wearables were presented in [25]. This work pioneered a portable monitoring system using accelerometers positioned on the ankles. This system employed fast Fourier transform (FFT) and amplitude analysis to distinguish between normal gait and FoG. In [26], authors further developed an ambulatory monitoring device using accelerometers on the shank, detecting FoG episodes with 78% sensitivity and 89% accuracy based on threshold-based approaches. Advancements continued in [27], where the authors integrated accelerometers and gyroscopes into a wearable system that detected FoG and provided metronome stimuli to assist patients in overcoming FoG episodes. In [28], authors introduced the first FoG detection system using machine learning (ML) algorithms in combination with wearable sensors and smartphones to provide an online FoG detection system. Furthermore, more recent approaches [29, 30] have leveraged smartphone sensors, threshold, and ML algorithms, such as convolutional neural networks (CNNs), to improve the accuracy and reliability of FoG detection. These systems have shown high sensitivity and specificity by automatically extracting discriminative features from sensor data.

Vision-Based Systems Vision-based systems use cameras to capture video data of the movements of PwPD, providing a noncontact method of FoG detection [31]. These systems often employ advanced image processing and ML techniques to analyze gait patterns. In [32], authors conducted early implementations of this approach using a video camera and a 3D motion analysis system with markers attached to various body parts (e.g., shoulders, ankles, heels, toes, and over vertebra T10). This method allowed precise identification (sensitivity up to 83% and specificity greater than 95%) of FoG episodes using a threshold and frequency analysis. Other studies used 3D motion capture systems, such as the Kinect depth camera, to extract motion information for detecting gait disorders [33, 34], achieving remarkable results (e.g., 90% accuracy rate for FoG prediction). Despite this performance, the assessment was restricted to indoor environments, such as laboratories or clinics, limiting its broader application. Recent innovations in vision-based systems have focused on DL approaches, for instance, in [35, 36], authors explored the use of CNNs and graph convolutional networks, respectively. These methods used video data to recognize FoG by analyzing spatial and temporal patterns of body movements, with promising results.

Localization and Radar-Based Systems Localization and radar-based systems offer another layer of unobtrusive monitoring by using Wi-Fi or radar signals to detect movement patterns indicative of FoG. Authors of [37] showed the use of Wi-Fi sensing systems to detect variations in channel state information (CSI) amplitude generated during FoG episodes. This approach allowed for continuous and noninvasive monitoring, which proved effective in distinguishing FoG from normal gait. On the other hand, a combination of Wi-Fi and radar technologies was proposed for FoG detection in [38]. Wi-Fi-based sensing leverages CSI to generate scalograms, while radar sensing utilizes micro-Doppler signatures from the reflected radar signals. Both methods produce distinctive patterns that correspond to different activities and movement anomalies, such as those observed in FoG episodes. This combined approach involves using DL models, specifically autoencoders, which have been shown to improve classification accuracy dramatically. The Wi-Fi-based system achieved an accuracy of 84.5%, and the radar-based system reached 85.5%, but when fused, the accuracy increased up to 98%.

2.2 Algorithms

Innovation in algorithmic solutions plays a pivotal role in the analysis and management of FoG. This section reviews cutting-edge algorithms developed to assess, monitor, and intervene in cases of FoG, enhancing clinical outcomes and quality of life. In particular, data analysis methods, including threshold-based approaches, ML techniques, and DL models, have been developed to address FoG detection. Although early methods laid the groundwork, the integration of advanced ML and DL algorithms has greatly enhanced the accuracy and reliability of FoG detection systems.

Threshold-Based Methods Early efforts in FoG detection predominantly used threshold-based methods. These methods rely on predefined criteria or thresholds to identify FoG episodes from sensor data, typically obtained from accelerometers and gyroscopes [21]. Early studies in FoG detection include the work presented in [25], which used accelerometers. They used the power of different frequency components obtained through a four-level Daubechies wavelet to distinguish between normal gait and FoG. Similarly, wearable sensors were used to monitor stride length and gait variability in PwPD [39]. By setting thresholds for these parameters, they could detect significant deviations from typical gait patterns, effectively identifying FoG. In [26], the authors proposed an ambulatory monitoring device that uses the frequency characteristics of vertical leg movements captured by an accelerometer on the left shank. Their threshold-based approach achieved a 78% detection rate and 89% accuracy. In [40], the authors further refined the threshold-based FoG detection method by incorporating multiple accelerometers and a metronome to assist patients during FoG episodes, resulting in improved performance. Instead, in [32], the authors conducted a study involving a 3D motion analysis system and video recording. Researchers used a threshold and frequency analysis on data from ten PwPD and ten healthy subjects walking on a treadmill with obstacles. These methods, while foundational, often struggled with sensitivity and specificity, necessitating the development of more sophisticated techniques.

Machine Learning Methods With technological advances, ML methods have been increasingly applied to FoG detection to improve aspects such as sensitivity and specificity. These methods use different sets of "handcrafted" features extracted from sensor data to train sophisticated classifiers that can distinguish between normal gait and FoG episodes. The first study introducing ML in FoG detection was proposed in [28]. This study introduced a wearable assistant system that combines a smartphone and wearable accelerometers. The researchers evaluated different temporal windows and various classification algorithms (e.g., random forest, decision trees, naive Bayes, and multilayer perceptron). The features used for classification were based on the work presented in [40], along with additional time and frequency domain features (e.g., mean, standard deviation, variance, entropy, energy, freezing index (freezing band power 3–8 Hz divided by locomotor band power 0.5–3 Hz), and the sum of the freezing band and locomotor band power). The proposed approach achieved high sensitivity (98.35%) and specificity (99.72%). Other studies using ML techniques include [41], where a combination of accelerometers, gyroscopes, and diverse classification algorithms (e.g., naive Bayes, random forests, decision trees, and random tree) was employed to detect FoG events with 81.9% sensitivity and 98.7% specificity. In [42], the authors found that spectral features are suitable for designing automatic FoG detection systems in combination with a support vector machine (SVM) classifier with data from a waist-worn accelerometer. Their system recorded activities of 20 PD patients in both medicated (ON) and non-medicated (OFF) states, achieving an accuracy of 98.7%. Similarly, the authors of [43] developed an ML algorithm based on SVM to detect FoG using a single three-axial accelerometer worn at the waist. This system aimed to detect FoG in real-life scenarios and accounted for the influence of medication on motor response, showing a sensitivity of 82.1% and a specificity of 97.2% in episodic evaluation. Furthermore, in [44], the authors improved (91.7% sensitivity and 87.4% specificity) the system proposed in [43] by systematically reducing the number of features, thus optimizing the classifier for real-time processing. Additionally, in [45], a wearable-based system was proposed for predicting FoG events before their actual occurrence. The authors used a k-nearest neighbor (k-NN) algorithm and data from triaxial accelerometer sensors (back, hip, and ankle) to classify three categories, namely, pre-FoG, no-FoG, and FoG. The proposed system successfully predicted FoG with a sensitivity of 94.1% and a specificity of 97.1%. Similarly, [46] showed that predicting freezing events by analyzing pre-FoG gait degradation is feasible using decision trees and SVM to achieve a sensitivity of 84.0% and a specificity of 88.3%. In addition, the researchers noted that the patients' motor status (ON-OFF) should be considered, as medication may attenuate pre-FoG periods, impacting FoG prediction.

Deep Learning Methods The advent of DL brought a new paradigm to FoG detection, with models capable of automatically learning relevant features from raw data without manual feature extraction. DL methods, particularly CNNs, have demonstrated superior performance due to their ability to automatically extract and learn features from raw data [18]. In [47], authors utilized CNNs to extract discriminative features from sensors embedded in smartphones. Their method showed a 20% higher sensitivity compared to traditional classifiers like RF. In [48], authors proposed an eight-layered CNN architecture for FoG detection using accelerometer data and spectral analysis through FFT. Their approach included complex activities such as cleaning a cup and carrying a glass of water, demonstrating the robustness of DL in real-life scenarios. Mohammadian Rad et al. [49] introduced a non-supervised method based on convolutional denoising autoencoders to detect unusual movement patterns. When learning the normal movement distributions from accelerometer data, deviations from the norm were quantified using a proposed normative probability map (NPM) [49].

In [50], authors evaluated a CNN with various data representations, including Mel-frequency cepstral coefficients (MFCC). This approach achieved an area under the curve (AUC) of 0.931. In particular, the authors found that a single accelerometer on the knee could match the performance of systems with multiple accelerometers. In the same manner, in [51], authors assessed a six-sensor setup (placed on feet, shanks, lumbar, and chest) along with a three-layer CNN. They observed high accuracy in detecting the freezing time and determined an optimal sensor configuration for FoG detection consisting of three sensors positioned on the lumbar region and both ankles. In the same vein of optimizing the previous methods and providing algorithms to be used in real time, lightweight DL algorithms were proposed based on multi-head CNNs [52] and Convmixer [53]. These models achieved a classification performance of 60–88% sensitivity and 88–93% specificity using the REMPARK dataset.

In addition to the extensive use of CNNs, recurrent networks such as gated recurrent unit (GRUs), long short-term memory (LSTMs), or transformer networks have been reported in several studies. In [54], the authors propose a combination of CNN and attention-enhanced LSTM, achieving an accuracy of 92% and an AUC of 0.945 in leave-one-subject-out cross validation (LOSO CV). Using only the back accelerometer data, the algorithm reached a sensitivity of 82.9% and a specificity of 90.8%. Furthermore, the combination of CNNs and LSTM processing layers was evaluated in [55] using three previous spectral windows, achieving 87.1% in both sensitivity and specificity using a LOSO CV. The previous system was improved by Sigcha et al. [56], by using an algorithm that integrates a CNN block with a transformer encoder in combination with the staked spectral windows. This method achieved 89.3% sensitivity and 89.3% specificity using a holdout evaluation.

Other novel approaches used a graph fusion neural network (GFN) [57] and temporal convolutional networks (TCN) [58]. The GFN was used to combine data from footstep pressure sensors and video recordings to create multimodal graphs to improve FoG detection. The GFN achieved promising results with an AUC of 0.882

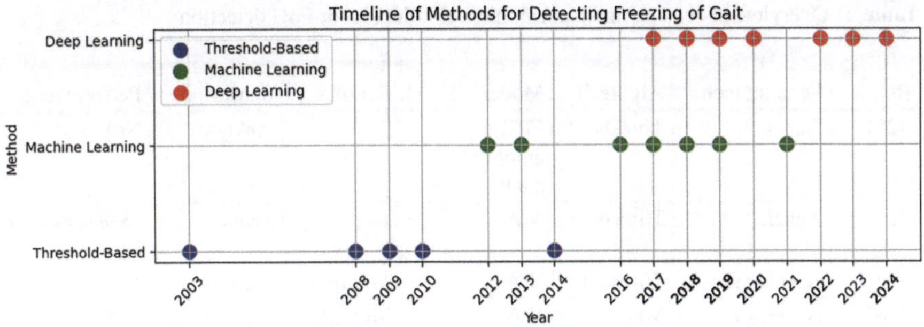

Fig. 5 Timeline of methods for detecting freezing of gait

on a multimodal FoG dataset. In addition, a TCN was trained using data collected using inertial measurement units (IMUs) with accelerometer and gyroscope in various FoG-provoking tasks and medication states. The results showed strong agreement between the model and expert video annotations in various task and medication conditions (intra-class correlation coefficient in the percentage of time frozen = 0.92).

On the other hand, vision-based approaches have been explored to reduce the obtrusiveness of wearable sensors. In [59], a system was proposed that exploited a spatial attention network combined with CNNs to learn and recognize FoG regions from video data from 45 PwPD. The results showed 68.2% sensitivity and 80.8% specificity. Similarly, in [60], a graph convolutional neural network architecture was introduced that constructs a directed graph of anatomical joints. Data were extracted from video recordings of 45 PwPD and used to characterize FoG and gait patterns. The results showed promising performance, with an AUC of 0.887. Regarding the use of Wi-Fi sensing systems, in [37], the authors used multiresolution time-frequency scalograms and a deep multiclass CNN for feature extraction and classification, achieving 97% sensitivity and 100% specificity. These DL approaches have established new benchmarks in FoG detection by leveraging the strengths of neural networks in handling large and complex datasets.

Figure 5 shows the timeline of the methods proposed for detecting FoG.

Table 1 summarizes the previously mentioned sensing and algorithmic technologies for FoG detection.

3 Public Datasets

The design of FoG detection algorithms requires access to high-quality datasets, which is fundamental for advancing research on FoG in PwPD. This section explores notable freely available datasets that provide valuable insights into the complexities of FoG. A comprehensive summary of each dataset is provided in

Table 1 Overview of sensing and algorithmic technologies for FoG detection

Ref.	Sensing tech.	Alg. tech.	Model	# of Sensors	Sensor pos.	Performance
[25]	Accel.	Threshold	FFT, amplitude analysis	2	Ankles	N/A
[26]	Accel.	Threshold	N/A	1	Shank	78% sens 89% accu
[27]	Accel, gyro	Threshold	N/A	Multiple	Various	N/A
[28]	Accel, gyro, smartphone	ML	N/A	Multiple	N/A	N/A
[29]	Smartphone sensors	Threshold	N/A	N/A	N/A	N/A
[30]	Smartphone sensors	ML	CNN	N/A	N/A	N/A
[32]	Video camera, 3D motion analysis	Threshold	Threshold, freq analysis	Multiple	Various body parts	83% sens >95% spec
[33, 34]	3D motion, Kinect	N/A	N/A	N/A	N/A	90% Accu
[35]	Video	DL	CNN	N/A	N/A	N/A
[36]	Video	DL	Graph conv. networks	N/A	N/A	N/A
[37]	Wi-Fi	DL	Autoencoders	N/A	N/A	84.5% Accu
[38]	Wi-Fi, radar	DL	Autoencoders	N/A	N/A	98% Accu
[28]	Accel., smartphone	ML	RF, DT, NB, MLP	Multiple	N/A	98.35% sens 99.72% spec
[42]	Accel.	ML	SVM	1	Waist	98.7% Accu
[43]	Accel.	ML	SVM	1	Waist	82.1% sens 97.2% spec
[44]	Accel.	ML	N/A	N/A	N/A	91.7% sens 87.4% spec
[45]	Accel.	ML	k-NN	Multiple	Back, hip, ankle	94.1% sens 97.1% spec
[46]	Accel.	ML	DT, SVM	N/A	N/A	84.0% sens 88.3% spec
[47]	Smartphone sensors	DL	CNN	N/A	N/A	93.8% sens 90.1% spec
[29]	Smartphone sensors	DL	CNN	N/A	N/A	89% sens 97% spec
[48]	Accel.	DL	CNN	N/A	N/A	91.9% sens 89.5% spec
[49]	Accel.	DL	Denoising autoencoders	N/A	N/A	N/A
[50]	Accel.	DL	CNN	1	Knee	AUC: 0.931

(continued)

Table 1 (continued)

Ref.	Sensing tech.	Alg. tech.	Model	# of Sensors	Sensor pos.	Performance
[51]	Accel.	DL	CNN	6	Feet, shanks, lumbar, chest	N/A
[52]	Accel.	DL	Multi-head CNNs	N/A	N/A	60–88% sens 88–93% spec
[53]	Accel.	DL	Convmixer	N/A	N/A	60–88% sens 88–93% spec
[54]	Accel.	DL	CNN, attn-enhanced LSTM	1	Back	91.9% Accu AUC: 0.945
[55]	Accel.	DL	CNN, LSTM	multiple	N/A	87.1% sens 87.1% spec
[56]	Accel.	DL	CNN, transformer encoder	Multiple	N/A	89.3% sens 89.3% spec
[57]	Footstep pressure, Video	DL	GFN	N/A	N/A	AUC: 0.882
[58]	IMUs (Accel, gyro)	DL	TCN	N/A	N/A	ICC: 0.92
[59]	Video	DL	Spatial attn network, CNN	N/A	N/A	68.2% sens 80.8% spec
[60]	Video	DL	graph conv. networks	N/A	N/A	AUC: 0.887
[37]	Wi-Fi	DL	CNN	N/A	N/A	97% sens 100% spec

Table 2 in terms of sample size, sensor settings, and FoG episode number and duration.

3.1 Multisensor

The dataset described in [51], henceforth referred to as multisensor, comprises data from seven PwPD. The sample consisted of four men and three women, with an average age of 58.4 ± 5.1 years and disease duration of 10.1 ± 2.4 years. Six IMUs were strapped on the top of both feet, the lateral side of both shanks, and the lumbar (L5) and chest regions. Data were recorded with a sampling frequency of $Fs = 128\,Hz$. Each participant provided different walking sessions through the turning and barrier course specifically designed to elicit FoG. Each walking trial (walk) consisted of two ellipses and two figures of eight around tall barriers. Participants completed all trials OFF medication and OFF deep brain stimulation. A video of each walk was synchronized with the IMU system. The experiments were carried out over two to six clinic visits separated by up to 44 months. A total of 89 minutes of data were collected, and 211 FoG episodes were identified, accounting for 24% of the total recording duration. The dataset can be found at this GitHub page.

3.2 Daphnet

The Daphnet dataset [61] comprises data from ten PwPD. The sample consisted of seven men and three women, with an average age of 66.4 ± 4.8 years, disease duration of 13.7 ± 9.7 years, and H&Y score of 2.6 ± 0.65 in ON condition. Data were recorded from three accelerometers placed on the shank, thigh, and lower back during the experiments. Participants were asked to complete three walking tasks that aimed to represent different aspects of daily walking. These tasks included walking back and forth in a straight line along the lab hallway and random walking in a reception hall space with initiated stops and 360-degree turns. In addition, walking while simulating ADL was considered in the protocol, including entry and exit of rooms and walking to the lab kitchen, getting a drink, and returning to the starting room with a cup of water. The experiments were carried out in the morning during the OFF stage of the medication cycle, which was more than 12 hours after their last drug intake. In total, 8.3 hours of data were recorded. Eight out of 10 participants manifested a total number of 237 FoG episodes (29 minutes), while 2 patients did not manifest any FoG. The dataset can be found at here.

3.3 tDCS

The tDCS dataset [62] comprises 71 PwPD with FoG. The sample consisted of 57 men and 14 women, with an average age of 69.9 ± 7.8 years and disease duration of 9.2 ± 5.7 years. Data were collected using an IMU attached to the participants' lower back, recording acceleration data at a sampling rate of 128 Hz. Participants undertook a FoG-provoking test in a controlled lab setting, involving three difficulty levels (i.e., single task, dual-motor task, and motor-cognitive task). The FoG-provoking test required participants to walk, turn in circles, enter a doorway, turn again, and return to a seated position. The study included assessments both ON and OFF. Each test was recorded and annotated offline to identify FoG episodes using specific criteria. Instances with short duration or annotation difficulties were reviewed by multiple evaluators. A total of 15.32 hours of acceleration data were recorded, including 1132 FoG episodes (4.75 hours), accounting for 31% of the total data. Data from 61 participants is available at this Kaggle page.

3.4 IMU

The IMU dataset [63] includes 35 idiopathic PwPD. The sample comprises 19 males and 16 females aged 44–84 years, H&Y of 2–4. Data were collected using an IMU attached to the ankle of the most affected side, recording acceleration, and angular velocity signals with a sampling rate of 128 Hz. Participants were asked to perform 3 trials of the 360° turning in place task for 2 minutes, both clockwise and anticlockwise. A total of 26.9 minutes of FoG were registered during the experiments from 23 subjects, with an average duration of FoG episodes of 3 s. The remaining 12 subjects did not manifest any FoG. The dataset can be found here.

3.5 Multimodal

This dataset described in [64], henceforth referred to as multimodal, comprises 12 subjects with PD. The sample consists of 6 men and 6 women, with an average age of 69.1 ± 7.9 years and a disease duration of 9.3 ± 6.8 years. Participants were equipped with four inertial modules on both shanks, lower back, and wrist, a wearable skin galvanic skin response (GSR) sensor on the wrist, three electromyography (EMG) sensors on both tibialis anterior and the right gastrocnemius muscle, and an electroencephalography (EEG) sensor with 25 electrodes. The sampling frequency was set to 1000 Hz for EEG and EMG data and 100 Hz for recording movement and skin conductance. Participants provided different gait tasks designed to trigger FoG, including rapid turning, approaching obstacles, and walking into narrow spaces. Participants completed all trials OFF medication. A video of each walk was synchronized with the IMU system. A total of 222 minutes of data were collected, and 334 FoG episodes were identified, accounting for 40% of the total recording duration. The dataset can be found here.

3.6 Limitations

As described above and summarized in Table 2, only five public FoG datasets exist. They differ in sample size and characteristics, number of sensors and placement, pharmacological conditions, and total FoG data and recorded episodes. All datasets include data collected in the laboratory during scripted gait and motor tasks and sometimes simulated activities of daily living. This aspect significantly limits the generalizability of the results to unconstrained remote environments (e.g., home). In fact, the range of activities performed in daily living and their inherent variability are difficult to represent in predefined motor tasks. The sample size is often reduced, as well as the total data recording time and the number and total duration of FoG episodes. The only exception is represented by the tDCS dataset, which comprises

Table 2 Characteristics of public FoG datasets

Dataset	# Subjects	# Sensors	Sensor type	Sensor position	Sampling frequency	Medication status	# FoG episodes	FoG duration (minutes)	Recording (minutes)
Multisensor	7	6	IMU	Feet, shanks, lower back, chest	64 Hz	OFF	211	21	90
Daphnet	10 (8 freezers)	3	ACC	Shank, thigh, lower back	50 Hz	OFF	237	29	498
tDCS	62 (41 freezers)	1	ACC	Lower back	128 Hz	ON, OFF	1132	285	918
Multimodal	12	3	IMU	Shanks, lower back	100 Hz	OFF	334	89	222
IMU	35 (22 freezers)	1	IMU	Shank	128 Hz	ON	NA	27	NA

more than 60 PwPD, producing 15 hours of recordings and more than a thousand FoG episodes, for a total FoG duration of about 5 hours. In addition, this is the only dataset that includes data recorded both in ON (under therapy) and OFF (not under therapy) conditions. All together, these make the tDCS dataset the largest and most heterogeneous public FoG dataset. Concerning sensor settings, there is a wide heterogeneity in the selection of sensors and their positioning. The lower back and shank/ankle represent the most common sensor locations, present in four of five datasets. These common positions may provide consistent data to evaluate the generalizability of developed FoG detection algorithms in different datasets. However, it is worth considering that while the accelerometer was employed in all studies, data from the gyroscope sensors are less common. Finally, all datasets were marked/labeled via direct visual inspection of video recordings by expert clinical raters. However, only binary labels (i.e., FoG, non-FoG) were provided in all cases. This hinders comprehensive evaluation of recognition performance, due to the impossibility to provide context for false negatives and false positives occurrence [65].

4 Open-Source FoG Detection Algorithms

Another central limitation in this research area concerns the lack of accessibility to the code of existing methodologies. On such purpose, this section discusses key algorithms that are freely available to the research community, fostering collaboration and innovation in addressing the challenges posed by FoG.

4.1 1D-CNN

The algorithm proposed in [51] is available open source on github. Raw acceleration signals are segmented into 2 s-windows. Windows are scaled using the z-score normalization, producing data with zero mean and unit variance. The final input data have shape 128×36, with 128 samples (2 s-window, Fs = 64 Hz) and 6 IMUs (each providing three acceleration and three angular velocity signals). The model is a 1D-CNN with two convolutional, one pooling, and one fully connected layers. The model was trained and evaluated on the multisensor dataset [51], comprising seven PwPD equipped with six sensors on the feet, ankles, lower back, and chest. A total of 89 minutes of walking and 211 FoG events were recorded. Performance was evaluated using a LOSO validation, providing an AUC of 0.80 for the full set of sensors and an AUC of 0.83 when using the IMUs on the lower back and ankles. An intra-class correlation coefficient (ICC) of 0.83 was found for the percent time spent with FoG, increasing to 0.95 for the estimation of the number of FoG events.

4.2 Threshold-Based

The algorithm proposed in [66] is available open source on github. Acceleration and angular velocity signals are segmented using 1s-long windows. Additional preprocessing procedures include filtering, cross-correlation computation, and FFT computation. The final input data have dimension 128 × 18, with 128 samples (1 s-window, Fs = 128 Hz) and 3 inertial modules, each providing 3 acceleration and 3 angular velocity signals. The detection algorithm is based on a threshold method. Specifically, acceleration and angular velocity data from the sensor on the back are used for gait detection. Then, windows are predicted as FoG if (a) the correlation between left and right angular velocity from the ankle-mounted sensors is below a certain threshold and (b) the freeze index of left and right acceleration data from the ankle-mounted sensors is above a certain threshold. Thresholds were optimized based on a dataset collected in the laboratory, comprising 27 freezers (79–150 FoG episodes) and 18 non-freezers. The detection method was tested on unsupervised home-based data from 23 freezers and 25 non-freezers. Performance was evaluated in terms of classification of freezers and non-freezers and ICC between algorithm and human rater prediction of the number of FoG episodes. An AUC in the range of 0.89–0.93 was obtained in discriminating freezers and non-freezers. From the home assessment monitoring period, the percent of time spent with FoG was higher ($p = 0.04$), while the variability of time spent with FoG was lower ($p = 0.02$) in freezers than in non-freezers.

4.3 Multi-head CNN

The algorithm proposed in [65] is available open source on github. Raw acceleration signals are segmented using windows of 2 s. The average value is removed from each acceleration component. No additional preprocessing is required. The final input shape is 80 × 3, with 80 samples (2 s-window, Fs = 40 Hz) and 3 acceleration components. The model is a multi-head 1D-CNN. The input layer has three heads in which data are analyzed using different spatial resolutions (kernels). The model has two convolutional, two pooling, and one fully connected layer. The detection algorithm was trained and optimized on the REMPARK dataset, comprising acceleration data from 21 PwPD that manifested a total of 1058 FoG episodes. Data were recorded in the home environment using a single IMU on the waist. Performance was calculated on the hold-out test set (five subjects) and on two external datasets including more than a hundred PwPD [2, 67]. The model provided an AUC of 0.946–0.953 on the REMPARK and external datasets, with 0.80–0.88 sensitivity and 0.88–0.98 specificity. In addition, 91–100% of FoG episodes were detected or predicted. Finally, inference time is 43 ms.

Table 3 Research query (2015–2024)

Operator	Keywords
	(Freezing of gait OR FoG)
AND	(Parkinson's disease)
AND	(Inertial sensors OR accelerometer OR gyroscope OR magnetometer OR IMU OR Wi-Fi sensing OR video-based systems OR computer vision)
AND	(Open source OR open-source)

4.4 Staying Updated on Open-Source Algorithms for FoG Detection

To keep abreast of the latest developments in open-source algorithms designed for detecting FoG in individuals with Parkinson's disease, it is essential to regularly search relevant academic and technical databases. Utilizing a comprehensive research query can help identify new studies and advancements in this field. Readers are encouraged to use the research query shown in Table 3 on databases such as Scopus, IEEE Xplore, PubMed, and ACM:

5 Discussion

A wide range of FoG recognition systems has been proposed over the past decades. However, the heterogeneity of sensor settings (number of sensors, type and position on the body) does not allow for robust conclusions on the best sensor configuration for FoG recognition. In this context, several sensor configurations were explored in [51]. Participants were equipped with up to 11 sensors placed on different parts of the body, including the lower and upper extremities, trunk, back, and head. This makes the dataset generated the most complete database in terms of the number of sensors. However, the sample size is small (seven subjects), and the experiments comprised only supervised laboratory tasks.

In addition to the heterogeneity of the sensor configuration, the experimental procedures show great variability. The FoG recognition algorithms were mostly evaluated on simple walking tasks, such as walking back and forth along a corridor [46], or the timed up-and-go test [68]. This generates a reduced set of tasks (walking and turning) in addition to FoG. For this reason, some studies included other activities and tasks that may generate false alarms, such as 360-degree turns [63], walking along a figure of eight [51], and pass through a doorway [46]. To further increase the likelihood of FoG occurrence, dual-motor or cognitive tasks were often employed [69]. Finally, some experimental protocols included the simulation of everyday life activities, which are essential for understanding the performance of the algorithm in daily life [70]. However, most of the studies were conducted in laboratory settings, which differ significantly from the home environment. This represents the most significant challenge, as algorithms trained on supervised data may not generalize to new, more complex environments and conditions. In this context, the REMPARK dataset [70] represents the first attempt to evaluate FoG in the home environment. However, despite several tasks being included in the data collection procedures, they were mostly supervised tasks. In addition, the dataset was carefully labeled by expert clinicians, but it is not publicly available. A further step toward the recognition of FoG in daily life is provided by the daily living dataset [72]. Participants were asked to wear a sensor on their lower back for seven consecutive days to monitor their daily activity. The sample includes 65 PwPD, and the dataset is publicly available; however, it is not labeled. In this case, unsupervised or one-shot learning approaches are required. As is evident, there is a lack of a comprehensive dataset that includes (a) a large number of subjects, (b) data collected using different sensors in different body positions, (c) unconstrained daily activities performed in the home or outdoors, (d) labeling by experienced evaluators, (e) both ON and OFF conditions, and (f) data over several days.

As shown, different approaches were proposed for FoG recognition, with different characteristics and results. Threshold-based methods are simple since they require thresholding on single relevant features (e.g., the freezing index) [21, 26, 35]. However, threshold tuning on each individual is essential for proper functioning, making them subject-specific (personalized) approaches that poorly perform when applied to new subjects. On the other hand, classic ML algorithms can be trained on a large amount of data to capture signal patterns common among different subjects. These algorithms proved effective, providing promising results [16]. However, they require manual feature engineering and extensive feature selection, and these choices significantly affect final performance.

On the other hand, DL methods provide state-of-the-art performance by capturing inherent and subtle data characteristics without the need for manual feature extraction. However, they require a large amount of data for proper network training. Moreover, these approaches share the need for labeled data used to train and optimize supervised classification models. This represents an essential aspect, as the recording of a large amount of data and the accurate labeling of such data performed by accurate visual inspection of video recordings require time and effort. This is a common problem in clinical movement analysis, and it is much more evident in FoG recognition tasks [17]. In fact, as discussed earlier, FoG is difficult to appreciate in supervised settings. For this reason, triggering situations, therapy suppression, and dual tasks are often used to elicit FoG. However, FoG is still an unpredictable, episodic phenomenon of short duration (in general). Thus, collecting a significant amount of FoG and comprehensively capturing all possible features and manifestations are rather difficult.

In addition to the quantity of data, the heterogeneity of such data is equally important. This aspect is particularly relevant when developing general models, which are trained on a set of subjects and can work on new, unseen subjects. Several studies showed that performance varies significantly among subjects and is particularly impaired in some patients due to their specific gait and motion patterns that are not well represented in the training set [53]. This is a common problem in PD, where a large variability in symptom manifestation is observed among PwPD. This suggests that more personalized approaches should be used that adapt to individual activity, gait, and FoG patterns. Some studies compared the performance of general and personalized models, proving that the latter provides significantly better results [70]. At the same time, personalized methods require training a specific model on each subject, and the need for adequate training data (in terms of quantity and heterogeneity) limits the applicability of these methods. In fact, the development of subject-specific models involves extensive data collection from that specific subject, data labeling by expert clinicians, optimization, and validation. This represents a huge effort, considering it should be done for each new subject. Instead, general models are preliminarily trained on a large amount of data from a set of subjects and then applied to new unseen subjects. However, they have limited accuracy and generalization capability [53].

At present, all public datasets provide only binary labels, i.e., FoG or non-FoG. This is enough to train binary supervised classification algorithms. However, accurate labeling of activities (e.g., walking, turning, postural transitions, standing, sitting) may provide extended detection capabilities, with improved performance and more efficient computing [52]. In this context, a first attempt can be represented by the tDCS dataset [62], where FoG is labeled according to the context (i.e., start a hesitation, turn FoG, FoG during straight walking). In this sense, a more comprehensive dataset is required to properly evaluate context-aware algorithms [71].

A final important limitation is the private algorithms developed for FoG detection. Although hundreds of approaches have been proposed, very few have been made available through public repositories. Sharing the source code provides a baseline, allows for a fair comparison between different approaches, and promotes reproducibility. It is therefore highly recommended.

6 Conclusions

FoG is one of the most intricate and debilitating symptoms of PD. To improve FoG monitoring and the quality of life of PwPD, significant efforts have been made to develop automatic FoG recognition systems. Several detection technologies and algorithms have been developed, optimized and validated using various experimental settings. In addition, public datasets shared with the research community have provided an essential resource for the evaluation of these methods. To date, some challenges remain, including the lack of a large public dataset that includes data recorded from a large sample, recorded through a comprehensive set of multimodal sensors, that includes careful labeling of contextual activities, representing different pharmacological conditions, and continuously monitoring subjects in an unconstrained daily life.

From the algorithmic point of view, significant efforts are needed to develop FoG recognition models that do not hardly rely on accurately labeled FoG data. Anomaly detection approaches may overcome the problem of having a large quantity of FoG by employing only gait data for model training. The binary (i.e., FoG vs non-FoG) class is thus transformed into a one-class problem (i.e., probability of being a normal gait instance). To the best of our knowledge, there is a single attempt to use anomaly detection for FoG detection. Specifically, in [73], a CNN-based deep gait anomaly detection model pre-trained with FoG and gait instances from a group of subjects was proposed and then fine-tuned on gait data from a different subject. This approach achieved an AUC of 0.82, a low sensitivity of 63%, and a high specificity of 98%. In addition, the algorithm was evaluated on a small sample of seven PwPD performing a simple walk back and forth task in a hallway, producing 154 FoG episodes. Further work is necessary to provide more robust results on a larger sample of PwPD. In general, a sound approach may consist of pre-training a general robust model on a large database comprising many PwPD and FoG and then fine-tuning (personalizing) the model using data from each specific subject. Nevertheless, limited work is available in this sense. Future studies are required to investigate the effect of fine-tuning and other unsupervised or semi-supervised strategies (e.g., active learning, one-shot learning) that provide robust results without extensive labeled data.

Finally, researchers are invited to evaluate the algorithms developed on different public datasets to enable a fair and comprehensive comparison of approaches. Furthermore, the calculation of computational complexity and memory usage, as well as inference time, is essential to assess the possibility of real-time data processing in stand-alone devices. Finally, the sharing of the proposed architecture, parameters, and implementation settings is crucial to accelerate progress in FoG recognition, with the aim of developing systems that are ready to function properly in real life.

References

1. Jankovic, J.: Parkinson's disease: clinical features and diagnosis. Journal of Neurology, Neurosurgery & Psychiatry **79**(4), 368–376 (2008). https://doi.org/10.1136/jnnp.2007.131045
2. Borzì, L., Olmo, G., Artusi, C.A., Fabbri, M., Rizzone, M.G., Romagnolo, A., Zibetti, M., Lopiano, L.: A new index to assess turning quality and postural stability in patients with Parkinson's disease. Biomedical Signal Processing and Control **62**, 102059 (2020). https://doi.org/10.1016/j.bspc.2020.102059
3. Chaudhuri, K., Healy, D., Schapira, A.: Non-motor symptoms of Parkinson's disease: diagnosis and management. The Lancet Neurology **5**(3), 235–245 (2006). https://doi.org/10.1016/S1474-4422(06)70373-8
4. Zhang, W.S., Gao, C., Tan, Y.Y., Chen, S.D.: Prevalence of freezing of gait in Parkinson's disease: a systematic review and meta-analysis. Journal of neurology **268**, 4138–4150 (2021). https://doi.org/10.1007/s00415-021-10685-5
5. Gao, C., Liu, J., Tan, Y., S, C.: Freezing of gait in Parkinson's disease: pathophysiology, risk factors and treatments. Translational Neurodegeneration **9** (2020). https://doi.org/10.1186/s40035-020-00191-5

6. Nutt, J.G., et al.: Freezing of gait: moving forward on a mysterious clinical phenomenon. The Lancet Neurology **10**(8), 734–744 (2011). https://doi.org/10.1016/S1474-4422(11)70143-0
7. Schaafsma, J.D., Balash, Y., Gurevich, T., Bartels, A.L., et al.: Characterization of freezing of gait subtypes and the response of each to levodopa in Parkinson's disease. European Journal of Neurology **10**(4), 391–398 (2003). https://doi.org/10.1046/j.1468-1331.2003.00611.x
8. Okuma, Y., Yanagisawa, N.: The clinical spectrum of freezing of gait in Parkinson's disease. Movement Disorders **23**(S2), S426–S430 (2008). https://doi.org/10.1002/mds.21934
9. Ehgoetz Martens, K.A., Ellard, C.G., Almeida, Q.J.: Does anxiety cause freezing of gait in Parkinson's disease? PLOS ONE **9**(9), 1–10 (2014). https://doi.org/10.1371/journal.pone.0106561
10. Demrozi, F., Bragoi, V., Tramarin, F., Pravadelli, G.: An indoor localization system to detect areas causing the freezing of gait in Parkinsonians. In: 2019 Design, Automation & Test in Europe Conference & Exhibition (DATE). pp. 952–955. IEEE (2019). https://doi.org/10.23919/DATE.2019.8715093
11. Nieuwboer, A., Rochester, L., Herman, T., Vandenberghe, W., Emil, G.E., Thomaes, T., Giladi, N.: Reliability of the new freezing of gait questionnaire: Agreement between patients with Parkinson's disease and their carers. Gait & Posture **30**(4), 459–463 (2009). https://doi.org/10.1016/j.gaitpost.2009.07.108
12. Zampogna, A., Borzì, L., Rinaldi, D., Artusi, C.A., Imbalzano, G., Patera, M., Lopiano, L., Pontieri, F., Olmo, G., Suppa, A.: Unveiling the unpredictable in Parkinson's disease: Sensor-based monitoring of dyskinesias and freezing of gait in daily life. Bioengineering **11**(5) (2024). https://doi.org/10.3390/bioengineering11050440
13. Barthel, C., Mallia, E., Debû, B., Bloem, B.R., Ferraye, M.U.: The practicalities of assessing freezing of gait. Journal of Parkinson's disease **6**(4), 667—674 (October 2016). https://doi.org/10.3233/jpd-160927
14. Amboni, M., et al.: Prevalence and associated features of self-reported freezing of gait in Parkinson disease: The deep fog study. Parkinsonism & Related Disorders **21**(6), 644–649 (2015). https://doi.org/10.1016/j.parkreldis.2015.03.028
15. Cui, C.K., Lewis, S.J.G.: Future therapeutic strategies for freezing of gait in Parkinson's disease. Frontiers in Human Neuroscience **15** (2021). https://doi.org/10.3389/fnhum.2021.741918
16. Sweeney, D., Quinlan, L.R., Browne, P., Richardson, M., et al.: A technological review of wearable cueing devices addressing freezing of gait in Parkinson's disease. Sensors **19**(6) (2019). https://doi.org/10.3390/s19061277
17. Cockx, H., Klaver, E., Tjepkema-Cloostermans, M., van Wezel, R., Nonnekes, J.: The gray area of freezing of gait annotation: A guideline and open-source practical tool. Movement Disorders Clinical Practice **9**(8), 1099–1104 (2022). https://doi.org/10.1002/mdc3.13556
18. Sigcha, L., Borzì, L., Amato, F., Rechichi, I., et al.: Deep learning and wearable sensors for the diagnosis and monitoring of Parkinson's disease: A systematic review. Expert Systems with Applications **229**, 120541 (2023). https://doi.org/10.1016/j.eswa.2023.120541
19. Pardoel, S., Kofman, J., Nantel, J., Lemaire, E.D.: Wearable-sensor-based detection and prediction of freezing of gait in Parkinson's disease: A review. Sensors **19**(23) (2019). https://doi.org/10.3390/s19235141
20. Zampogna, A., Borzì, L., Soares, C., Demrozi, F.: Editorial: High-tech personalized healthcare in movement disorders. Frontiers in Neurology **15** (2024). https://doi.org/10.3389/fneur.2024.1452612
21. Zhang, W., Sun, H., Huang, D., Zhang, Z., Li, J., Wu, C., Sun, Y., Gong, M., Wang, Z., Sun, C., et al.: Detection and prediction of freezing of gait with wearable sensors in Parkinson's disease. Neurological Sciences **45**(2), 431–453 (2024). https://doi.org/10.1007/s10072-023-07017-y
22. Martínez-Villaseñor, L., Ponce, H., Miralles-Pechuán, L.: A survey on freezing of gait detection and prediction in Parkinson's disease. In: Advances in Soft Computing: 19th Mexican International Conference on Artificial Intelligence, MICAI 2020, Mexico City, Mexico, October 12–17, 2020, Proceedings, Part I 19. pp. 169–181. Springer (2020). https://doi.org/10.1007/978-3-030-60884-2_13

23. Demrozi, F., Borzì, L., Olmo, G.: Wearable sensors for supporting diagnosis, prognosis, and monitoring of neurodegenerative diseases. Electronics **12**(6) (2023). https://doi.org/10.3390/electronics12061269
24. Polvorinos-Fernández, C., Sigcha, L., Borzì, L., Olmo, G., Asensio, C., López, J.M., de Arcas, G., Pavón, I.: Evaluating motor symptoms in Parkinson's disease through wearable sensors: A systematic review of digital biomarkers. Applied Sciences **14**(22) (2024). https://doi.org/10.3390/app142210189
25. Han, J.H., Lee, W.J., Ahn, T.B., Jeon, B.S., Park, K.S.: Gait analysis for freezing detection in patients with movement disorder using three dimensional acceleration system. In: Proceedings of the 25th Annual International Conference of the IEEE Engineering in Medicine and Biology Society (IEEE Cat. No. 03CH37439). vol. 2, pp. 1863–1865. IEEE (2003). https://doi.org/10.1109/IEMBS.2003.1279781
26. Moore, S.T., MacDougall, H.G., Ondo, W.G.: Ambulatory monitoring of freezing of gait in Parkinson's disease. Journal of neuroscience methods **167**(2), 340–348 (2008). https://doi.org/10.1016/j.jneumeth.2007.08.023
27. Bachlin, M., Roggen, D., Troster, G., Plotnik, M., Inbar, N., Meidan, I., Herman, T., Brozgol, M., Shaviv, E., Giladi, N., et al.: Potentials of enhanced context awareness in wearable assistants for Parkinson's disease patients with the freezing of gait syndrome. In: 2009 International Symposium on Wearable Computers. pp. 123–130. IEEE (2009). https://doi.org/10.1109/ISWC.2009.14
28. Mazilu, S., Hardegger, M., Zhu, Z., Roggen, D., Tröster, G., Plotnik, M., Hausdorff, J.M.: Online detection of freezing of gait with smartphones and machine learning techniques. In: 2012 6th International Conference on Pervasive Computing Technologies for Healthcare (PervasiveHealth) and Workshops. pp. 123–130. IEEE (2012)
29. Pepa, L., Ciabattoni, L., Verdini, F., Capecci, M., Ceravolo, M.G.: Smartphone based fuzzy logic freezing of gait detection in Parkinson's disease. In: 2014 IEEE/ASME 10th International Conference on Mechatronic and Embedded Systems and Applications (MESA). pp. 1–6. IEEE (2014)
30. Borzì, L., Varrecchia, M., Olmo, G., Artusi, C.A., Fabbri, M., Rizzone, M.G., Romagnolo, A., Zibetti, M., Lopiano, L.: Home monitoring of motor fluctuations in Parkinson's disease patients. Journal of Reliable Intelligent Environments **5**(3), 145–162 (2019)
31. Singh, J.P., Jain, S., Arora, S., Singh, U.P.: Vision-based gait recognition: A survey. Ieee Access **6**, 70497–70527 (2018)
32. Delval, A., Snijders, A.H., Weerdesteyn, V., Duysens, J.E., Defebvre, L., Giladi, N., Bloem, B.R.: Objective detection of subtle freezing of gait episodes in Parkinson's disease. Movement Disorders **25**(11), 1684–1693 (2010)
33. Amini, A., Banitsas, K., Young, W.R.: Kinect4fog: monitoring and improving mobility in people with Parkinson's using a novel system incorporating the microsoft kinect v2. Disability and Rehabilitation: Assistive Technology **14**(6), 566–573 (2019)
34. Soltaninejad, S., Cheng, I., Basu, A.: Kin-fog: Automatic simulated freezing of gait (fog) assessment system for Parkinson's disease. Sensors **19**(10), 2416 (2019)
35. Sun, R., Wang, Z., Martens, K.E., Lewis, S.: Convolutional 3d attention network for video based freezing of gait recognition. In: 2018 Digital Image Computing: Techniques and Applications (DICTA). pp. 1–7. IEEE (2018)
36. Hu, K., Wang, Z., Mei, S., Martens, K.A.E., Yao, T., Lewis, S.J., Feng, D.D.: Vision-based freezing of gait detection with anatomic directed graph representation. IEEE journal of biomedical and health informatics **24**(4), 1215–1225 (2019)
37. Tahir, A., Ahmad, J., Shah, S.A., Morison, G., Skelton, D.A., Larijani, H., Abbasi, Q.H., Imran, M.A., Gibson, R.M.: Wifreeze: Multiresolution scalograms for freezing of gait detection in Parkinson's leveraging 5g spectrum with deep learning. Electronics **8**(12), 1433 (2019)
38. Shah, S.A., Tahir, A., Ahmad, J., Zahid, A., Pervaiz, H., Shah, S.Y., Ashleibta, A.M.A., Hasanali, A., Khattak, S., Abbasi, Q.H.: Sensor fusion for identification of freezing of gait episodes using wi-fi and radar imaging. IEEE Sensors Journal **20**(23), 14410–14422 (2020)

39. Salarian, A., Russmann, H., Vingerhoets, F.J., Dehollain, C., Blanc, Y., Burkhard, P.R., Aminian, K.: Gait assessment in Parkinson's disease: toward an ambulatory system for long-term monitoring. IEEE transactions on biomedical engineering **51**(8), 1434–1443 (2004)
40. Bächlin, M., Hausdorff, J.M., Roggen, D., Giladi, N., Plotnik, M., Tröster, G.: Online detection of freezing of gait in Parkinson's disease patients: A performance characterization. In: Proceedings of the Fourth International Conference on Body Area Networks. pp. 1–8 (2009)
41. Tripoliti, E.E., Tzallas, A.T., Tsipouras, M.G., Rigas, G., Bougia, P., Leontiou, M., Konitsiotis, S., Chondrogiorgi, M., Tsouli, S., Fotiadis, D.I.: Automatic detection of freezing of gait events in patients with Parkinson's disease. Computer methods and programs in biomedicine **110**(1), 12–26 (2013)
42. Ahlrichs, C., Samà, A., Lawo, M., Cabestany, J., Rodríguez-Martín, D., Pérez-López, C., Sweeney, D., Quinlan, L.R., Laighin, G.Ò., Counihan, T., et al.: Detecting freezing of gait with a tri-axial accelerometer in Parkinson's disease patients. Medical & biological engineering & computing **54**, 223–233 (2016)
43. Rodríguez-Martín, D., Pérez-López, C., Samà, A., Català, A., Moreno Arostegui, J.M., Cabestany, J., Mestre, B., Alcaine, S., Prats, A., Cruz Crespo, M.d.l., et al.: A waist-worn inertial measurement unit for long-term monitoring of Parkinson's disease patients. Sensors **17**(4), 827 (2017)
44. Samà, A., Rodríguez-Martín, D., Pérez-López, C., Català, A., Alcaine, S., Mestre, B., Prats, A., Crespo, M.C., Bayés, À.: Determining the optimal features in freezing of gait detection through a single waist accelerometer in home environments. Pattern Recognition Letters **105**, 135–143 (2018)
45. Demrozi, F., Bacchin, R., Tamburin, S., Cristani, M., Pravadelli, G.: Toward a wearable system for predicting freezing of gait in people affected by Parkinson's disease. IEEE journal of biomedical and health informatics **24**(9), 2444–2451 (2019)
46. Borzì, L., Mazzetta, I., Zampogna, A., Suppa, A., Olmo, G., Irrera, F.: Prediction of freezing of gait in Parkinson's disease using wearables and machine learning. Sensors **21**(2), 614 (2021)
47. Kim, H.B., Lee, H.J., Lee, W.W., Kim, S.K., Jeon, H.S., Park, H.Y., Shin, C.W., Yi, W.J., Jeon, B., Park, K.S.: Validation of freezing-of-gait monitoring using smartphone. Telemedicine and e-Health **24**(11), 899–907 (2018)
48. Camps, J., Samà, A., Martín, M., Rodriguez-Martin, D., Pérez-López, C., Arostegui, J.M.M., Cabestany, J., Catala, A., Alcaine, S., Mestre, B., et al.: Deep learning for freezing of gait detection in Parkinson's disease patients in their homes using a waist-worn inertial measurement unit. Knowledge-Based Systems **139**, 119–131 (2018)
49. Mohammadian Rad, N., Van Laarhoven, T., Furlanello, C., Marchiori, E.: Novelty detection using deep normative modeling for imu-based abnormal movement monitoring in Parkinson's disease and autism spectrum disorders. Sensors **18**(10), 3533 (2018)
50. San-Segundo, R., Navarro-Hellín, H., Torres-Sánchez, R., Hodgins, J., De la Torre, F.: Increasing robustness in the detection of freezing of gait in Parkinson's disease. Electronics **8**(2), 119 (2019)
51. O'Day, J., Lee, M., Seagers, K., Hoffman, S., et al.: Assessing inertial measurement unit locations for freezing of gait detection and patient preference. Journal of NeuroEngineering and Rehabilitation **19**(20) (2022). https://doi.org/10.1186/s12984-022-00992-x
52. Borzì, L., Sigcha, L., Olmo, G.: Context recognition algorithms for energy-efficient freezing-of-gait detection in Parkinson's disease. Sensors **23**(9) (2023). https://doi.org/10.3390/s23094426
53. Sigcha, L., Borzi, L., Olmo, G.: Deep learning algorithms for detecting freezing of gait in Parkinson's disease: A cross-dataset study. Expert Systems with Applications **255**, 124522 (2024). https://doi.org/10.1016/j.eswa.2024.124522
54. Li, B., Yao, Z., Wang, J., Wang, S., Yang, X., Sun, Y.: Improved deep learning technique to detect freezing of gait in Parkinson's disease based on wearable sensors. Electronics **9**(11), 1919 (2020)

55. Sigcha, L., Costa, N., Pavón, I., Costa, S., Arezes, P., López, J.M., De Arcas, G.: Deep learning approaches for detecting freezing of gait in Parkinson's disease patients through on-body acceleration sensors. Sensors **20**(7), 1895 (2020)
56. Sigcha, L., Borzì, L., Pavón, I., Costa, N., Costa, S., Arezes, P., López, J.M., De Arcas, G.: Improvement of performance in freezing of gait detection in Parkinson's disease using transformer networks and a single waist-worn triaxial accelerometer. Engineering Applications of Artificial Intelligence **116**, 105482 (2022)
57. Hu, K., Wang, Z., Martens, K.A.E., Hagenbuchner, M., Bennamoun, M., Tsoi, A.C., Lewis, S.J.: Graph fusion network-based multimodal learning for freezing of gait detection. IEEE Transactions on Neural Networks and Learning Systems **34**(3), 1588–1600 (2021)
58. Yang, P.K., Filtjens, B., Ginis, P., Goris, M., Nieuwboer, A., Gilat, M., Slaets, P., Vanrumste, B.: Freezing of gait assessment with inertial measurement units and deep learning: effect of tasks, medication states, and stops. Journal of NeuroEngineering and Rehabilitation **21**(1), 24 (2024)
59. Sun, R., Wang, Z., Martens, K.E., Lewis, S.: Convolutional 3d attention network for video based freezing of gait recognition. In: 2018 Digital Image Computing: Techniques and Applications (DICTA). pp. 1–7. IEEE (2018)
60. Hu, K., Wang, Z., Mei, S., Martens, K.A.E., Yao, T., Lewis, S.J., Feng, D.D.: Vision-based freezing of gait detection with anatomic directed graph representation. IEEE journal of biomedical and health informatics **24**(4), 1215–1225 (2019)
61. Bächlin, M., Plotnik, M., Roggen, D., Maidan, I., et al.: Wearable assistant for Parkinson's disease patients with the freezing of gait symptom. IEEE Trans Inf Technol Biomed **14**(2), 436–446 (2010). https://doi.org/10.1109/TITB.2009.2036165
62. Reches, T., Dagan, M., Herman, T., Gazit, E., et al.: Using wearable sensors and machine learning to automatically detect freezing of gait during a fog-provoking test. Sensors **20**(16) (2020). https://doi.org/10.3390/s20164474
63. Ribeiro De Souza, C., Miao, R., Ávila De Oliveira, J., Cristina De Lima-Pardini, A., Fragoso De Campos, D., Silva-Batista, C., Teixeira, L., Shokur, S., Mohamed, B., Coelho, D.B.: A public data set of videos, inertial measurement unit, and clinical scales of freezing of gait in individuals with Parkinson's disease during a turning-in-place task. Frontiers in Neuroscience **16** (2022). https://doi.org/10.3389/fnins.2022.832463
64. Zhang, W., Yang, Z., Li, H., Huang, D., Wang, L., Wei, Y., Zhang, L., Ma, L., Feng, H., Pan, J., et al.: Multimodal data for the detection of freezing of gait in Parkinson's disease. Scientific data **9**(1), 606 (2022)
65. Borzì, L., Sigcha, L., Rodríguez-Martín, D., Olmo, G.: Real-time detection of freezing of gait in Parkinson's disease using multi-head convolutional neural networks and a single inertial sensor. Artificial Intelligence in Medicine **135**, 102459 (2023). https://doi.org/10.1016/j.artmed.2022.102459
66. Mancini, M., Shah, V.V., Stuart, S., Curtze, C., Horak, F.B., Safarpour, D., Nutt, J.G.: Measuring freezing of gait during daily-life: an open-source, wearable sensors approach. Journal of neuroengineering and rehabilitation **18**, 1–13 (2021)
67. Borzì, L., Olmo, G., Artusi, C., Lopiano, L.: Detection of freezing of gait in people with Parkinson's disease using smartphones. 2020 IEEE 44th Annual Computers, Software, and Applications Conference (COMPSAC) pp. 625–635. (2020). https://doi.org/10.1109/COMPSAC48688.2020.0-186
68. Shi, B., Tay, A., Au, W.L., Tan, D.M.L., et al.: Detection of freezing of gait using convolutional neural networks and data from lower limb motion sensors. IEEE Transactions on Biomedical Engineering **69**(7), 2256–2267 (2022). https://doi.org/10.1109/TBME.2022.3140258
69. D'Cruz, N., Seuthe, J., De Somer, C., Hulzinga, F., et al.: Dual task turning in place: A reliable, valid, and responsive outcome measure of freezing of gait. Movement Disorders **37**(2), 269–278 (2022). https://doi.org/10.1002/mds.28887
70. Rodríguez-Martín, D., Samà, A., Pérez-López, C., Català, A., et al.: Home detection of freezing of gait using support vector machines through a single waist-worn triaxial accelerometer. PLOS ONE **12**(2), 1–26 (02 2017). https://doi.org/10.1371/journal.pone.0171764

71. Borzí, L., Demrozi, F., Bacchin, R.A. et al. A multi-level annotated sensor dataset of gait freezing manifestations and severity in Parkinson's disease. Scientific Data (2026). https://doi.org/10.1038/s41597-026-06645-1
72. Salomon, A., Gazit, E., Ginis, P., et al.: A machine learning contest enhances automated freezing of gait detection and reveals time-of-day effects. Nature Communications **15**, 4853 (2024). https://doi.org/10.1038/s41467-024-49027-0
73. Naghavi, N., Wade, E.: Towards real-time prediction of freezing of gait in patients with Parkinson's disease: a novel deep one-class classifier. IEEE Journal of Biomedical and Health Informatics **26**(4), 1726–1736 (2021)

Computation in Memory for Reliable and Energy-Efficient Diabetic Retinopathy Screening

Sumit Diware, Koteswararao Chilakala, and Rajendra Bishnoi

> *"Innovation is taking two things that exist and putting them together in a new way."*
>
> *—Tom Freston*

Abstract Diabetic retinopathy (DR) refers to permanent vision loss resulting from high glucose and blood pressure. As it is a major cause of blindness worldwide, regular screening for DR is crucial to its timely treatment. Neural network-based classifiers offer significant potential to automate this process. Such classifiers must ensure consistent performance throughout development and deployment, provide explanatory feedback for predictions, and support usability for widespread adoption. Energy-efficient deployment on edge devices is also essential for global accessibility. In this chapter, we present a reliable and energy-efficient hardware for DR screening on edge devices. We first develop a DR classification model using custom training data that incorporates diverse image quality and image sources along with improved class balance. This enables effective handling of both on-field variations in retinal images and minority DR classes, enhancing post-deployment reliability. We then present a pseudo-binary classification scheme to further improve the model performance and provide supplementary information about the predictions. Additionally, we introduce an energy-efficient hardware design for our model using memristor-based computation-in-memory for deployment on edge devices.

Keywords Diabetic retinopathy screening · Computation-in-memory (CIM) · Memristors · Edge AI · Energy-efficient hardware · Multiclass medical classification

S. Diware (✉) · K. Chilakala · R. Bishnoi
Computer Engineering Laboratory, Delft University of Technology, Delft, The Netherlands
e-mail: S.S.Diware@tudelft.nl

B. Farahani et al. (eds.), *Smart and Connected Health: Applications and Real-World Use Cases*, https://doi.org/10.1007/978-3-032-09253-3_8

1 Introduction

Diabetic retinopathy (DR) is a medical condition characterized by irreversible damage to the retina due to high glucose levels and elevated blood pressure. It stands as a major cause of permanent vision loss worldwide, with projections indicating that 70 million individuals will be affected by 2045 [1]. Also, all individuals with diabetes are especially at risk of developing DR [2]. Since vision loss due to DR cannot be recovered, early detection is vital for timely interventions to avert further retinal damage. Consequently, regular screenings for DR are critical for identifying this condition at early stages. Recent breakthroughs in artificial intelligence have paved the way for developing automated systems for quick, effective, and convenient DR screening. These systems utilize neural networks due to their inherent proficiency in classification tasks to effectively identify DR categories from retinal images.

Publicly available DR datasets exhibit inherent image inconsistencies to pose a tougher classification challenge than private ones, resulting in more robust and adaptive models. Moreover, their wider accessibility is valuable for driving further innovation in automated DR classification. Hence, we focus on DR classification literature based on publicly available datasets. Such works face reliability issues, where the model performs well during development but exhibits poor performance upon deployment. This can arise due to several factors such as small training data size [2–8], absence of external test data [9–13], and lack of diversity in training data [14, 15]. Furthermore, the inherent class imbalance in public datasets can bias the model performance toward majority classes. This can hinder the identification of minority DR classes (indicating retinal damage), further aggravating the reliability concern. Additionally, supplementary information about model prediction is crucial for widespread adoption of automated DR classification [13, 16], for instance, when a DR model assists human specialists in double reading [17], supplementary information bolsters specialist's confidence when human diagnosis matches the model prediction and helps resolve conflicts when these two differ. However, the aforementioned works do not address the inclusion of additional information about the prediction of the model. Lastly, deploying automated DR classification on portable edge devices can address the global scarcity of DR screening facilities [18, 19]. This requires energy-efficient hardware design of DR classifiers to achieve uninterrupted operation despite limited energy resources, enabling large-scale screening programs even in remote regions. Nevertheless, the aforementioned works focus only on software model development.

In this chapter, we present a reliable and energy-efficient DR screening hardware targeting deployment on edge devices. We first develop a reliable DR classification model via training on a newly created custom dataset. This dataset encompasses image quality inconsistencies, diverse image sources, and reduced class imbalance. This enables our trained model to effectively handle real-world retinal images and perform well on minority classes, ensuring post-deployment reliability. Furthermore, we introduce a pseudo-binary classification scheme that internally uses multiclass classification to achieve binary screening. This enhances our model's classification performance and also provides supplementary information to aid its wider adoption. We then present energy-efficient hardware design of our model based on computation-in-memory (CIM) paradigm. It uses emerging memory devices known as memristors to perform computations directly within the memory. This eliminates the data transfer bottleneck and provides superior energy efficiency suitable for edge device deployment. Results show that we achieve reliable DR classification while consuming three orders of magnitude less energy compared to the state-of-the-art hardware platforms.

The rest of the chapter is structured as follows: Sects. 2 and 3 present the basics of diabetic retinopathy and memristor-based computation in memory, while Sect. 4 provides a review of related existing literature. A new methodology is described in Sect. 5, followed by simulation setup details in Sect. 6 and simulation results in Sect. 7. Finally, Sect. 8 concludes the chapter.

2 Diabetic Retinopathy Basics

Diabetic retinopathy (DR) is a permanent condition triggered by high glucose levels and hypertension. It damages the blood vessels in the retina and leads to irreversible vision loss. There exist five distinct severity levels (classes) of DR according to international guidelines [21]:

- No DR (DR-0)
- Mild nonproliferative (DR-1)
- Moderate nonproliferative (DR-2)
- Severe nonproliferative (DR-3)
- Proliferative (DR-4)

These classes are distinguished from each other based on certain features present in the retina, known as lesions. Figure 1 depicts the four most common lesions: microaneurysms, hemorrhages, hard exudates, and soft exudates. They can be described as follows [2, 20, 22]:

- Microaneurysms (MA): These are the earliest visible signs of retinal damage. They manifest as tiny red dots arising due to capillary dilation.
- Hemorrhages (HM): These are red spots with irregular margins and/or uneven density. They are bigger than MA and occur due to the leakage of weak capillaries.
- Hard exudates (HE): These are yellow-white deposits in outer layers of the retina caused by leakage of plasma.
- Soft exudates (SE): These are grayish oval or round-shaped patches arising due to the swelling of the nerve fiber. They are also called cotton wool spots.

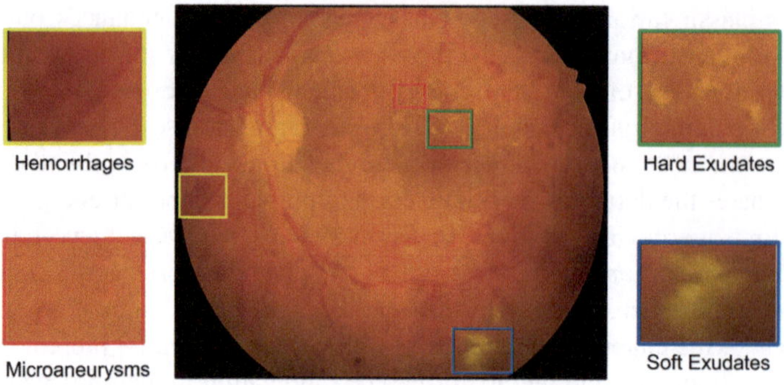

Fig. 1 Retinal image annotated with the four most common lesions: microaneurysms, hemorrhages, hard exudates, and soft exudates [20]

Table 1 Lesion-based diagnosis of DR classes

DR severity level	DR class	Lesion-based diagnosis
No DR	DR-0	No lesions
Mild nonproliferative	DR-1	Only MA present
Moderate nonproliferative	DR-2	MA and other lesions present Less prominence than DR-3
Severe nonproliferative	DR-3	At least one of the following present: HM (>20 in each quadrant) Blood spillage (>2 quadrants) No indicators of DR-4
Proliferative	DR-4	Vitreous/preretinal HM and/or NV

- Neovascularization (NV): It refers to the abnormal growth of new blood vessels on the inner surface of the retina. Such blood vessels often bleed into the vitreous cavity and lead to obscured vision.

Table 1 shows how various DR classes are diagnosed based on the composition of these lesions. Conventional DR detection typically begins with pre-capture medical procedures on patients to enlarge their pupils and facilitate better coverage of the retinal area during image capture. Skilled operators then employ specialized fundus cameras and meticulously adjust settings, such as focus, exposure, alignment, etc., to capture high-quality retinal images. Subsequently, the severity of DR is evaluated through visual inspection of various lesions within the captured retinal image. The advancements in artificial intelligence have opened avenues to employ automated systems to achieve DR identification from retinal images. They leverage neural networks, which inherently excel at extracting crucial lesion information from retinal images and autonomously categorize them into distinct DR classes. Thus, neural network-based DR classification systems offer an effective and efficient approach to DR detection. Deploying these classifiers on edge devices can significantly improve the accessibility of DR screening. However, this requires the classifier module to cope with the limited energy resources on these battery-powered devices. Memristor-based computation-in-memory (CIM) paradigm can overcome this challenge to facilitate energy-efficient hardware design for neural networks. We will delve into the fundamentals of memristor-based CIM in the next subsection.

3 Fundamentals of Computation in Memory (CIM)

3.1 CIM Paradigm

Neural networks have conventionally been implemented on hardware platforms including CPUs [23], GPUs [24], and dedicated ASICs like TPUs [25], which are based on the von Neumann architecture and utilize CMOS technology. This architecture features separate units for memory and computation, resulting in frequent data transfers during vector-matrix multiplication (VMM) operations in neural networks. Such transfers contribute to high energy consumption, given that VMM tasks represent a significant portion of overall computations in these networks [26–30]. Additionally, the challenges of CMOS technology include significant subthreshold leakage, and scalability issues [31–40]. *Computation-in-memory* (CIM) paradigm leverages emerging memory technologies, such as resistive random access memories (RRAMs), also called memristors [41–51], to address these limitations. It performs computations directly within the memory, eliminating the data transfer bottleneck. Furthermore, memristors are nonvolatile, offer high scalability, and are compact in size. Hence, CIM emerges as a highly promising alternative to conventional hardware for neural network implementation [52–58].

3.2 CIM System Architecture

A typical CIM system architecture depicted in Fig. 2 uses a tiled structure to exploit the inherent parallelism in neural network computations. It consists of tiles, functional units, global buffer, and global control logic. Tiles are responsible for performing vector-matrix multiplication (VMM) operations. Functional units perform other computations such as summing up partial outputs from tiles, pooling, or activation function. Global buffer serves as a storage for input/output feature maps, while global control logic governs the interactions among all system components.

Each tile is made up of CIM-based VMM units, input buffer, output buffer, tile aggregation units, and control logic. This structure allows distribution of the workload across multiple VMM units for parallel execution. Control logic regulates the operation of all the tile components. Additionally, it synchronizes control sequences with the global control logic. VMM units receive input data from the input buffer. Their outputs are post-processed by tile aggregation unit through

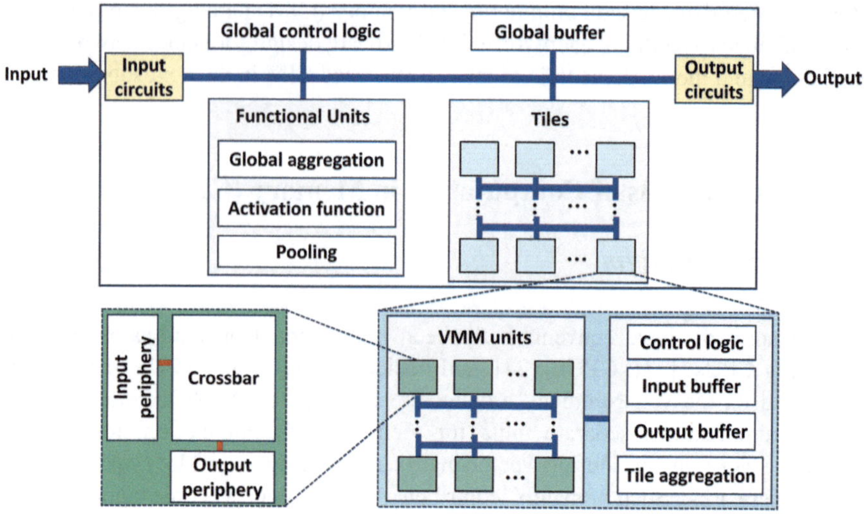

Fig. 2 System architecture of a typical memristor-based computation-in-memory (CIM) hardware

appropriate shift-and-add operations. The final aggregated outputs are stored in the output buffer. Each VMM unit includes control logic, a memristive memory array called crossbar and peripheral circuits. The crossbar holds the network weights, enabling VMM between weights and inputs (activations) in the analog domain. The peripheral circuits act as interface between the crossbar and the rest of the system. They first convert digital inputs from input buffer into analog domain to perform VMM within the crossbar. Subsequently, they convert analog VMM outputs back into digital domain for further processing in the system. This approach allows the VMM unit to benefit from energy efficiency of analog computations while maintaining compatibility with surrounding digital system.

The operation of this system can be described as follows: The global control unit first distributes the input data from global buffer across all tiles. Each tile stores the assigned inputs in its input buffer and feeds it into its VMM units. The partial outputs generated by the VMM units within each tile are then combined and stored in the tile output buffers. After all tiles finish their VMM computations, their output buffer data is consolidated by global aggregation units. This combined data is then passed through activation and/or pooling units, and the resulting outputs are stored back in the global buffer.

3.3 *CIM-Based Vector-Matrix Multiplication (VMM) Unit*

We will now explore the working of CIM-based VMM unit shown in Fig. 3, which is the core component of CIM system architecture. It utilizes memristor-based memory cells for data storage. These cells are arranged in a grid-like structure known as crossbar to enable dense data storage and parallel computation. Within the crossbar, weight matrix is stored as conductance values (Gs) within each memory cell. Due to the limited bit capacity of memristor devices, multi-bit weights are typically split across multiple memory cells in a single row, for instance, a 2-bit weight is distributed across two 1-bit memristors. This technique is known as bit-slicing [59–61]. The input vector is converted into voltages (Vs) using digital-to-analog converters (DACs) and applied to crossbar rows. Multi-bit inputs are often divided into smaller chunks and applied to crossbar using time-division multiplexing, due to the limitations associated with input voltage range and high-

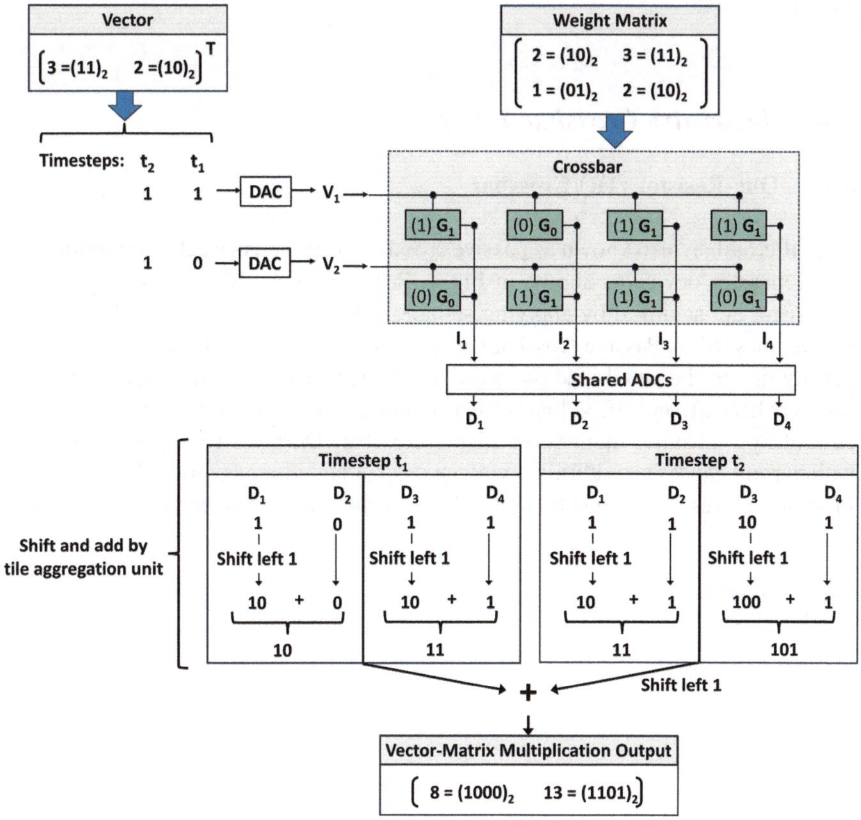

Fig. 3 Illustration of vector-matrix multiplication operation on CIM hardware

resolution DACs, for example, a 2-bit input is divided into single bit chunks that are converted to voltages and applied to the crossbar rows across two timesteps.

At each timestep, the current flowing through each G is equivalent to element-wise multiplication of Vs and Gs as per Ohm's law. Currents from all the G's in a column get accumulated according to Kirchhoff's law to produce output currents (Is). These aggregated currents represent multiply-and-accumulate (MAC) operations in analog domain. The MAC operations across all the columns together represent a partial VMM operation (for that specific timestep). These partial VMM outputs are converted to digital domain by analog-to-digital converters (ADCs) [59, 62–67]. The ADCs are typically shared across multiple columns due to their large pitch size compared to the crossbar. The ADC outputs undergo shift-and-add operation to account for weight slicing across multiple columns. An additional round of shift-and-add operations then merges this output with that of previous timestep to account for time-division multiplexing of input bits. The final full precision output is obtained by repeating this process till the last timestep of input vector. We now present three types of crossbar designs used in CIM-based VMM unit next.

3.4 *Memristive Crossbar Types*

3.4.1 One-Resistor (1R) Crossbar

The 1R crossbar, also known as passive crossbar, uses memory cells containing only the memristor device as shown in Fig. 4. Programming and reading of a specific memristor are achieved by applying suitable voltages at the bit lines (BLs) and the source lines (SLs). Passive crossbar can achieve the best area efficiency and memory density due to the small size of memristor devices. However, it suffers sneak path issues, where SL and BL voltages to program/read a specific memristor also induce current flow in other memristors of the crossbar. During write operations, sneak path current causes an additional voltage drop along wire parasitics. This can result in write failures due to insufficient voltage at the intended memristor. During read

Fig. 4 One-resistor (1R) crossbar design

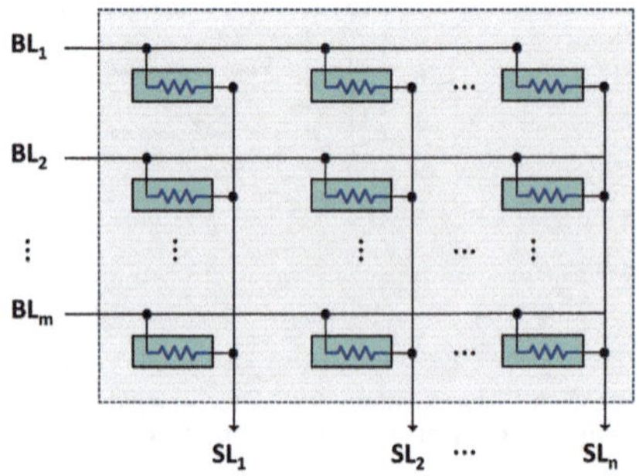

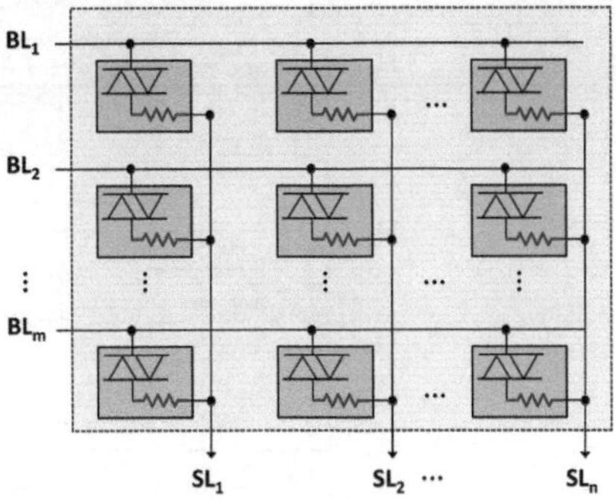

Fig. 5 One-selector and one-resistor (1S1R) crossbar design

operations, sneak currents from other memristors introduce deviations in the read current of the selected memristor. This can cause read failure due to incorrect interpretation of the target memristor's conductance state. Thus, sneak paths present a significant challenge for widespread adoption of 1R crossbars. This issue can be mitigated by integrating an additional selector device in series with the memristor, giving rise to the two crossbar types discussed next.

3.4.2 One-Selector and One-Resistor (1S1R) Crossbar

The selector is basically a device with bidirectional diode-like I–V characteristics [68]. The memory cells in 1S1R crossbar integrate such a selector device in series with each memristor as shown in Fig. 5. This eliminates the sneak path issue. Moreover, the selector device can be stacked on top of the memristor, potentially yielding a high memory density and small area footprint. However, development of selector devices that meet the desired performance requirements remains a significant challenge. This limitation hampers the widespread adoption of 1S1R arrays.

3.4.3 One-Transistor and One-Resistor (1T1R) Crossbar

This crossbar incorporates memory cells composed of transistors in series with the memristors as shown in Fig. 6. It requires additional control lines, known as word lines (WLs), to selectively activate or deactivate the transistors. This approach effectively mitigates the sneak paths, ensuring accurate programming and reading of the memristors. Although the inclusion of a transistor in each memory cell leads to a larger area footprint for 1T1R crossbar, it still remains the most preferred choice in CIM designs. This widespread adoption is driven by maturity CMOS fabrication processes, enabled by high-yield production of large-scale 1T1R crossbars. The next subsection presents the device technologies used for implementing memristors in the crossbar memory cells and their commercial availability.

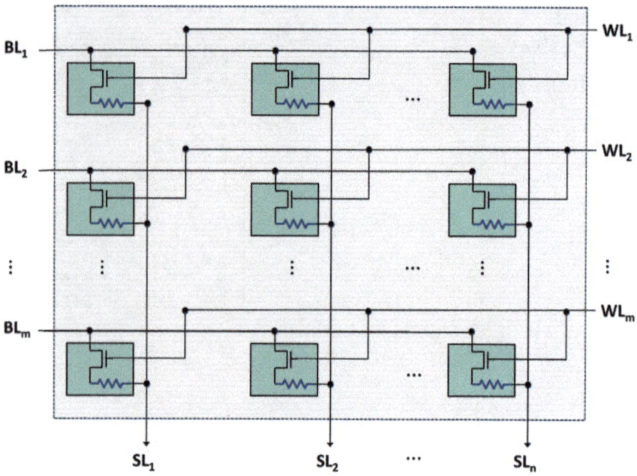

Fig. 6 One-transistor and one-resistor (1T1R) crossbar design

3.5 *Memristor Device Technologies*

3.5.1 Resistive Random Access Memory (RRAM)

The RRAM device is made up of an oxide material sandwiched between two metal electrodes [69–72], as shown in Fig. 7. Its conductance can be modulated by creating and disrupting the conductive filament (CF) composed of oxygen vacancies within the oxide layer. RRAM device exhibits a high conductance (logic 1) when CF connects the electrodes, while a disrupted CF leads to low conductance (logic 0). The SET process achieves high conductance state through a high electric field that forces oxygen ions to drift toward one electrode. This leaves behind vacant oxygen sites in oxide layer, which form CF to increase its conductivity. The RESET process leads to low conductance state using an electric field with polarity opposite to that in SET process. This causes oxygen ions to migrate back into the oxide layer. These ions combine with oxygen vacancies to disrupt the CF, reducing the conductivity of oxide layer. To read data from RRAM device without altering the

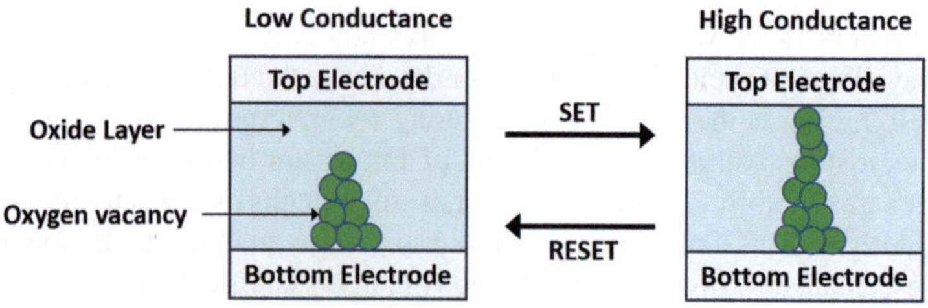

Fig. 7 Resistive random access memory (RRAM) device

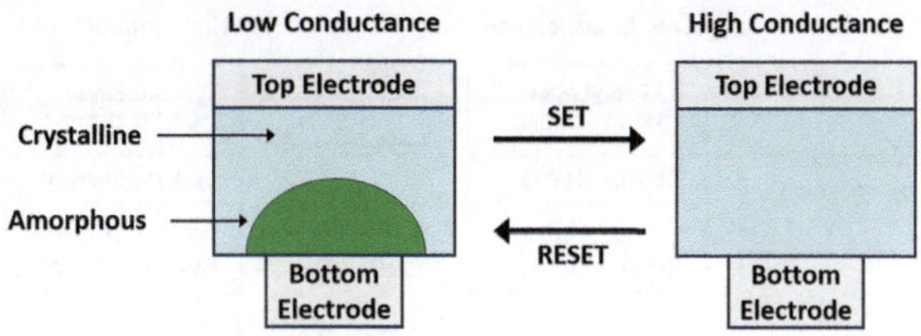

Fig. 8 Phase-change memory (PCM) device

stored information, a small voltage is applied across it to detect its conductance state.

3.5.2 Phase-Change Memory (PCM)

The PCM device consists of a top electrode, a phase-change material layer (e.g., $Ge_2Sb_2Te_5$), and a bottom electrode [73], as shown in Fig. 8. The phase-change material exhibits high conductance in crystalline phase (logic 1) and low conductance in amorphous phase (logic 0).

The transition from low to high conductance state is termed the SET process. Here, a current pulse heats the amorphous material above its crystallization threshold but below its melting point for a duration sufficient enough for crystallization to occur. Conversely, transitioning from high to low conductance state involves a high-current pulse with an abrupt trailing edge. This pulse melts the phase-change material through Joule heating, followed by rapid cooling to solidify it in the amorphous state. To read the state of the PCM device without disturbing its existing state, a small electrical current is passed through it for detecting its conductance.

3.5.3 Spin-Transfer Torque Magnetic Random Access Memory (STT-MRAM)

The STT-MRAM device consists of an oxide barrier sandwiched between two ferromagnetic layers [74–81], as shown in Fig. 9. One ferromagnetic layer (called reference layer) maintains a fixed magnetic field direction, while the other ferromagnetic layer (called free layer) has a programmable magnetic field direction. A parallel (P) alignment of magnetic fields in these layers yields high conductance (logic 1), while an antiparallel (AP) alignment results in low conductance (logic 0).

Data is written to the STT-MRAM device using current, which must exceed a minimum threshold value to switch the magnetic field direction of the free layer. For the SET operation (changing from AP to P orientation), a write current surpassing this threshold flows from the reference layer to the free layer [82–89]. The reference layer acts as a spin filter, only allowing electrons aligned with its magnetic field to pass through. These filtered electrons then apply spin-transfer torque to the free

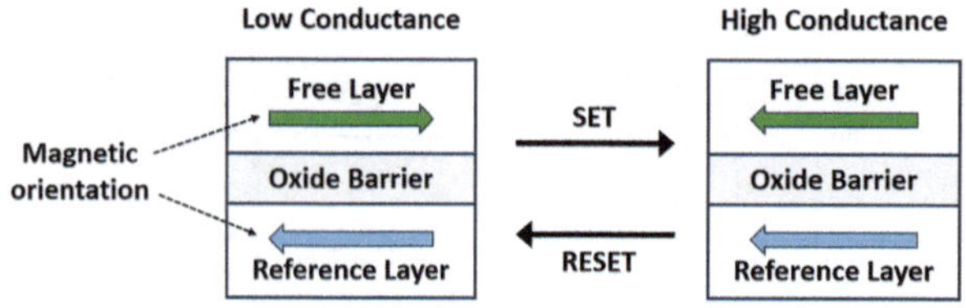

Fig. 9 Spin-transfer torque magnetic random access memory (STT-MRAM) device

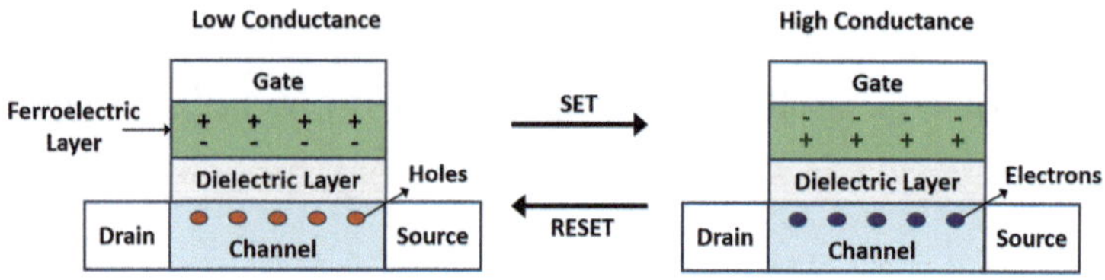

Fig. 10 Ferroelectric field-effect transistor (FeFET) device

layer, altering its magnetic field direction. This results in parallel orientation of the two magnetic layers, which leads to high conductance [90–99]. In the RESET operation (changing from P to AP orientation), a write current above the threshold flows from the free layer to the reference layer. Out of the electrons emerging from the free layer, reference layer reflects back those with spins opposite to its magnetic field [100–103]. Such reflected electrons apply spin-transfer torque to the free layer. This alters free layer magnetic field direction toward antiparallel orientation, resulting in low conductance [90, 91, 104–108]. Reading an STT-MRAM device without disturbing its existing state involves applying a small current to detect its conductance.

3.5.4 Ferroelectric Field-Effect Transistor (FeFET)

FeFET device is basically a transistor with an additional ferroelectric (FE) layer in MOS structure [109], as shown in Fig. 10. The polarization of FE layer can be modulated by applying gate-to-source voltage (V_{GS}) exceeding a coercive voltage (V_C). It manifests high conductance (logic 1) if FE layer polarization helps in channel formation while exhibits low conductance (logic 0) if it opposes channel formation.

In an nMOS-based FeFET (n-FeFET), SET process involves a positive V_{GS} ($V_{GS} > V_C$) to polarize FE layer in the direction pointing to the channel. This assists the electrons in the substrate to form a channel, leading to high conductance state. Conversely, RESET process uses a negative V_{GS} ($|V_{GS}| > V_C$) to polarize the FE layer in the direction pointing to the gate terminal. This obstructs the electrons from forming a channel, resulting in low conductance state. Similar concept can be directly extended to a pMOS-based FeFET (p-FeFET). Reading the data from

Table 2 Examples of commercial fabrication processes for memristor device technologies

Memristor device	Fabrication process
RRAM	TSMC 40/28/22 nm [114–116], Intel 22 nm [117]
PCM	TSMC 40 nm [118], STMicroelectronics 28 nm [119]
STT-MRAM	TSMC 22 nm [120], Globalfoundries 22 nm [121], Samsung 28 nm [122]
FeFET	Globalfoundries 28/22 nm [123, 124]

FeFET device without disturbing its state involves applying a small read voltage at its gate to sense the drain-to-source current (I_{DS}) [110].

3.5.5 Memristor Fabrication

Memristor fabrication processes are now available from leading manufacturers like TSMC, Intel, STMicroelectronics, Globalfoundries, and Samsung as shown in Table 2. These processes typically offer 1-bit storage per memristor. Multi-level cell capability, where a single memristor stores multiple bits using intermediate conductance states, has drawn particular interest from the research community. Efforts in this direction have shown promising results for RRAM and PCM technologies [111–113], signifying a pivotal step toward advancement of CIM architectures.

4 Existing Works

The effectiveness of neural network-based DR classification solutions can be significantly enhanced by publicly available DR datasets. These dataset exhibit image inconsistencies due to factors like varying equipment quality and operator expertise, making classification more difficult compared to private datasets collected under controlled conditions. This complexity also mimics the real-world data characteristics, ultimately leading to the development of more robust and adaptive models. Furthermore, the broad availability of these datasets to the research community can facilitate the creation of innovative DR classification solutions. Hence, we focus on the literature related to DR classification that utilizes publicly available datasets. These works suffer from several challenges like model reliability issue, lack of supplementary information, and scarcity of DR screening facilities. These challenges are discussed in detail next, with a summary provided in Table 3.

Model reliability issue arises when the model performs well during development phase but struggles after deployment. Existing works that are susceptible to reliability issue can be organized into three categories: The first category involves works using small datasets for model development. This hinders model's generalization ability leading to reliability concern. Datasets like APTOS [125], Messidor-2 [126–

Table 3 Summary of the related works on neural network-based DR classification

Paper	Large dataset	Diverse dataset	External testing	Supplementary information	Hardware design consideration
Feng et al. [3]	×	×	×	×	×
Khan et al. [4]	×	×	×	×	×
Wong et al. [5]	×	×	×	×	×
Islam et al. [6]	×	×	×	×	×
Mohammadi et al. [7]	×	×	×	×	×
Zhou et al. [2]	×	×	×	×	×
Sikder et al. [8]	×	×	×	×	×
Poranki et al. [9]	✓	×	×	×	×
Sadeghzadeh et al. [10]	✓	×	×	×	×
Sait et al. [11]	✓	×	×	×	×
Alyoubi et al. [12]	✓	×	×	×	×
Li et al. [13]	✓	×	×	×	×
Korot et al. [14]	✓	×	✓	×	×
Ludwig et al. [15]	✓	×	✓	×	×
This work	✓	✓	✓	✓	✓

128], HRF [129], small-scale merger of EyePACS with Messidor-2 [130], and FGADR [2] are small-sized. Several studies exemplify this issue, for instance, Feng et al. [3] present a cascaded convolutional and graph neural networks using APTOS and Messidor-2 datasets. Khan et al. [4] perform transfer learning using HRF and APTOS datasets. Wong et al. [5] propose transfer learning with optimized feature weights on APTOS dataset and small-scale merger of EyePACS and Messidor-2 datasets. Islam et al. [6] evaluate supervised contrastive learning on APTOS and Messidor-2 datasets. Mohammadi et al. [7] use Messidor-2 dataset to explore automated machine learning platforms. Zhou et al. [2] introduce FGADR dataset and its benchmarking. Sikder et al. [8] present decision tree-based ensemble learning using APTOS dataset.

The second category involves works that use large training datasets like EyePACS [131] or DDR [13] but do not use an external test data. External test data with a different origin is necessary to expose the model to truly unseen characteristics, mimicking real-world deployment. The internal test data is not enough as it shares common origin and characteristics with training/validation data. As a result, generalization ability of the model is not properly assessed leading to reliability issue. Many works suffer from this challenge. Poranki et al. [9] use EyePACS dataset to develop hybrid learning model combining graph learning and deep learning. Sadeghzadeh et al. [10] propose a hybrid of convolutional and transformer network evaluated on EyePACS dataset. Sait et al. [11] train MobileNet V3-based network on EyePACS dataset. Alyoubi et al. [12] use DDR dataset for convolutional neural network-based classification. Li et al. [13] present DDR dataset and its benchmarking.

The third category includes works that incorporate both large training dataset and external test dataset but lack diversity in training data. This affects the reliability by hindering the model's adaptability to post-deployment data, for instance, Korot et al. [14] use EyePACS datasets for model development and IDRiD dataset [132] for external validation. However, relying on a single training dataset restricts the data diversity. Similarly, Ludwig et al. [15] use EyePACS and APTOS datasets for model development while employing Messidor-2 dataset for external validation. However, a significantly larger number of EyePACS images compared to APTOS ones dominate the training data, thus restricting its diversity. Moreover, the inherent class imbalance in all public datasets can lead to the model excelling at handling majority classes while struggling with minority DR classes. This can aggravate the reliability issue by reducing model's effectiveness in identifying instances of retinal damage.

Double reading is a process where two specialists (readers) independently analyze the same recording (e.g., retinal image) and exchange their insights to reach a unified diagnosis [133, 134]. Neural network models can streamline double reading for DR screening by replacing one of the human specialists [17]. However, the remaining human specialist has no insight into the rationale behind model's diagnosis due to the lack of supplementary information. This black box nature of model predictions hinders collaborative reasoning and building trust with the specialist, reducing effectiveness of its assistance [16]. Moreover, supplementary information is also crucial for instilling confidence in patients [13]. Despite such importance, none of the aforementioned works incorporate supplementary information in their output.

There exists a global scarcity of DR screening facilities due to the shortage of specialists, inadequate medical infrastructure, and economic constraints [18]. Deploying automated DR classification on portable edge devices offers a promising solution to this challenge. These devices can integrate portable imaging technology [19] with a dedicated DR classification chip. This chip must exhibit energy efficiency to ensure prolonged operation, even with limited battery resources and unstable power grids. Sharing such edge device among multiple regions with similar healthcare resource limitations can significantly expand the screening outreach and serve a larger population over time. Hence, there is a need for a hardware solution that can facilitate reliable and energy-efficient DR screening at the edge.

5 Methodology

5.1 Overview

Developing neural network-based DR screening solutions involves two phases, as shown in Fig. 11. The first phase is the development (pre-deployment) phase, where the model is trained and translated into a corresponding hardware design. The

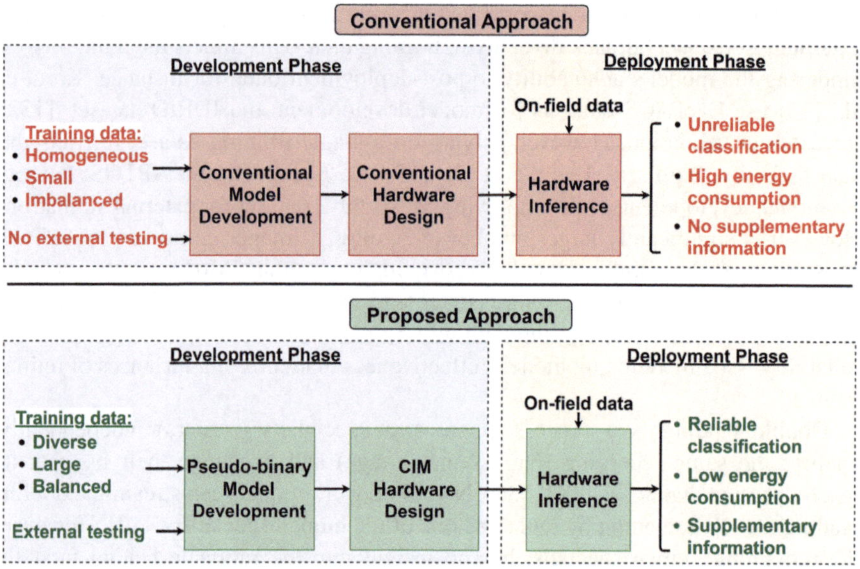

Fig. 11 Overview of conventional and proposed approaches for neural network-based DR screening

second phase is the post-deployment phase, which involves using on-field images to perform inference on this hardware. Models developed using the conventional approach face reliability issues, where they perform well during the development phase but fail in the post-deployment phase. This is a consequence of model's poor generalization ability arising from small-sized, nondiverse, and imbalanced training data, coupled with the absence of external test data. Moreover, they do not provide supplementary information about the model prediction. This limits their widespread adoption by both specialists and patients. Additionally, there has been almost no effort directed toward energy-efficient hardware design for DR classification models targeting edge device deployment, which is critical for their global accessibility.

To overcome these challenges, we first create a large, diverse, and balanced custom dataset by combining data from multiple sources and taking measures to reduce class imbalance. We then train our DR classification model with this dataset and also assess its reliability with external test data. Moreover, we introduce a pseudo-binary classification scheme that improves the model performance and also provides supplementary information to facilitate its widespread adoption. Furthermore, we present energy-efficient hardware design for our model using memristor-based CIM for deployment on edge devices to improve accessibility. Thus, we provide a solution that offers reliable DR classification, supplementary information, and energy-efficient edge deployment. We will describe the details of our approach in the next subsections.

5.2 *Model Development*

5.2.1 Dataset Creation

Inconsistent post-deployment retinal image quality presents a challenge for DR classification models. Factors like improper exposure, misalignment, and incomplete retinal coverage lead to this issue. Moreover, data distribution after deployment can also vary widely from that during training. Hence, training dataset must incorporate these inconsistencies and capture post-deployment diversity. We build such a comprehensive training dataset by leveraging the following publicly available datasets, whose samples are shown in Fig. 12:

- EyePACS dataset [131]: It is provided by EyePACS Inc. and contains images collected from different parts of the USA.
- DDR dataset [13]: It contains images collected across 147 hospitals in China.
- APTOS dataset [125]: It is provided by Aravind Eye Hospital in India, as a part of competition organized by Asia Pacific Tele-Ophthalmology Society in 2019.

After acquiring these datasets, we filter out corrupt images and merge them in varying proportions to create three merged datasets. We then undersample the majority classes in each merged dataset to limit class imbalance to 10× or less. This is because 10× or less imbalance suffices for neural networks perform well on minority classes [135] and achieving perfect class balance is impractical due to the huge number of healthy retina images. As a result, we end up with three new proposed datasets: small (S), medium (M), and large (L). The classwise distributions of the original and new datasets are shown in Table 4. It can be seen that the new M and L datasets exhibit better class balance than original EyePACS and DDR datasets. Moreover, merging enhances the data diversity in M and L datasets compared to EyePACS and DDR datasets. As a result, models developed using M and L datasets can potentially exhibit better reliability in handling both on-field image inconsistencies and minority DR classes. Furthermore, the three datasets can be used to obtain crucial insights into how dataset size influences classification performance.

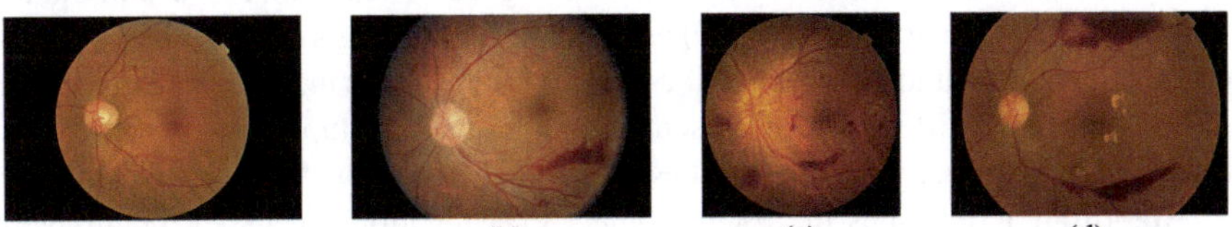

(a) (b) (c) (d)

Fig. 12 Samples from EyePACS [131], DDR [13], and APTOS [125] public datasets. (**a**) DR-0 (EyePACS). (**b**) DR-4 (EyePACS). (**c**) DR-4 (DDR). (**d**) DR-4 (APTOS)

Table 4 Overview of the original public datasets and newly proposed datasets. Imbalance is the ratio of the sizes of the largest and smallest class

	Original datasets			Proposed datasets		
	APTOS [125]	DDR [13]	EyePACS [131]	Small (S): APTOS	Medium (M): APTOS & DDR	Large (L): APTOS, DDR & EyePACS
DR-0	1805	6266	65,343	1798	3000	10,000
DR-1	370	630	6205	365	991	7180
DR-2	999	4477	13,153	991	3000	10,000
DR-3	193	236	2087	188	424	2504
DR-4	295	913	1914	292	1204	3117
Total	3662	12,522	88,702	3634	8619	32,801
Imbalance	9.4×	26.5×	34.1×	9.6×	7.1×	4.0×

5.2.2 Pseudo-binary Classification

Recommended management guidelines [136] suggest annual screening for DR-0/DR-1, follow-up every 6 months for DR-2, and referral to an ophthalmologist for DR-3/DR-4. Thus, the recommended DR management approach shifts from annual screening to more frequent monitoring as the severity reaches DR-2. Consequently, grouping the five DR classes into the following two categories suffices for screening [15]: non-referable DR (consisting of DR-0 and DR-1) and referable DR (consisting of DR-2, DR-3, and DR-4). Thus, DR screening becomes a binary classification task involving referable DR and non-referable DR as its two classes.

A straightforward approach to binary screening involves relabeling the five original classes into these two broad categories and developing a binary classification model. However, this leads to a hard decision between the two categories which is more susceptible to misclassifications. It also weakens the interpretability by hindering the derivation of supplementary information. To alleviate this problem, we introduce a pseudo-binary classification approach. It internally employs a multiclass DR classifier and uses additional decision-making logic to ultimately produce a binary classification outcome, as shown in Fig. 13. It capitalizes on cumulative probabilities within non-referable (0) and referable (1) categories instead of hinging on a single maximum probability for decision-making, reducing susceptibility to misclassifications. Moreover, it presents the outcome as a tuple containing prediction, confidence level, and referable DR probability, providing better interpretability.

Algorithm 1 describes the pseudo-binary classification process. It begins with multiclass classification to obtain prediction probabilities for the five original DR classes. Subsequently, it calculates a score for the non-referable (0) class by adding the probabilities of DR-0 and DR-1. Similarly, a score for the referable (1) class is computed by adding the probabilities of DR-2, DR-3, and DR-4. The broad class (referable 0 or non-referable 1) with the higher score is selected as the prediction value. Furthermore, if the scores differ by more than a predefined confidence threshold, we indicate high confidence (“H”), otherwise low confidence

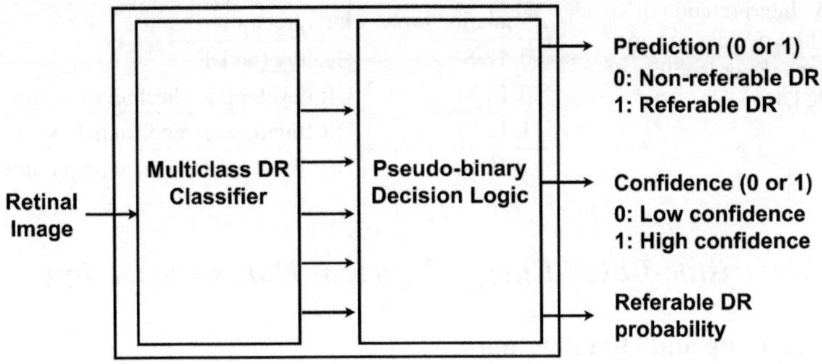

Fig. 13 Pseudo-binary classification concept

Algorithm 1: Pseudo-binary DR classification

```
input : Confidence threshold C_th, retinal image I
output: Prediction tuple P
softmax ← multiclass_inference(I);
non_ref_score ← softmax(DR-0) + softmax(DR-1);
ref_score ← softmax(DR-2) + softmax(DR-3) + softmax(DR-4);
Δ ← ref_score - non_ref_score;
if Δ > 0 then
    prediction ← 1;
    if Δ > C_th then
        confidence ← H;
    else
        confidence ← L;
else
    prediction ← 0;
    if |Δ| > C_th then
        confidence ← H;
    else
        confidence ← L;
P ← (prediction, confidence, ref_score);
return P;
```

("L"). Additionally, the referable class score indicates the probability of referable DR, for example, consider a scenario with confidence threshold 0.25 and softmax probabilities as [0.20 (DR-0), 0.33 (DR-1), 0.4 (DR-2), 0.03 (DR-3), 0.04 (DR-4)]. This leads to a pseudo-binary prediction tuple as (0, L, 47%), indicating that the patient has non-referable DR (class 0), detected with low confidence (L), and a 47% likelihood of referable DR. Thus, the patient appears to be developing DR-1 and is recommended to have a checkup in the near future. The interpretation of various output tuples of pseudo-binary classification is summarized in Table 5.

Table 5 Interpretation of pseudo-binary classification predictions (S: probability of referable DR)

Prediction tuple	Interpretation
(0, H, S)	Healthy (no DR)
(0, L, S)	DR developing, checkup recommended
(1, L, S)	DR found, seek medical help soon
(1, H, S)	DR found, seek medical help immediately

5.3 *Memristor-Based Energy-Efficient Hardware Design*

5.3.1 Pruning and Quantization

We optimize our pseudo-binary DR classification model for CIM hardware deployment via pruning and quantization to minimize hardware resource usage. Pruning removes a specific portion of low-magnitude weights from each layer. As this can lead to accuracy loss, we retrain the model after pruning to restore the accuracy. A critical factor during retraining phase is hyperparameter selection, particularly the learning rate. Hence, we adjust the learning rate dynamically within a small range around its initial value during retraining. This iterative cycle continues until we achieve the desired level of pruning while preserving the network's original accuracy. We then quantize pruned model to further reduce hardware resource requirements. To avoid accuracy degradation due to aggressive quantization, we perform design space exploration to minimize bit sizes without impacting classification performance.

5.3.2 Mapping to CIM Architecture

We map our optimized model to the memristor-based CIM architecture in [59], whose fundamental building block is depicted in Fig. 14. It divides the full-precision neural network weights and inputs into smaller slices. This is because memristors have limited bit capacity and high-resolution data converters (DACs and ADCs) consume significant energy and area. We transform 2-bit slices of the weights into conductance values, which are then mapped onto distinct columns within the memristor crossbar. We also convert 1-bit slices of the inputs into voltages that are applied to the crossbar at different timesteps, for instance, with 1-bit DACs for 32-bit digital inputs, the DACs are fed with 1-bit at a time across 32 timesteps. The DACs convert the bits at each timestep into voltages, generating a current in each column of the crossbar. These currents are captured by sample-and-hold circuits (S&H) and then converted into digital outputs by ADCs. To account for the slicing of weights across crossbar columns, a shift-and-add operation is performed across the columns for the ADC outputs. Furthermore, another round of shift-and-add operations merges partial outputs from various timesteps to produce the final full-precision digital output.

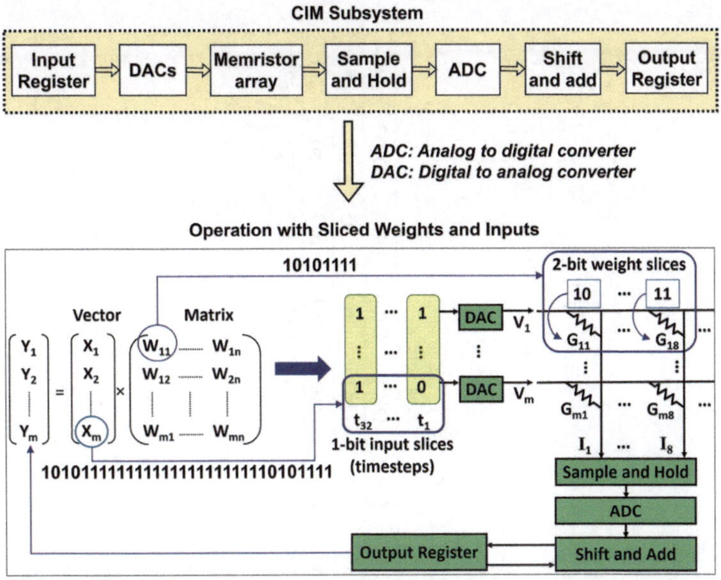

Fig. 14 CIM hardware architecture for DR model implementation

6 Simulation Setup

6.1 Simulation Platform

For developing our DR classification model, we use TensorFlow [137] deep learning framework. We select Inception-v3 (IV3) [138] and DenseNet121 (DN121) [139] neural network architectures for exploration due to their strong performance on complex image datasets [140]. These architectures are customized for DR classification by adding three additional fully connected layers (IV3: 2048×128, 128×128, 128×5 and DN121: 1024×128, 128×128, and 128×5) and dropout layers (probability 0.5). We train these networks with each of our new S/M/L datasets, following the train-validation-test split shown in Table 6. Employing transfer learning, we only train the newly added fully connected layers while freezing the pre-trained weights from the ImageNet dataset for all the other layers. During training, we perform grid search followed by manual fine-tuning to establish optimal values for the hyperparameters. Post-training model performance is evaluated with the corresponding S/M/L test set.

We evaluate the reliability of trained models using publicly accessible Messidor-2 dataset [126–128] as an external test set. It contains 1748 images where 1058 images are provided by the Messidor program partners [126] and the remaining are collected at Brest University Hospital in France. The labels for Messidor-2 dataset are sourced from [149], following the study in [150]. This labeling

Table 6 Data splitting for the new proposed datasets: small (S), medium (M), and large (L)

Dataset	Training set					Validation set					Test set				
	DR-0	DR-1	DR-2	DR-3	DR-4	DR-0	DR-1	DR-2	DR-3	DR-4	DR-0	DR-1	DR-2	DR-3	DR-4
Small (S)	1059	229	589	112	191	382	67	197	33	48	357	69	205	43	53
Medium (M)	1775	598	1808	257	733	620	202	565	87	250	605	191	627	80	221
Large (L)	6021	4314	5992	1465	1887	2034	1432	1975	517	603	1945	1434	2033	522	627

Table 7 Simulation platform details

Deep learning framework	TensorFlow [137]
Network architectures	Modified Inception-v3 (IV3), modified densenet121 (DN121)
DR datasets	New datasets (Table 4): small (S), medium (M), large (L)
Hardware platforms	CIM:ISAAC [59], CPU: Intel Core i7-9750H [141] GPU: NVIDIA GeForce GTX 1650 [142] mTPU: Google Edge TPU on Coral dev board [143]
Power profiling tools	CPU: s-tui [144], GPU: nvidia-smi [145] mTPU: datasheet [146], CIM: data in [59]
Latency profiling tools	CPU and GPU: TensorFlow profiler [147] mTPU: Python datetime package [148], CIM: data in [59]

process has deemed four images ungradable, leaving 1744 usable images with classwise distribution as follows: DR-0, 1017; DR-1, 270; DR-2, 347; DR-3, 75; and DR-4, 35. As Messidor-2 embodies a data characteristics distinct from S/M/L train-validation-test sets, model performance on Messidor-2 serves as an indicator of its reliability.

To map the trained reliable model onto the memristor-based CIM hardware, we employ the architecture described in Sect. 5.3.2. We leverage power consumption and latency data as presented in [59] to assess CIM energy consumption. This energy consumption is then compared against three conventional state-of-the-art hardware platforms: CPU (Intel Core i7-9750H [141]), GPU (NVIDIA GeForce GTX 1650 [142]), and mTPU (Google Edge TPU on Coral development board [143]). CPU and GPU represent general-purpose conventional hardware, while the mTPU embodies AI-optimized conventional hardware. To quantify energy consumption across these conventional hardware platforms, we first measure their latency and power consumption and then calculate energy consumption as a product of these two values. The latency for both CPU and GPU is obtained via TensorFlow profiler [147], while that for mTPU is measured using Python datetime package [148]. The power consumption of the CPU is recorded using s-tui [144], while nvidia-smi [145] is used to record GPU power consumption. We use mTPU's datasheet to obtain its power consumption [146]. The simulation platform details are summarized in Table 7.

6.2 Performance Metrics

The key performance metrics for the evaluation of the proposed DR classification include accuracy, F1-score, and energy consumption. They are described as follows:

- **Accuracy** It is defined as the ratio of the number of correctly classified retinal images to the total number of input retinal images, expressed as a percentage.

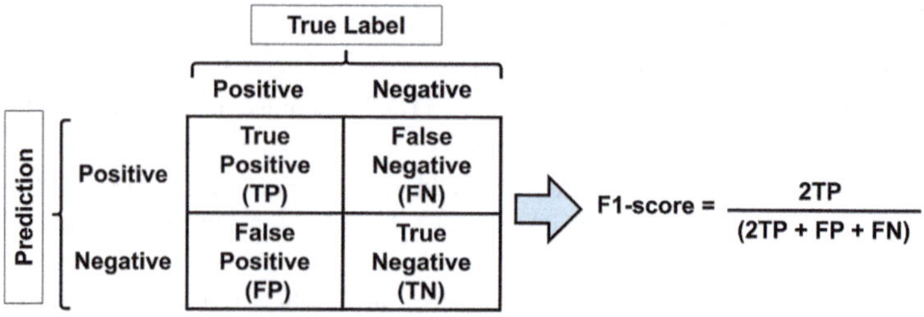

Fig. 15 F1-score calculation from confusion matrix representation

- **F1-score** The F1-score provides a deeper assessment of the model's behavior than accuracy. It reflects the model's ability to make correct predictions while keeping false alarms to a minimum. It is calculated using a table called confusion matrix as shown in Fig. 15.
- **Energy consumption** Deployment of automated DR screening on portable edge devices can significantly improve its global accessibility, even in remote areas. To achieve this, such devices must be able to operate with limited and interrupted energy availability. Hence, energy consumed by a DR classification hardware is an important performance metric.

7 Simulation Results

7.1 Model Reliability Assessment

The reliability assessment of our modified IV3 and DN121 networks is shown in Figs. 16 and 17. We train these networks across our three new datasets (S, M, and L in Table 4) by employing the pseudo-binary classification approach. This yields six distinct models: IV3-S (IV3 network trained on S dataset), IV3-M (IV3 network trained on M dataset), IV3-L (IV3 network trained on L dataset), DN121-S (DN121 network trained on S dataset), DN121-M (DN121 network trained on M dataset), and DN121-L (DN121 network trained on L dataset). The development phase performance of these models is assessed as their accuracy and F1-score on the corresponding S/M/L test set. To emulate post-deployment scenarios, we evaluate their accuracy and F1-score on Messidor-2 as an external test dataset. A reliable neural network model should exhibit consistent accuracy and F1-score, both during the development phase (on S/M/L test sets) and in post-deployment situations (on Messidor-2 dataset). This criterion forms the basis to assess the reliability of these six models in Figs. 16 and 17.

Models developed using S dataset (IV3-S and DN121-S) show commendable performance on S/M/L test sets but struggle on Messidor-2, indicating low reliability. The other models trained with M and L datasets exhibit a marked improvement

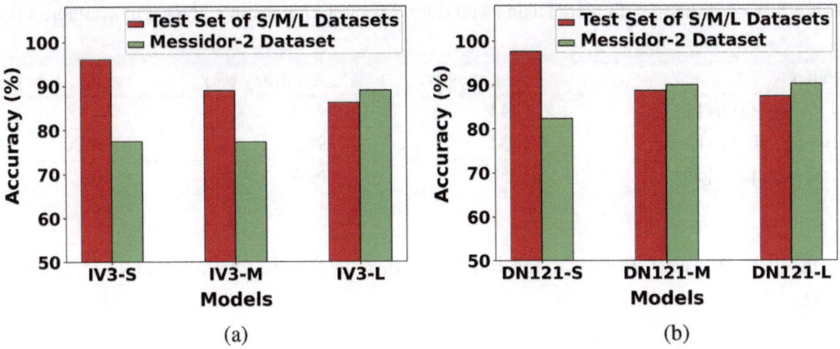

Fig. 16 Reliability assessment of IV3 and DN121 models in terms of accuracy. (**a**) IV3 accuracy. (**b**) DN121 accuracy

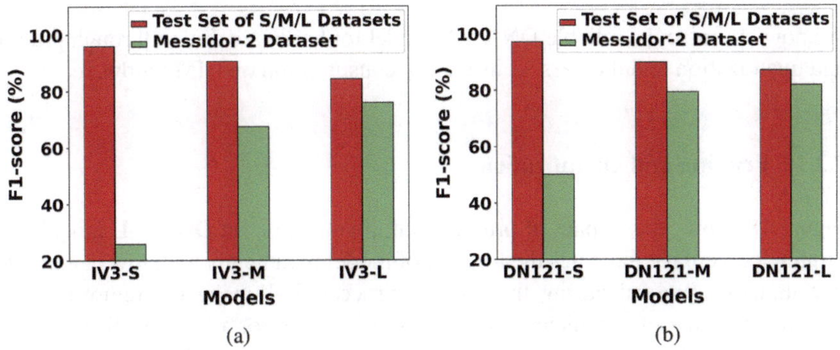

Fig. 17 Reliability assessment of IV3 and DN121 models in terms of F1-score. (**a**) IV3 F1-score. (**b**) DN121 F1-score

in reliability, consistently maintaining robust performance on both S/M/L test sets and Messidor-2 data. Also, model reliability improves as we transition from the M to the L dataset. This can be observed as IV3-L and DN121-L models outperform their M dataset counterparts IV3-M and DN121-M. This also highlights the pivotal role of large datasets in ensuring model reliability. As DN121-L exhibits better reliability than IV3-L in terms of both accuracy and F1-score, it becomes our final choice.

We now compare the reliability of our DN121-L model with other works from the literature. While several works [3, 6, 7, 9] have conducted evaluations on Messidor-2 dataset, they incorporate it within the training data rather than exclusively reserving it for testing. Consequently, a fair comparison with such studies is not possible. Additionally, works like [14] use datasets other than Messidor-2 for external testing and cannot be directly compared with our work. Therefore, we compare our approach with [15, 151] which use Messidor-2 dataset only for external testing. As shown in Table 8, DN121-L achieves competitive accuracy and F1-score despite testing on 65% more Messidor-2 images compared to [15, 151], validating its

Table 8 Comparison with works using Messidor-2 for external testing. "N/A": unavailable values

Paper	Test images	Accuracy (%)	F1-score (%)
This work (DN121-L)	1744	90.3	81.8
Blair et al. [151]	1054	93.5	N/A
Ludwig et al. [15]	1058	N/A	83

reliability. We will discuss CIM hardware design for this reliable DN121-L model next.

7.2 Memristor-Based CIM Design of Reliable Model

We now optimize the reliable DN121-L model for hardware design through pruning and quantization and then assess its energy consumption on CIM hardware.

7.2.1 Pruning and Quantization

Figure 18 shows the impact of pruning and quantization on DN121-L model. The pruning percentage indicates the fraction of model parameters with low magnitudes that undergo removal during the pruning process. Following the removal of the specified fraction of parameters, we conduct retraining for a few epochs to recover the accuracy loss incurred during pruning. Observing the classification performance across various pruning percentages, it becomes evident that the 50% pruned version of DN121-L exhibits the best classification performance. We select this version, denote it as DN121-L-P50, and subject it to weight quantization. We observe that we can use 4-bit weights with almost no accuracy loss. Hence, we select the 4-bit

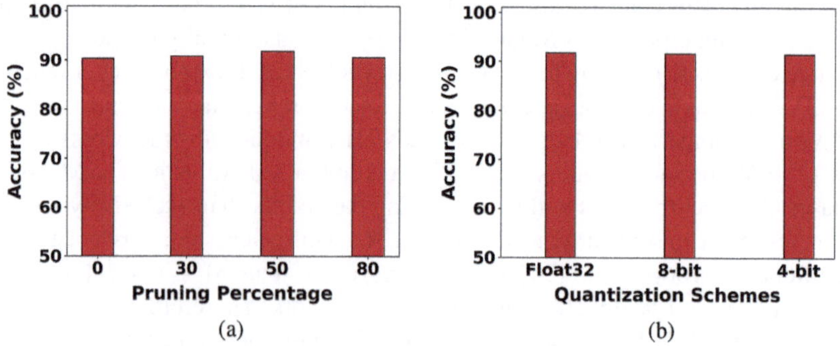

Fig. 18 Impact of pruning and quantization on DN121 accuracy. (**a**) Pruning of DN121-L model. (**b**) Quantization of DN121-L-P50 model

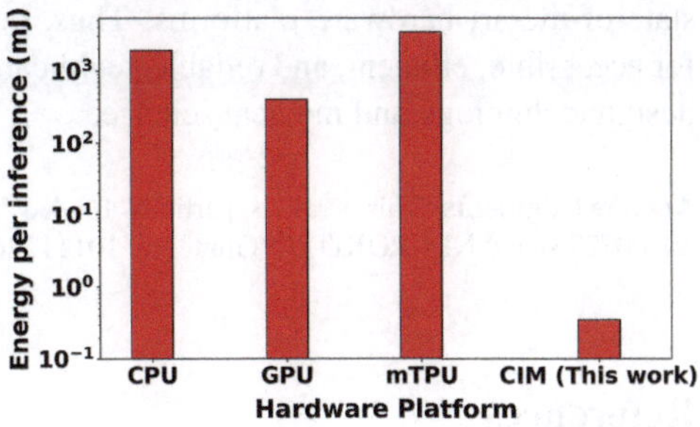

Fig. 19 Energy per image inference for various hardware platforms

quantized version of DN2121-L-P50 and denote it as CIM-DN121. We will analyze its energy consumption on CIM hardware next.

7.2.2 Energy Efficiency Assessment

The energy consumption for CIM-DN121 model, along with that of DN121 on state-of-the-art conventional hardware platforms like CPU, GPU, and edge TPU (mTPU), is depicted in Fig. 19. It quantifies the energy required for executing inference on a single retinal image. The mTPU turns out to be the least energy-efficient despite being designed for AI applications. This is because mTPU's efficiency is limited by other resources on Coral dev board which handle tasks such as code context management and input/output data processing. For a large neural network model like DN121, these resources become the bottleneck, leading to significantly longer execution latency and increased energy consumption compared to CPUs or GPUs. Therefore, mTPU dev board may not be the best choice for energy-efficient execution of large neural network models. On the other hand, CIM-DN121 demonstrates 5441× reduction in energy consumption compared to CPU. Furthermore, it consumes 1144× and 9686× less energy compared to GPU and mTPU, respectively. This highlights the tremendous potential of memristor-based CIM for developing energy-efficient hardware for DR screening.

8 Conclusion

This chapter presented a reliable and energy-efficient hardware for DR screening. We achieved reliable classification by using diverse and inconsistent data quality for model development while also addressing the class imbalance issue. We then introduced a pseudo-binary classification scheme to further improve the model performance and provide supplementary information. Furthermore, we presented energy-efficient hardware design for our reliable DR model, targeting deployment on edge devices for enhanced healthcare accessibility. We achieved reliable classification performance while consuming three orders of magnitude less energy than state-of-the-art hardware platforms. Thus, our work has established a foundation for accessible, efficient, and reliable healthcare through the intersection of hardware design technology and medical science.

Acknowledgments This work is partially funded by the European Union, DAIS (Grant No. 101007273) and NEUROKIT2E (Grant No. 101112268).

References

1. "Sight for all." https://sightforall.org/diabetic-retinopathy-initiative/.
2. Y. Zhou, B. Wang, L. Huang, S. Cui, and L. Shao, "A benchmark for studying diabetic retinopathy: Segmentation, grading, and transferability," *IEEE Transactions on Medical Imaging*, vol. 40, no. 3, pp. 818–828, 2021.
3. M. Feng, J. Wang, K. Wen, and J. Sun, "Grading of diabetic retinopathy images based on graph neural network," *IEEE access: practical innovations, open solutions*, vol. 11, pp. 98391–98401, 2023.
4. M. B. Khan, M. Ahmad, S. B. Yaakob, R. Shahrior, M. A. Rashid, and H. Higa, "Automated diagnosis of diabetic retinopathy using deep learning: On the search of segmented retinal blood vessel images for better performance," *Bioengineering*, vol. 10, no. 4, 2023.
5. W. K. Wong, F. H. Juwono, and C. Apriono, "Diabetic retinopathy detection and grading: A transfer learning approach using simultaneous parameter optimization and feature-weighted ECOC ensemble," *IEEE access: practical innovations, open solutions*, vol. 11, pp. 83004–83016, 2023.
6. M. R. Islam, L. F. Abdulrazak, M. Nahiduzzaman, M. O. F. Goni, M. S. Anower, M. Ahsan, J. Haider, and M. Kowalski, "Applying supervised contrastive learning for the detection of diabetic retinopathy and its severity levels from fundus images," *Computers in Biology and Medicine*, vol. 146, p. 105602, 2022.
7. S. S. Mohammadi and Others, "A user-friendly approach for the diagnosis of diabetic retinopathy using ChatGPT and automated machine learning," *Ophthalmology Science*, p. 100495, 2024.
8. N. Sikder, M. Masud, A. K. Bairagi, A. S. M. Arif, A. A. Nahid, and H. A. Alhumyani, "Severity classification of diabetic retinopathy using an ensemble learning algorithm through analyzing retinal images," *Symmetry*, vol. 13, no. 4, 2021.
9. V. K. R. Poranki and B. Srinivasarao, "Computer-aided diagnosis-based grading classification of diabetic retinopathy using deep graph correlation network with IRF," *SN Computer Science*, vol. 5, no. 2, 2024.
10. A. Sadeghzadeh, M. S. Junayed, T. Aydin, and M. B. Islam, "Hybrid CNN+Transformer for diabetic retinopathy recognition and grading," in *2023 Innovations in Intelligent Systems and Applications Conference, ASYU 2023*, pp. 1–6, 2023.
11. A. R. Wahab Sait, "A lightweight diabetic retinopathy detection model using a deep-learning technique," *Diagnostics*, vol. 13, no. 19, 2023.
12. W. L. Alyoubi, M. F. Abulkhair, and W. M. Shalash, "Diabetic retinopathy fundus image classification and lesions localization system using deep learning," *Sensors*, vol. 21, no. 11, 2021.
13. T. Li, Y. Gao, K. Wang, S. Guo, H. Liu, and H. Kang, "Diagnostic assessment of deep learning algorithms for diabetic retinopathy screening," *Information Sciences*, vol. 501, pp. 511–522, 2019.

14. E. Korot, Z. Guan, D. Ferraz, S. K. Wagner, G. Zhang, X. Liu, L. Faes, N. Pontikos, S. G. Finlayson, H. Khalid, G. Moraes, K. Balaskas, A. K. Denniston, and P. A. Keane, "Code-free deep learning for multi-modality medical image classification," *Nature Machine Intelligence*, vol. 3, no. 4, pp. 288–298, 2021.
15. C. A. Ludwig, C. Perera, D. Myung, M. A. Greven, S. J. Smith, R. T. Chang, and T. Leng, "Automatic identification of referral-warranted diabetic retinopathy using deep learning on mobile phone images," *Translational Vision Science and Technology*, vol. 9, no. 2, pp. 1–7, 2020.
16. G. Quellec, K. Charrière, Y. Boudi, B. Cochener, and M. Lamard, "Deep image mining for diabetic retinopathy screening," *Medical Image Analysis*, vol. 39, pp. 178–193, 2017.
17. R. Sayres, A. Taly, E. Rahimy, K. Blumer, D. Coz, N. Hammel, J. Krause, A. Narayanaswamy, Z. Rastegar, D. Wu, S. Xu, S. Barb, A. Joseph, M. Shumski, J. Smith, A. B. Sood, G. S. Corrado, L. Peng, and D. R. Webster, "Using a deep learning algorithm and integrated gradients explanation to assist grading for diabetic retinopathy," *Ophthalmology*, vol. 126, no. 4, pp. 552–564, 2019.
18. Z. L. Teo, Y. C. Tham, M. Yu, M. L. Chee, T. H. Rim, N. Cheung, M. M. Bikbov, Y. X. Wang, Y. Tang, Y. Lu, I. Y. Wong, D. S. W. Ting, G. S. W. Tan, J. B. Jonas, C. Sabanayagam, T. Y. Wong, and C. Y. Cheng, "Global prevalence of diabetic retinopathy and projection of burden through 2045: Systematic review and meta-analysis," *Ophthalmology*, vol. 128, no. 11, pp. 1580–1591, 2021.
19. N. Panwar, P. Huang, J. Lee, P. A. Keane, T. S. Chuan, A. Richhariya, S. Teoh, T. H. Lim, and R. Agrawal, "Fundus photography in the 21st century -a review of recent technological advances and their implications for worldwide healthcare," *Telemedicine and e-Health*, vol. 22, no. 3, pp. 198–208, 2016.
20. L. Lin, M. Li, Y. Huang, P. Cheng, H. Xia, K. Wang, J. Yuan, and X. Tang, "The SUSTech-SYSU dataset for automated exudate detection and diabetic retinopathy grading," *Scientific Data*, vol. 7, no. 1, p. 409, 2020.
21. S. D. Solomon and M. F. Goldberg, "ETDRS grading of diabetic retinopathy: Still the gold standard?," *Ophthalmic Research*, vol. 62, no. 4, pp. 190–195, 2019.
22. S. Diware, K. Chilakala, R. V. Joshi, S. Hamdioui, and R. Bishnoi, "Reliable and Energy-Efficient Diabetic Retinopathy Screening Using Memristor-Based Neural Networks," *IEEE Access*, vol. 12, pp. 47469–47482, 2024.
23. E. Rotem, A. Yoaz, L. Rappoport, S. J. Robinson, J. Y. Mandelblat, A. Gihon, E. Weissmann, R. Chabukswar, V. Basin, R. Fenger, M. Gupta, and A. Yasin, "Intel alder lake CPU architectures," *IEEE Micro*, vol. 42, no. 3, pp. 13–19, 2022.
24. J. Choquette, W. Gandhi, O. Giroux, N. Stam, and R. Krashinsky, "NVIDIA A100 tensor core GPU: Performance and innovation," *IEEE Micro*, vol. 41, no. 2, pp. 29–35, 2021.
25. N. P. Jouppi, G. Kurian, S. Li, P. Ma, R. Nagarajan, L. Nai, N. Patil, S. Subramanian, A. Swing, B. Towles, C. Young, X. Zhou, Z. Zhou, and D. Patterson, "TPU v4: An optically reconfigurable supercomputer for machine learning with hardware support for embeddings," in *Proceedings - International Symposium on Computer Architecture*, pp. 1147–1160, 2023.
26. S. Jain, A. Sengupta, K. Roy, and A. Raghunathan, "RxNN: A framework for evaluating deep neural networks on resistive crossbars," *IEEE Transactions on Computer-Aided Design of Integrated Circuits and Systems*, vol. 40, no. 2, pp. 326–338, 2021.
27. S. Diware, A. Singh, A. Gebregiorgis, R. V. Joshi, S. Hamdioui, and R. Bishnoi, "Accurate and energy-efficient bit-slicing for rram-based neural networks," *IEEE Transactions on Emerging Topics in Computational Intelligence*, vol. 7, no. 1, pp. 164–177, 2023.
28. L. Huijbregts, H.-H. Liu, P. Detterer, S. Hamdioui, A. Yousefzadeh, and R. Bishnoi, "Energy-efficient snn architecture using 3nm finfet multiport sram-based cim with online learning," in *Design Automation Conference*, pp. 1–6, 2024.
29. S. Diware, M. A. Yaldagard, and R. Bishnoi, "Hardware-aware quantization for accurate memristor-based neural networks," in *International Conference on Computer Aided Design (ICCAD)*, 2024.

30. S. Diware, Y. Dong, M. A. Yaldagard, and R. Bishnoi, "Adaptive multi-threshold encoding for energy-efficient ecg classification architecture using spiking neural network," in *Design, Automation and Test in Europe (DATE)*, 2025.
31. "International Roadmap for Devices and Systems." standards.ieee.org/develop/indconn/irds/index.html.
32. S. M. Nair, R. Bishnoi, M. B. Tahoori, H. Zahedmanesh, K. Croes, K. Garello, G. S. Kar, and F. Catthoor, "Physics based modeling of bimodal electromigration failure distributions and variation analysis for vlsi interconnects," in *International Reliability Physics Symposium (IRPS)*, pp. 1–5, IEEE, 2020.
33. A. T. Erozan, D. D. Weller, F. Rasheed, R. Bishnoi, J. Aghassi-Hagmann, and M. B. Tahoori, "A novel printed-lookup-table-based programmable printed digital circuit," *IEEE Transactions on Very Large Scale Integration (VLSI) Systems*, vol. 28, no. 6, pp. 1496–1504, 2020.
34. A. T. Erozan, G. Y. Wang, R. Bishnoi, J. Aghassi-Hagmann, and M. B. Tahoori, "A compact low-voltage true random number generator based on inkjet printing technology," *IEEE Transactions on Very Large Scale Integration (VLSI) Systems*, vol. 28, no. 6, pp. 1485–1495, 2020.
35. F. Rasheed, M. Hefenbrock, R. Bishnoi, M. Beigl, J. Aghassi-Hagmann, and M. B. Tahoori, "Crossover-aware placement and routing for inkjet printed circuits," *ACM Journal on Emerging Technologies in Computing Systems (JETC)*, vol. 16, no. 2, pp. 1–22, 2020.
36. F. Rasheed, M. Hefenbrock, R. Bishnoi, M. Beigl, J. Aghassi-Hagmann, and M. B. Tahoori, "Predictive modeling and design automation of inorganic printed electronics," in *Design, Automation & Test in Europe Conference & Exhibition (DATE)*, pp. 30–35, IEEE, 2019.
37. A. T. Erozan, R. Bishnoi, J. Aghassi-Hagmann, and M. B. Tahoori, "Inkjet-printed true random number generator based on additive resistor tuning," in *Design, Automation & Test in Europe Conference & Exhibition (DATE)*, pp. 1361–1366, IEEE, 2019.
38. S. M. Nair, R. Bishnoi, M. B. Tahoori, H. Zahedmanesh, K. Croes, K. Garello, G. S. Kar, and F. Catthoor, "Variation-aware physics-based electromigration modeling and experimental calibration for vlsi interconnects," in *2019 IEEE International Reliability Physics Symposium (IRPS)*, pp. 1–6, 2019.
39. G. Tshagharyan, G. Harutyunyan, Y. Zorian, A. Gebregiorgis, M. S. Golanbari, R. Bishnoi, and M. B. Tahoori, "Modeling and testing of aging faults in finfet memories for automotive applications," in *International Test Conference (ITC)*, pp. 1–10, 2018.
40. A. T. Erozan, G. C. Marques, M. S. Golanbari, R. Bishnoi, S. Dehm, J. Aghassi-Hagmann, and M. B. Tahoori, "Inkjet-printed egfet-based physical unclonable function–design, evaluation, and fabrication," *IEEE Transactions on Very Large Scale Integration (VLSI) Systems*, vol. 26, no. 12, pp. 2935–2946, 2018.
41. S. Yu, W. Shim, X. Peng, and Y. Luo, "RRAM for compute-in-memory: From inference to training," *IEEE Transactions on Circuits and Systems I: Regular Papers*, vol. 68, no. 7, pp. 2753–2765, 2021.
42. M. A. Yaldagard, S. Diware, R. V. Joshi, S. Hamdioui, and R. Bishnoi, "Read-disturb detection methodology for RRAM-based computation-in-memory architecture," in *AICAS 2023 – IEEE International Conference on Artificial Intelligence Circuits and Systems*, pp. 1–5, 2023.
43. S. Diware, M. A. Yaldagard, A. Gebregiorgis, R. V. Joshi, S. Hamdioui, and R. Bishnoi, "Dynamic detection and mitigation of read-disturb for accurate memristor-based neural networks," in *2024 IEEE 6th International Conference on AI Circuits and Systems (AICAS)*, pp. 393–397, 2024.
44. M. Fieback, G. C. Medeiros, L. Wu, H. Aziza, R. Bishnoi, M. Taouil, and S. Hamdioui, "Defects, fault modeling, and test development framework for rrams," *ACM Journal on Emerging Technologies in Computing Systems (JETC)*, vol. 18, no. 3, pp. 1–26, 2022.
45. A. Singh, S. Diware, A. Gebregiorgis, R. Bishnoi, F. Catthoor, R. V. Joshi, and S. Hamdioui, "Low-power memristor-based computing for edge-ai applications," in *International Symposium on Circuits and Systems (ISCAS)*, pp. 1–5, 2021.

46. M. Fieback, S. Nagarajan, R. Bishnoi, M. Tahoori, M. Taouil, and S. Hamdioui, "Testing scouting logic-based computation-in-memory architectures," in *European Test Symposium (ETS)*, pp. 1–6, IEEE, 2020.
47. A.R. Bommana, R. Bishnoi, N. Karimi, F. Firouzi, and K. Chakrabarty, "PHANTOM: Power Hammering Attack and Countermeasure on Multi-tenant ReRAM Compute-in-Memory Accelerators", *IEEE Transactions on Information Forensics and Security*, 2026. IEEE
48. A.R. Bommana, F. Firouzi, C. Ogbogu, B. Joardar, J. Doppa, P.P. Pande, and K. Chakrabarty, "SEC-CiM+: Selective and Reconfigurable Error Compensation for ReRAM-Based Compute-in-Memory", *ACM Transactions on Design Automation of Electronic Systems*, 2026. ACM New York, NY
49. A.R. Bommana, F. Firouzi, and K. Chakrabarty, "COMET-3D: Compute-in-Memory Based Transformer Accelerator with Optimized Pipeline and 3D Heterogeneous Integration", *IEEE Transactions on Computer-Aided Design of Integrated Circuits and Systems*, 2025. IEEE
50. S.S.R. Nakkilla, A. Datta, A.R. Bommana, J. Zhang, F. Firouzi, and K. Chakrabarty, "LiM Pruner: Efficient Pruning for Large Language-in-Memory Models", in *2025 IEEE International Conference on Omni-layer Intelligent Systems (COINS)*, pp. 1–7, 2025. IEEE
51. A.R. Bommana, F. Firouzi, C. Ogbogu, B.K. Joardar, J.R. Doppa, P.P. Pande, and K. Chakrabarty, "SEC-CiM: Selective Error Compensation for ReRAM-based Compute-in-Memory", in *2024 IEEE International Test Conference (ITC)*, pp. 177–186, 2024. IEEE
52. X. Yang, B. Taylor, A. Wu, Y. Chen, and L. O. Chua, "Research progress on memristor: From synapses to computing systems," *IEEE Transactions on Circuits and Systems I: Regular Papers*, vol. 69, no. 5, pp. 1845–1857, 2022.
53. R. Bishnoi, S. Diware, A. Gebregiorgis, S. Thomann, S. Mannaa, B. Deveautour, C. Marchand, A. Bosio, D. Deleruyelle, I. O'Connor, H. Amrouch, and S. Hamdioui, "Energy-efficient computation-in-memory architecture using emerging technologies," in *Proceedings of the International Conference on Microelectronics, ICM*, pp. 325–334, 2023.
54. S. Diware, A. Gebregiorgis, R. V. Joshi, S. Hamdioui, and R. Bishnoi, "Unbalanced Bit-slicing Scheme for Accurate Memristor-based Neural Network Architecture," in *International Conference on Artificial Intelligence Circuits and Systems (AICAS)*, pp. 1–4, 2021.
55. A. Singh, M. Zahedi, T. Shahroodi, M. Gupta, A. Gebregiorgis, M. Komalan, R. V. Joshi, F. Catthoor, R. Bishnoi, and S. Hamdioui, "Cim-based robust logic accelerator using 28 nm stt-mram characterization chip tape-out," in *International Conference on Artificial Intelligence Circuits and Systems (AICAS)*, pp. 451–454, 2022.
56. A. Singh, R. Bishnoi, R. V. Joshi, and S. Hamdioui, "Referencing-in-array scheme for rram-based cim architecture," in *Design, Automation & Test in Europe Conference & Exhibition (DATE)*, pp. 1413–1418, IEEE, 2022.
57. C. Münch, R. Bishnoi, and M. B. Tahoori, "Tolerating retention failures in neuromorphic fabric based on emerging resistive memories," in *2020 25th Asia and South Pacific Design Automation Conference (ASP-DAC)*, pp. 393–400, IEEE, 2020.
58. R. Bishnoi, L. Wu, M. Fieback, C. Münch, S. M. Nair, M. Tahoori, Y. Wang, H. Li, and S. Hamdioui, "Special session–emerging memristor based memory and cim architecture: Test, repair and yield analysis," in *VLSI Test Symposium (VTS)*, pp. 1–10, IEEE, 2020.
59. A. Shafiee, A. Nag, N. Muralimanohar, R. Balasubramonian, J. P. Strachan, M. Hu, R. S. Williams, and V. Srikumar, "ISAAC: A convolutional neural network accelerator with in-situ analog arithmetic in crossbars," in *Proceedings - 2016 43rd International Symposium on Computer Architecture, ISCA 2016*, pp. 14–26, 2016.
60. A. Ankit, I. El Hajj, S. Rahul Chalamalasetti, G. Ndu, M. Foltin, R. S. Williams, P. Faraboschi, W. M. Hwu, J. Paul Strachan, K. Roy, and D. S. Milojicic, "PUMA: A programmable ultra-efficient memristor-based accelerator for machine learning inference," in *International Conference on Architectural Support for Programming Languages and Operating Systems - ASPLOS*, pp. 715–731, 2019.
61. A. Ankit, I. E. Hajj, S. R. Chalamalasetti, S. Agarwal, M. Marinella, M. Foltin, J. P. Strachan, D. Milojicic, W. M. Hwu, and K. Roy, "PANTHER: A programmable architecture for neural network training harnessing energy-efficient ReRAM," *IEEE Transactions on Computers*, vol. 69, no. 8, pp. 1128–1142, 2020.

62. A. Singh, R. Bishnoi, A. Kaichouhi, S. Diware, R. V. Joshi, and S. Hamdioui, "A 115.1 tops/w, 12.1 tops/mm 2 computation-in-memory using ring-oscillator based adc for edge ai," in *International Conference on Artificial Intelligence Circuits and Systems (AICAS)*, pp. 1–5, 2023.
63. A. Gebregiorgis, A. Singh, S. Diware, R. Bishnoi, and S. Hamdioui, "Dealing with non-idealities in memristor based computation-in-memory designs," in *International Conference on Very Large Scale Integration (VLSI-SoC)*, pp. 1–6, 2022.
64. A. Singh, M. Fieback, R. Bishnoi, F. Bradarić, A. Gebregiorgis, R. V. Joshi, and S. Hamdioui, "Accelerating rram testing with a low-cost computation-in-memory based dft," in *International Test Conference (ITC)*, pp. 400–409, 2022.
65. M. Mayahinia, A. Singh, C. Bengel, S. Wiefels, M. A. Lebdeh, S. Menzel, D. J. Wouters, A. Gebregiorgis, R. Bishnoi, R. Joshi, *et al.*, "A voltage-controlled, oscillation-based adc design for computation-in-memory architectures using emerging rerams," *ACM Journal on Emerging Technologies in Computing Systems (JETC)*, vol. 18, no. 2, pp. 1–25, 2022.
66. A. Singh, M. A. Lebdeh, A. Gebregiorgis, R. Bishnoi, R. V. Joshi, and S. Hamdioui, "Srif: Scalable and reliable integrate and fire circuit adc for memristor-based cim architectures," *IEEE Transactions on Circuits and Systems I: Regular Papers*, vol. 68, no. 5, pp. 1917–1930, 2021.
67. C. Münch, N. Sayed, R. Bishnoi, and M. Tahoori, "A novel oscillation-based reconfigurable in-memory computing scheme with error correction," *IEEE Transactions on Magnetics*, vol. 57, no. 2, pp. 1–10, 2020.
68. Y. Deng, P. Huang, B. Chen, X. Yang, B. Gao, J. Wang, L. Zeng, G. Du, J. Kang, and X. Liu, "RRAM Crossbar Array With Cell Selection Device: A Device and Circuit Interaction Study," *IEEE Transactions on Electron Devices*, vol. 60, no. 2, pp. 719–726, 2013.
69. A. Sebastian, M. Le Gallo, R. Khaddam-Aljameh, and E. Eleftheriou, "Memory devices and applications for in-memory computing," *Nature Nanotechnology*, vol. 15, no. 7, pp. 529–544, 2020.
70. Rajendran, Gokulnath and Banerjee, Writam and Chattopadhyay, Anupam and Aly, Mohamed M. Sabry, "Application of resistive random access memory in hardware security: A review," *Advanced Electronic Materials*, vol. 7, no. 12, p. 2100536, 2021.
71. S. Diware, S. Dash, A. Gebregiorgis, R. V. Joshi, C. Strydis, S. Hamdioui, and R. Bishnoi, "Severity-based hierarchical ecg classification using neural networks," *IEEE Transactions on Biomedical Circuits and Systems*, vol. 17, no. 1, pp. 77–91, 2023.
72. C. Bengel, J. Mohr, S. Wiefels, A. Singh, A. Gebregiorgis, R. Bishnoi, S. Hamdioui, R. Waser, D. Wouters, and S. Menzel, "Reliability aspects of binary vector-matrix-multiplications using reram devices," *Neuromorphic computing and engineering*, vol. 2, no. 3, p. 034001, 2022.
73. M. L. Gallo and A. Sebastian, "An overview of phase-change memory device physics," *Journal of Physics D: Applied Physics*, vol. 53, no. 21, p. 213002, 2020.
74. A. M. Hosseini Monazzah, A. M. Rahmani, A. Miele, and N. Dutt, "CAST: Content-Aware STT-MRAM Cache Write Management for Different Levels of Approximation," *IEEE Transactions on Computer-Aided Design of Integrated Circuits and Systems*, vol. 39, no. 12, pp. 4385–4398, 2020.
75. P. Girard, Y. Cheng, A. Virazel, W. Zhao, R. Bishnoi, and M. B. Tahoori, "A survey of test and reliability solutions for magnetic random access memories," *Proceedings of the IEEE*, vol. 109, no. 2, pp. 149–169, 2020.
76. N. Sayed, R. Bishnoi, and M. B. Tahoori, "Approximate spintronic memories," *ACM Journal on Emerging Technologies in Computing Systems (JETC)*, vol. 16, no. 4, pp. 1–22, 2020.
77. S. M. Nair, R. Bishnoi, and M. B. Tahoori, "Mitigating read failures in stt-mram," in *VLSI Test Symposium (VTS)*, pp. 1–6, IEEE, 2020.
78. M. Tahoori, S. M. Nair, R. Bishnoi, L. Torres, S. Senni, G. Patrigeon, P. Benoit, G. Di Pendina, and G. Prenat, "A universal spintronic technology based on multifunctional standardized stack," in *Design, Automation & Test in Europe Conference & Exhibition (DATE)*, pp. 394–399, IEEE, 2020.

79. S. M. Nair, R. Bishnoi, A. Vijayan, and M. B. Tahoori, "Dynamic faults based hardware trojan design in stt-mram," in *2020 Design, Automation & Test in Europe Conference & Exhibition (DATE)*, pp. 933–938, 2020.
80. S. B. Dodo, R. Bishnoi, and M. B. Tahoori, "Secure stt-mram bit-cell design resilient to differential power analysis attacks," *IEEE Transactions on Very Large Scale Integration (VLSI) Systems*, vol. 28, no. 1, pp. 263–272, 2019.
81. S. M. Nair, R. Bishnoi, and M. B. Tahoori, "A comprehensive framework for parametric failure modeling and yield analysis of stt-mram," *IEEE Transactions on Very Large Scale Integration (VLSI) Systems*, vol. 27, no. 7, pp. 1697–1710, 2019.
82. N. Sayed, L. Mao, R. Bishnoi, and M. B. Tahoori, "Compiler-assisted and profiling-based analysis for fast and efficient stt-mram on-chip cache design," *ACM Transactions on Design Automation of Electronic Systems (TODAES)*, vol. 24, no. 4, pp. 1–25, 2019.
83. N. Sayed, R. Bishnoi, and M. B. Tahoori, "Fast and reliable stt-mram using nonuniform and adaptive error detecting and correcting scheme," *IEEE Transactions on Very Large Scale Integration (VLSI) Systems*, vol. 27, no. 6, pp. 1329–1342, 2019.
84. C. Münch, R. Bishnoi, and M. B. Tahoori, "Reliable in-memory neuromorphic computing using spintronics," in *Asia and South Pacific design automation conference*, pp. 230–236, 2019.
85. S. M. Nair, R. Bishnoi, M. B. Tahoori, G. Tshagharyan, H. Grigoryan, G. Harutyunyan, and Y. Zorian, "Defect injection, fault modeling and test algorithm generation methodology for stt-mram," in *International Test Conference (ITC)*, pp. 1–10, 2018.
86. N. Sayed, R. Bishnoi, F. Oboril, and M. B. Tahoori, "A cross-layer adaptive approach for performance and power optimization in stt-mram," in *2018 Design, Automation & Test in Europe Conference & Exhibition (DATE)*, pp. 791–796, IEEE, 2018.
87. C. Münch, R. Bishnoi, and M. B. Tahoori, "Multi-bit non-volatile spintronic flip-flop," in *Design, Automation & Test in Europe Conference & Exhibition (DATE)*, pp. 1229–1234, 2018.
88. S. M. Nair, R. Bishnoi, and M. B. Tahoori, "Parametric failure modeling and yield analysis for stt-mram," in *2018 Design, Automation & Test in Europe Conference & Exhibition (DATE)*, pp. 265–268, IEEE, 2018.
89. A. Gebregiorgis, R. Bishnoi, and M. B. Tahoori, "Spintronic normally-off heterogeneous system-on-chip design," in *2018 Design, Automation & Test in Europe Conference & Exhibition (DATE)*, pp. 113–118, IEEE, 2018.
90. N. Mahdavi, F. Razaghian, and H. Farbeh, "An Architectural-Level Reliability Improvement Scheme in STT-MRAM Main Memory," *Microprocessors and Microsystems*, vol. 90, p. 104462, 2022.
91. T. Kawahara, K. Ito, R. Takemura, and H. Ohno, "Spin-transfer torque RAM technology: Review and prospect," *Microelectronics Reliability*, vol. 52, no. 4, pp. 613–627, 2012.
92. M. Tahoori, S. M. Nair, R. Bishnoi, S. Senni, J. Mohdad, F. Mailly, L. Torres, P. Benoit, A. Gamatié, P. Nouet, *et al.*, "Using multifunctional standardized stack as universal spintronic technology for iot," in *Design, Automation & Test in Europe Conference & Exhibition (DATE)*, pp. 931–936, 2018.
93. N. Sayed, S. M. Nair, R. Bishnoi, and M. B. Tahoori, "Process variation and temperature aware adaptive scrubbing for retention failures in stt-mram," in *2018 23rd Asia and South Pacific Design Automation Conference (ASP-DAC)*, pp. 203–208, IEEE, 2018.
94. S. M. Nair, R. Bishnoi, M. S. Golanbari, F. Oboril, and M. B. Tahoori, "Vaet-stt: A variation aware estimator tool for stt-mram based memories," in *Design, Automation & Test in Europe Conference & Exhibition (DATE), 2017*, pp. 1456–1461, 2017.
95. S. M. Nair, R. Bishnoi, M. S. Golanbari, F. Oboril, F. Hameed, and M. B. Tahoori, "Vaet-stt: Variation aware stt-mram analysis and design space exploration tool," *IEEE Transactions on Computer-Aided Design of Integrated Circuits and Systems*, vol. 37, no. 7, pp. 1396–1407, 2017.
96. M. Tahoori, S. M. Nair, R. Bishnoi, S. Senni, J. Mohdad, F. Mailly, L. Torres, P. Benoit, P. Nouet, R. Ma, *et al.*, "Great: heterogeneous integrated magnetic technology using multi-

functional standardized stack," in *Computer Society Annual Symposium on VLSI (ISVLSI)*, pp. 344–349, IEEE, 2017.
97. S. Mittal, R. Bishnoi, F. Oboril, H. Wang, M. Tahoori, A. Jog, and J. S. Vetter, "Architecting sot-ram based gpu register file," in *Computer Society Annual Symposium on VLSI (ISVLSI)*, pp. 38–44, IEEE, 2017.
98. N. Sayed, F. Oboril, A. Shirvanian, R. Bishnoi, and M. B. Tahoori, "Exploiting stt-mram for approximate computing," in *IEEE European Test Symposium (ETS)*, pp. 1–6, IEEE, 2017.
99. N. Sayed, M. Ebrahimi, R. Bishnoi, and M. B. Tahoori, "Opportunistic write for fast and reliable stt-mram," in *Design, Automation & Test in Europe Conference & Exhibition (DATE), 2017*, pp. 554–559, IEEE, 2017.
100. R. Bishnoi, F. Oboril, and M. B. Tahoori, "Design of defect and fault-tolerant nonvolatile spintronic flip-flops," *IEEE Transactions on Very Large Scale Integration (VLSI) Systems*, vol. 25, no. 4, pp. 1421–1432, 2016.
101. F. Oboril, F. Hameed, R. Bishnoi, A. Ahari, H. Naeimi, and M. Tahoori, "Normally-off stt-mram cache with zero-byte compression for energy efficient last-level caches," in *Proceedings of the 2016 International Symposium on Low Power Electronics and Design*, pp. 236–241, 2016.
102. R. Bishnoi, M. Ebrahimi, F. Oboril, and M. B. Tahoori, "Improving write performance for stt-mram," *IEEE Transactions on Magnetics*, vol. 52, no. 8, pp. 1–11, 2016.
103. R. Bishnoi, F. Oboril, and M. B. Tahoori, "Low-power multi-port memory architecture based on spin orbit torque magnetic devices," in *Proceedings of the 26th edition on Great Lakes Symposium on VLSI*, pp. 409–414, 2016.
104. R. Bishnoi, F. Oboril, and M. B. Tahoori, "Fault tolerant non-volatile spintronic flip-flop," in *2016 Design, Automation & Test in Europe Conference & Exhibition (DATE)*, pp. 261–264, IEEE, 2016.
105. R. Bishnoi, F. Oboril, and M. B. Tahoori, "Non-volatile non-shadow flip-flop using spin orbit torque for efficient normally-off computing," in *2016 21st Asia and South Pacific Design Automation Conference (ASP-DAC)*, pp. 769–774, IEEE, 2016.
106. R. Bishnoi, F. Oboril, M. Ebrahimi, and M. B. Tahoori, "Self-timed read and write operations in stt-mram," *IEEE Transactions on Very Large Scale Integration (VLSI) Systems*, vol. 24, no. 5, pp. 1783–1793, 2015.
107. R. Bishnoi, M. Ebrahimi, F. Oboril, and M. B. Tahoori, "Asynchronous asymmetrical write termination (aawt) for a low power stt-mram," in *2014 Design, Automation & Test in Europe Conference & Exhibition (DATE)*, pp. 1–6, IEEE, 2014.
108. R. Bishnoi, F. Oboril, M. Ebrahimi, and M. B. Tahoori, "Avoiding unnecessary write operations in stt-mram for low power implementation," in *Fifteenth International Symposium on Quality Electronic Design*, pp. 548–553, IEEE, 2014.
109. S. K. Kim, T. W. Oh, S. Lim, D. H. Ko, and S.-O. Jung, "High-Performance and Area-Efficient Ferroelectric FET-Based Nonvolatile Flip-Flops," *IEEE Access*, vol. 9, pp. 35549–35561, 2021.
110. S. Kumar, S. Chatterjee, S. Thomann, Y. S. Chauhan, and H. Amrouch, "Cross-Layer Reliability Modeling of Dual-Port FeFET: Device-Algorithm Interaction," *IEEE Transactions on Circuits and Systems I: Regular Papers*, vol. 70, no. 7, pp. 2891–2903, 2023.
111. P. Yao, H. Wu, B. Gao, J. Tang, Q. Zhang, W. Zhang, J. J. Yang, and H. Qian, "Fully hardware-implemented memristor convolutional neural network," *Nature*, vol. 577, no. 7792, pp. 641–646, 2020.
112. W. Kim, R. Bruce, T. Masuda, G. Fraczak, N. Gong, P. Adusumilli, S. Ambrogio, H. Tsai, J. Bruley, J.-P. Han, M. Longstreet, F. Carta, K. Suu, and M. BrightSky, "Confined PCM-based Analog Synaptic Devices offering Low Resistance-drift and 1000 Programmable States for Deep Learning," in *2019 Symposium on VLSI Technology*, pp. T66–T67, 2019.
113. S. Diware, A. Gebregiorgis, R. V. Joshi, S. Hamdioui, and R. Bishnoi, "Mapping-aware biased training for accurate memristor-based neural networks," in *2023 IEEE 5th International Conference on Artificial Intelligence Circuits and Systems (AICAS)*, pp. 1–5, 2023.

114. C.-C. Chou, Z.-J. Lin, P.-L. Tseng, C.-F. Li, C.-Y. Chang, W.-C. Chen, Y.-D. Chih, and T.-Y. J. Chang, "An N40 256K×44 embedded RRAM macro with SL-precharge SA and low-voltage current limiter to improve read and write performance," in *2018 IEEE International Solid-State Circuits Conference - (ISSCC)*, pp. 478–480, 2018.
115. C.-F. Yang, C.-Y. Wu, M.-H. Yang, W. Wang, M.-T. Yang, T.-C. Chien, V. Fan, S.-C. Tsai, Y.-H. Lee, W.-T. Chu, and A. Hung, "Industrially Applicable Read Disturb Model and Performance on Mega-Bit 28nm Embedded RRAM," in *2020 IEEE Symposium on VLSI Technology*, pp. 1–2, 2020.
116. C.-C. Chou, Z.-J. Lin, C.-A. Lai, C.-I. Su, P.-L. Tseng, W.-C. Chen, W.-C. Tsai, W.-T. Chu, T.-C. Ong, H. Chuang, Y.-D. Chih, and T.-Y. J. Chang, "A 22nm 96KX144 RRAM Macro with a Self-Tracking Reference and a Low Ripple Charge Pump to Achieve a Configurable Read Window and a Wide Operating Voltage Range," in *2020 IEEE Symposium on VLSI Circuits*, pp. 1–2, 2020.
117. P. Jain, U. Arslan, M. Sekhar, B. C. Lin, L. Wei, T. Sahu, J. Alzate-vinasco, A. Vangapaty, M. Meterelliyoz, N. Strutt, A. B. Chen, P. Hentges, P. A. Quintero, C. Connor, O. Golonzka, K. Fischer, and F. Hamzaoglu, "13.2 A 3.6 Mb 10.1 Mb/mm^2 Embedded Non-Volatile ReRAM Macro in 22 nm FinFET Technology with Adaptive Forming/Set/Reset Schemes Yielding Down to 0.5 V with Sensing Time of 5 ns at 0.7 V," in *2019 IEEE International Solid-State Circuits Conference – (ISSCC)*, pp. 212–214, 2019.
118. J. Wu, Y. Chen, W. S. Khwa, S. M. Yu, T. Y. Wang, J. Tseng, Y. Chih, and C. H. Diaz, "A 40 nm Low-Power Logic Compatible Phase Change Memory Technology," in *2018 IEEE International Electron Devices Meeting (IEDM)*, pp. 27.6.1–27.6.4, 2018.
119. F. Arnaud, P. Zuliani, J. Reynard, A. Gandolfo, F. Disegni, P. Mattavelli, E. Gomiero, G. Samanni, C. Jahan, R. Berthelon, O. Weber, E. Richard, V. Barral, A. Villaret, S. Kohler, J. Grenier, R. Ranica, C. Gallon, A. Souhaite, D. Ristoiu, L. Favennec, V. Caubet, S. Delmedico, N. Cherault, R. Beneyton, S. Chouteau, P. Sassoulas, A. Vernhet, Y. Le Friec, F. Domengie, L. Scotti, D. Pacelli, J. Ogier, F. Boucard, S. Lagrasta, D. Benoit, L. Clement, P. Boivin, P. Ferreira, R. Annunziata, and P. Cappelletti, "Truly Innovative 28nm FDSOI Technology for Automotive Micro-Controller Applications embedding 16MB Phase Change Memory," in *2018 IEEE International Electron Devices Meeting (IEDM)*, pp. 18.4.1–18.4.4, 2018.
120. Y.-D. Chih, Y.-C. Shih, C.-F. Lee, Y.-A. Chang, P.-H. Lee, H.-J. Lin, Y.-L. Chen, C.-P. Lo, M.-C. Shih, K.-H. Shen, H. Chuang, and T.-Y. J. Chang, "13.3 A 22 nm 32 Mb Embedded STT-MRAM with 10 ns Read Speed, 1 M Cycle Write Endurance, 10 Years Retention at 150 °C and High Immunity to Magnetic Field Interference," in *2020 IEEE International Solid-State Circuits Conference - (ISSCC)*, pp. 222–224, 2020.
121. V. B. Naik, K. Lee, K. Yamane, R. Chao, J. Kwon, N. Thiyagarajah, N. L. Chung, S. H. Jang, B. Behin-Aein, J. H. Lim, T. Y. Lee, W. P. Neo, H. Dixit, S. K, L. C. Goh, T. Ling, J. Hwang, D. Zeng, J. W. Ting, E. H. Toh, L. Zhang, R. Low, N. Balasankaran, L. Y. Zhang, K. W. Gan, L. Y. Hau, J. Mueller, B. Pfefferling, O. Kallensee, S. L. Tan, C. S. Seet, Y. S. You, S. T. Woo, E. Quek, S. Y. Siah, and J. Pellerin, "Manufacturable 22nm FD-SOI Embedded MRAM Technology for Industrial-grade MCU and IOT Applications," in *2019 IEEE International Electron Devices Meeting (IEDM)*, pp. 2.3.1–2.3.4, 2019.
122. Y. J. Song, J. H. Lee, S. H. Han, H. C. Shin, K. H. Lee, K. Suh, D. E. Jeong, G. H. Koh, S. C. Oh, J. H. Park, S. O. Park, B. J. Bae, O. I. Kwon, K. H. Hwang, B. Seo, Y. Lee, S. H. Hwang, D. S. Lee, Y. Ji, K. Park, G. T. Jeong, H. S. Hong, K. P. Lee, H. K. Kang, and E. S. Jung, "Demonstration of Highly Manufacturable STT-MRAM Embedded in 28nm Logic," in *2018 IEEE International Electron Devices Meeting (IEDM)*, pp. 18.2.1–18.2.4, 2018.
123. M. Trentzsch, S. Flachowsky, R. Richter, J. Paul, B. Reimer, D. Utess, S. Jansen, H. Mulaosmanovic, S. Müller, S. Slesazeck, J. Ocker, M. Noack, J. Müller, P. Polakowski, J. Schreiter, S. Beyer, T. Mikolajick, and B. Rice, "A 28nm HKMG super low power embedded NVM technology based on ferroelectric FETs," in *2016 IEEE International Electron Devices Meeting (IEDM)*, pp. 11.5.1–11.5.4, 2016.

124. S. Dünkel, M. Trentzsch, R. Richter, P. Moll, C. Fuchs, O. Gehring, M. Majer, S. Wittek, B. Müller, T. Melde, H. Mulaosmanovic, S. Slesazeck, S. Müller, J. Ocker, M. Noack, D.-A. Löhr, P. Polakowski, J. Müller, T. Mikolajick, J. Höntschel, B. Rice, J. Pellerin, and S. Beyer, "A FeFET based super-low-power ultra-fast embedded NVM technology for 22 nm FDSOI and beyond," in *2017 IEEE International Electron Devices Meeting (IEDM)*, pp. 19.7.1–19.7.4, 2017.
125. "APTOS 2019 Blindness Detection." https://www.kaggle.com/c/aptos2019-blindness-detection/data.
126. "Messidor." https://www.adcis.net/en/third-party/messidor/.
127. E. Decencière, X. Zhang, G. Cazuguel, B. Laÿ, B. Cochener, C. Trone, P. Gain, J. R. Ordóñez-Varela, P. Massin, A. Erginay, B. Charton, and J. C. Klein, "Feedback on a publicly distributed image database: The Messidor database," *Image Analysis and Stereology*, vol. 33, no. 3, pp. 231–234, 2014.
128. M. D. Abràmoff, J. C. Folk, D. P. Han, J. D. Walker, D. F. Williams, S. R. Russell, P. Massin, B. Cochener, P. Gain, L. Tang, M. Lamard, D. C. Moga, G. Quellec, and M. Niemeijer, "Automated analysis of retinal images for detection of referable diabetic retinopathy," *JAMA Ophthalmology*, vol. 131, no. 3, pp. 351–357, 2013.
129. A. Budai, R. Bock, A. Maier, J. Hornegger, and G. Michelson, "Robust vessel segmentation in fundus images," *International Journal of Biomedical Imaging*, vol. 2013, no. 154860, 2013.
130. "Diabetic Retinopathy Messidor EyePac Pre Processed." https://www.kaggle.com/datasets/mohammadasimbluemoon/.
131. "Diabetic Retinopathy Detection, Kaggle.." https://www.kaggle.com/competitions/diabetic-retinopathy-detection/overview.
132. P. Porwal, S. Pachade, R. Kamble, M. Kokare, G. Deshmukh, V. Sahasrabuddhe, and F. Meriaudeau, "Indian diabetic retinopathy image dataset (IDRiD): A database for diabetic retinopathy screening research," *Data*, vol. 3, no. 3, 2018.
133. P. Taylor and H. W. Potts, "Computer aids and human second reading as interventions in screening mammography: Two systematic reviews to compare effects on cancer detection and recall rate," *European Journal of Cancer*, vol. 44, no. 6, pp. 798–807, 2008.
134. H. Geijer and M. Geijer, "Added value of double reading in diagnostic radiology, a systematic review," *Insights into Imaging*, vol. 9, no. 3, pp. 287–301, 2018.
135. "Determining whether a dataset is imbalanced or not." https://datascience.stackexchange.com/questions/122571/determining-whether-a-dataset-is-imbalanced-or-not.
136. R. Chakrabarti, c. A. Harper, and J. E. Keeffe, "Diabetic retinopathy management guidelines," *Expert Review of Ophthalmology*, vol. 7, no. 5, pp. 417–439, 2012.
137. M. Abadi and Others, "TensorFlow: A system for large-scale machine learning," in *USENIX Conference on Operating Systems Design and Implementation*, pp. 265–283, 2016.
138. C. Szegedy, V. Vanhoucke, S. Ioffe, J. Shlens, and Z. Wojna, "Rethinking the inception architecture for computer vision," in *Proceedings of the IEEE Computer Society Conference on Computer Vision and Pattern Recognition*, vol. 2016-Decem, pp. 2818–2826, 2016.
139. G. Huang, Z. Liu, L. Van Der Maaten, and K. Q. Weinberger, "Densely connected convolutional networks," in *Proceedings - 30th IEEE Conference on Computer Vision and Pattern Recognition, CVPR 2017*, vol. 2017-Janua, pp. 2261–2269, 2017.
140. "ImageNet Large Scale Visual Recognition Challenge." https://www.image-net.org/challenges/LSVRC/.
141. "Intel Core i7-9750H Processor." https://ark.intel.com/content/www/us/en/ark/products/191045/intel-core-i7-9750h-processor-12m-cache-up-to-4-50-ghz.html.
142. "GTX 1650 Gaming Laptops." https://www.nvidia.com/en-eu/geforce/gaming-laptops/gtx-1650/.
143. "Dev Board." https://coral.ai/products/dev-board/.
144. "The Stress Terminal UI: s-tui." https://github.com/amanusk/s-tui.
145. "System Management Interface SMI." https://developer.nvidia.com/nvidia-system-management-interface.
146. "Coral dev board datasheet." https://coral.ai/docs/dev-board/datasheet/.

147. "Optimize TensorFlow performance using the Profiler." https://www.tensorflow.org/guide/profiler.
148. "datetime – Basic date and time types." https://docs.python.org/3/library/datetime.html.
149. "MESSIDOR-2 DR Grades." https://www.kaggle.com/datasets/google-brain/messidor2-dr-grades.
150. J. Krause, V. Gulshan, E. Rahimy, P. Karth, K. Widner, G. S. Corrado, L. Peng, and D. R. Webster, "Grader variability and the importance of reference standards for evaluating machine learning models for diabetic retinopathy," *Ophthalmology*, vol. 125, no. 8, pp. 1264–1272, 2018.
151. J. P. Blair, J. N. Rodriguez, R. M. Lasagni Vitar, M. A. Stadelmann, R. Abreu-González, J. Donate, C. Ciller, S. Apostolopoulos, C. Bermudez, and S. De Zanet, "Development of LuxIA, a cloud-based AI diabetic retinopathy screening tool using a single color fundus image," *Translational Vision Science and Technology*, vol. 12, no. 11, p. 38, 2023.

AI in Diabetes Management: Neural Architecture Search for Blood Glucose Prediction

Aritra Ray, Peter Domanski, Farshad Firouzi, Dirk Pflüger, and Krishnendu Chakrabarty

"The science of today is the technology of tomorrow."

– Edward Teller

Abstract For individuals with type 1 diabetes mellitus, accurate prediction of future blood glucose values is crucial to aid its regulation with insulin administration, tailored to the individual's specific needs. In this chapter, we propose a novel approach for the integration of a neural architecture search framework with deep reinforcement learning to autonomously generate and train architectures, optimized for each subject over model size and analytical prediction performance, for the blood glucose prediction task in individuals with type 1 diabetes. We evaluate the proposed approach on the OhioT1DM dataset, which includes blood glucose monitoring records at 5-minute intervals over 8 weeks for 12 patients with type 1 diabetes mellitus. Prior work focused on predicting blood glucose levels in 30- and 45-minute prediction horizons, equivalent to six and nine data points, respectively. The proposed method demonstrates significant improvements on average for mean

This chapter is adapted from Domanski et al. [1].

A. Ray (✉)
Department of Electrical and Computer Engineering, Duke University, Durham, NC, USA
e-mail: aritra.ray@duke.edu

F. Firouzi
Department of Electrical and Computer Engineering, Duke University, Durham, NC, USA

Arizona State University, Tempe, AZ, USA

P. Domanski · D. Pflüger
Institute of Parallel and Distributed Systems Scientific Computing, University of Stuttgart, Stuttgart, Germany

K. Chakrabarty
Department of Electrical and Computer Engineering, Duke University, Durham, NC, USA

Arizona State University, Tempe, AZ, USA

B. Farahani et al. (eds.), *Smart and Connected Health: Applications and Real-World Use Cases*, https://doi.org/10.1007/978-3-032-09253-3_9

absolute error in the 30-minute and 45-minute prediction horizons, respectively, through the proposed deep reinforcement learning framework, in comparison to the state of the art. Using the deep reinforcement learning framework, the best-case and worst-case analytical performance measured over root mean square error and mean absolute error was obtained for subject ID 570 and subject ID 584, respectively. Models optimized for performance on the prediction task and model size were obtained after implementing neural architecture search in conjunction with deep reinforcement learning on these two extreme cases. We demonstrate improvements of 4.8% using long short-term memory-based architectures and 5.7% with gated recurrent unit-based architectures for patient ID 570 on the analytical prediction performance by integrating neural architecture search with deep reinforcement learning framework. The patient with the lowest performance (ID 584) on the deep reinforcement learning method had an even greater performance boost, with improvements of 10.0% and 12.6% observed for the long short-term memory and gated recurrent units, respectively. The subject-specific optimized models over performance and model size from the neural architecture search in conjunction with deep reinforcement learning had a reduction in model size which ranged from 20 to 150 times compared to the model obtained using only the deep reinforcement learning method. The smaller size, indicating a reduction in model complexity in terms of the number of trainable network parameters, was achieved without a loss in the prediction performance.

1 Introduction

Diabetes mellitus is a medical condition marked by high blood glucose (BG) levels [2, 3]. If left untreated, individuals experiencing uncontrolled high BG levels are susceptible to severe health complications, escalating the risk of heart disease, diabetic retinopathy, or kidney damage [4]. The International Diabetes Federation (IDF) estimates that approximately 537 million adults aged 20 to 79 are affected by diabetes, with the year 2021 alone witnessing 6.7 million deaths attributed to diabetes [5]. Diabetes manifests in various forms, including type 1, type 2, gestational, prediabetes, monogenic, and cystic fibrosis-related diabetes [6]. In this chapter, we present methodologies for predicting future BG values based on historical BG values for individuals with type 1 diabetes mellitus (T1DM). T1DM is characterized as an organ-specific autoimmune disease affecting pancreatic β cells [2], wherein the β cells eventually lose their capacity for insulin synthesis.

The application of recurrent neural networks (RNNs) for BG prediction has been extensively explored in the literature [7, 8, 9, 10, 11, 12, 13, 16, 17, 18, 19, 20, 21] [22]. In the healthcare domain, where insulin delivery relies on predicted BG values, surpassing the state of the art in analytical and clinical performance is of paramount importance for the advancement of T1DM care. However, RNNs such as long short-term memory (LSTM) and gated recurrent units (GRUs) exhibit inherent numerical instability in time-series prediction tasks. Table 1 provides

Table 1 Performance analysis of simple recurrent networks on OhioT1DM dataset while predicting all BG values throughout the 30-minute prediction horizon, with respect to RMSE and MAE metrics (Adapted from [7])

Reference	Subjects from OhioT1DM dataset	Features used	Methods	Analytical performance RMSE (PH = 30min)
[11]	Testing: 584, 567, 596, 552, 544, 540. Training: all subjects	CGM, finger stick, basal rate, galvanic skin response, skin temp., bolus dose	LSTM	25.0
			BiLSTM	24.4
			Convolutional LSTMs	23.2
			Seq-2-Seq CNN-LSTM	21.0
			Seq-2-Seq LSTM	20.8
			Seq-2-Seq BiLSTM	20.6
[10]	All subjects	CGM	LSTM	22.13
			GRU	22.00
			WaveNet	22.49
			WaveNet+GRU	22.21
			WaveNet+LSTM	22.35
			GRU+LSTM	21.98
[21]	All subjects	CGM	Regression	19.85
			Vanilla LSTM	19.83
			BiLSTM	20.05
			Ensemble via stacking	19.63
			Ensemble via multivariate	19.64
			Ensemble via subsequence	19.62

insights into the performance of simple RNNs on the OhioT1DM dataset [23] for a prediction horizon (PH) of 30 minutes. OhioT1DM [23] includes continuous glucose monitoring (CGM) data for 12 patients with T1DM, recorded at 5-minute intervals over an 8-week duration. Bhimireddy et al. [11] evaluated LSTM, BiLSTM, convolutional LSTM, sequence-to-sequence (Seq-2-Seq) LSTM, Seq-2-Seq BiLSTM, and Seq-2-Seq CNN-LSTM, considering six features for testing on a portion of the OhioT1DM dataset. Consequently, Dudukcu et al. [10] considered one feature and trained the simple LSTM to achieve a better performance than Bhimireddy et al. [11]. Thus, in this chapter, we also defined the BG prediction task as a univariate time-series forecasting problem. Performances on other simple RNNs like GRUs and WaveNets were also evaluated by Dudukcu et al. [10], wherein combining simple RNNs like GRU with LSTM led to a performance boost. Nemat et al. [21] trained a simple regression model to have comparable performance to Vanilla LSTM and BiLSTM models, while ensembling the regression, Vanilla LSTM, and BiLSTM models led to a performance boost. While the use of RNNs in time-series prediction is predominantly studied [24], the role of RNNs in combination with feature extractors [25, 26] has also shown promising results for time-series forecasting tasks. Additionally, Fu et al. [27] discusses the application of various reinforcement learning (RL) algorithms, including actor-critic methods, in controlling energy consumption in buildings, highlighting their relevance to time-series prediction problems.

We present a network architecture that combines RNNs with densely connected encoding-decoding networks as feature encoders within a deep reinforcement learning (DRL) framework. Combining RNNs with encoding-decoding networks offers several advantages, such as improving the handling of long-term dependencies or allowing continuous, real-time learning with customization for specific individuals. Additionally, RL is robust to noise and anomalies in the data as it learns from reward signals rather than direct supervision. Furthermore, the presented design can effectively manage varying sequence lengths, maintaining performance over extended periods where simpler or smaller models might fail. The DRL framework, introduced in [14], trains an architecture with a default number of densely connected encoding-decoding layers, LSTM layers, and RL, as outlined in Fig. 1. To elaborate, the input encoding network consists of densely connected layers that harness their representational capabilities to extract informative features for the subsequent LSTM network followed by an output encoding network. The output

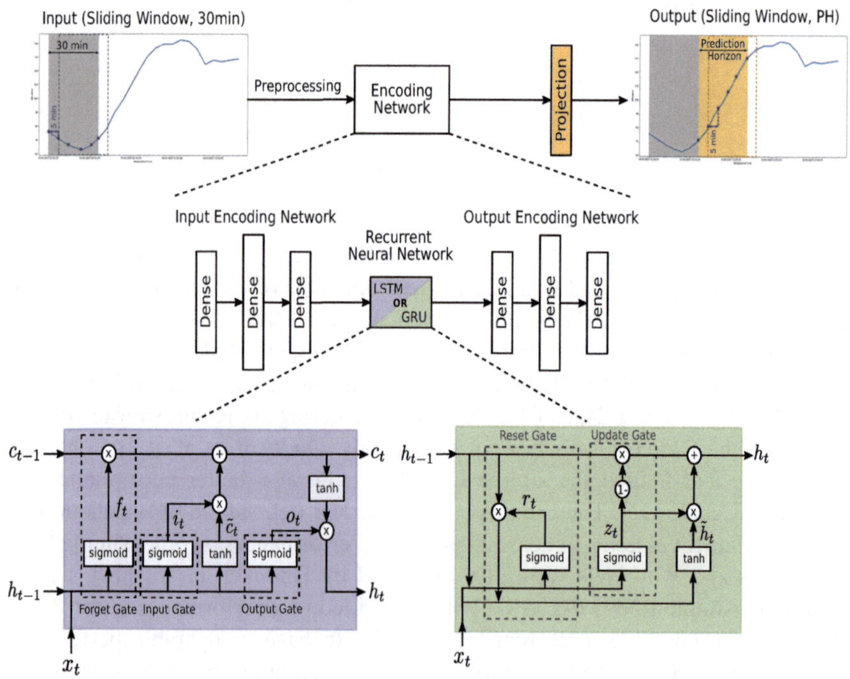

Fig. 1 Overview of the presented RNNs with densely connected encoding networks as feature extractors and the projection layer. The BG prediction task is formulated as a DRL learning task. The DRL-based approach requires the need of human experts to manually craft network architectures based on domain knowledge and intuition (Adapted from [7])

of this encoding network, in conjunction with the LSTM, is fed into a projection layer that generates a probability distribution across prediction values, ensuring a nonzero mean and bounded output to improve numerical stability. We formulated BG prediction as a DRL learning task and integrated the Soft Actor-Critic (SAC) in the DRL framework. The presented DRL method surpassed the state-of-the-art BG prediction approach in both root mean square error (RMSE) and mean absolute error (MAE) within 30-minute and 45-minute PHs. Compared to the previous best error, when the presented method was evaluated over the entire OhioT1DM dataset [23] and used only one feature for the prediction task, the presented DRL method demonstrated improvements of 18.4% and 22.5% on average and for MAE in the 30-minute and 45-minute PH, respectively, without additional optimization of the network's architectural structure. In the 30-minute PH, patient ID 570 exhibited the best-case prediction performance. In contrast, patient ID 584 represents the worst-case scenario regarding the analytical performance measured over RMSE on the prediction task using only the DRL method.

The default setting of the DRL framework is derived from human intuition, which might lead to non-optimized network architectures. In this chapter, we present a neural architecture search (NAS) framework, integrated with the DRL framework, which is designed to autonomously generate optimal or highly efficient neural network structures for the given BG prediction task. NAS with DRL eliminates the necessity for manual intervention by human experts in constructing network architectures based on domain knowledge and intuition. We present the introduction of the NAS framework with Bayesian optimization (BO). This provides a systemic approach for (a) designing networks optimized for analytical prediction performance on the BG prediction task, and model size for each patient, and (b) searching through a large search space of network hyperparameters to design optimized subject-specific models in terms of model complexity or the number of trainable network parameters. We iteratively generate candidate architectures using the NAS framework, train the architectures using DRL, and assess their performance in the BG prediction task. To accomplish this, we employ Optuna, a BO-based framework [28], to pinpoint architectures with low complexity that demonstrate high performance in BG prediction. In addition to the NAS, we incorporate GRU into the NAS space, an extension of prior work [14] which only utilized LSTM over the DRL framework. The efficacy of the presented method is evaluated using the OhioT1DM dataset [23]. The candidate architecture, generated by the NAS framework, is evaluated on the BG prediction task. The performance of the candidate architecture, or the prediction error, on the BG prediction task for the particular subject is used as a reward signal for the NAS controller. The reward signal helps the NAS controller decide on how well the generated candidate architecture performed on the prediction task, thereby helping it generate a new candidate architecture to meet the specific objective. Using a subject's past six BG values as inputs, this chapter predicts future BG values in the forecasting prediction horizon of 30 and 45 minutes, corresponding to six and nine BG values, respectively.

Evaluation results prove that the presented method achieved additional improvements of 4.8% using LSTM-based architectures and 5.7% with GRU-based architectures for patient ID 570 by integrating BO-based NAS to the DRL framework. In the 30-minute prediction horizon, patient ID 570 exhibited the best-case prediction performance, while patient ID 584 represents the worst-case scenario in terms of the analytical performance measured over RMSE on the prediction task using only the DRL method. Patient ID 584 had an even greater performance boost, with improvements of 10.0% and 12.6% observed for the LSTM and GRU, respectively, with the NAS combined with DRL approach. Moreover, the BO approach excels in identifying more suitable hyperparameter sets. With these sets, we achieve competitive results with recent methods [14] while utilizing a fraction of trainable network parameters. Through the NAS approach, we achieved a decrease in model size compared to solely employing the DRL method. The subject-specific optimized models over performance and model size from the NAS in conjunction DRL had a reduction in model size which ranged from 20 to 150 times compared to the model obtained using only the DRL method. The size reduction, indicating a reduction in model complexity in terms of the number of trainable network parameters, was achieved without a loss in the prediction performance compared to the state of the art [14]. The reduction varied depending on the specific patient and architecture. Compared with the state-of-the-art results [10, 14, 21], wherein the same number of features and all subjects were considered for training and testing, the presented NAS based on Bayesian optimization framework, built on top of DRL, expressed a very comparable performance in clinical settings, as elucidated in Sect. 5. While evaluating clinically using surveillance error grid (SEG) [29], the NAS with DRL expressed only marginal changes in the none and slight risk zones, within a range of ±3%, compared to the state of the art [10].

The rest of the chapter is organized as follows: Sect. 2 talks about the state-of-the-art approaches presented in the literature. In Sect. 3, we delve deeper into the presented methodology. The experimental settings are detailed in Sect. 4. The evaluation results are discussed in Sect. 5, while Sect. 6 concludes the chapter. The experimental results can be reproduced from the GitHub repository.[1]

2 Related Prior Work

Numerous studies have been conducted to predict BG for patients diagnosed with T1DM [10, 11, 12, 13, 16, 17, 18, 19, 20, 21, 22, 30]. Several external factors like insulin and carbohydrate intake, exhaustion from physical activity, and stress influence the fluctuation of BG levels in individuals with T1DM [31]. Accurate

[1] https://github.com/SC-SGS/Optimization-of-DRL-based-BG-Prediction.git

prediction of future BG levels helps in regulating insulin intake and thus helps to avoid serious health complications. To predict BG in T1DM, several techniques have been presented in the literature that involve both machine learning (ML) and deep learning (DL) approaches [11, 32, 33]. The effectiveness of the devised techniques is evaluated using datasets obtained from individuals with T1DM, as well as through dataset from simulators. Development of simulators for T1DM is also an active area of research. The subsequent subsections discuss simulators, datasets, and concentrate on pertinent literature concerning the OhioT1DM [23] dataset.

2.1 *Simulators and Datasets for Type 1 Diabetes*

Simulation environments such as the Diabetes Mellitus Metabolic Simulator for Research (DMMS.R) [34] are designed for diabetes research. DMMS.R facilitates clinical studies on virtual subjects with type 1, type 2, or prediabetes, providing in silico environments for testing interventions in diabetes treatment and monitoring, including modeling devices and exploring treatment protocols. It can be leveraged to create device models, apply comprehensive test protocols, analyze subject responses to drug treatments, and evaluate sensor and pump performance in the in silico environment before clinical trials. Moving on to the next, the Padova T1DM Simulator [35] was developed to model the glucose-insulin system during a meal based on closed-loop control [36]. It relates plasma concentrations of glucose and insulin to various glucose and insulin fluxes, incorporating a two-compartment model for the glucose subsystem. This simulator has been further enhanced to account for nonlinear increases in insulin-dependent utilization during hypoglycemia and includes models for glucagon kinetics and secretion [35]. Additional simulation tools for modeling T1DM BG levels include the web-based simulation tool [37], Simglucose [38], and GluCoEnv [39].

Predictions on *data from T1DM* like [23, 40, 41, 42, 43, 44, 45, 46] also contribute to a significant portion of the relevant literature. OhioT1DM [23] includes an 8-week dataset capturing the CGM values, insulin records, physiological sensor data, and self-reported life events for each of 12 individuals with T1DM. The OAPS [40] data collection, initiated in 2015, strives to improve accessibility and transparency of artificial pancreas system technology for individuals with T1DM [19]. Participants voluntarily share their data, encompassing CGM-recorded glucose levels, insulin rates, carbohydrate intake, physical activity, and other physiological details. Tidepool [41, 47] is another substantial collection of CGM values and insulin pump data, encompassing around 100 participants with a total of 152 million data instances [48].

2.2 Relevant Literature on the OhioT1DM Dataset

In 2018, Zhu et al. [33] explored the combination of causal dilated CNN layers with fast WaveNet for predicting BG on the OhioT1DM dataset [23], for six subjects, using four different features. In 2020, online Autoregressive Integrated Moving Average (ARIMA) with residual compensation network was presented by Ma et al. [49]. Daniels et al. [15] presented the use of deep multitask networks and convolutional RNNs, while dilated RNNs in the context of BG prediction for T1DM individuals were explored by Zhu et al. [12]. In the same year, deep residual networks, latent variable-based statistical modelling, and shallow neural networks were explored by Rubin-Falcone et al. [48], Sun et al. [50], and Pavan et al. [51], respectively, on the OhioT1DM dataset [23]. Multi-scale LSTM with multi-lag structures, convolutional RNN (CRNN), neural networks (NN), and RNN in conjunction with attention mechanisms were presented by Yang et al. [13], Freiburghaus et al. [16], and Bevan et al. [17], respectively. In the same year, a neural physiological encoder with LSTM, Seq-2-Seq models with RNNs, knowledge distillation via RNN, and generative adversarial networks (GAN) for time-series prediction on the OhioT1DM [23] were explored by Gu et al. [18], Bhimireddy et al. [11], Hameed et al. [19], and Zhu et al. [52], respectively.

In the following year, combinations of LSTM, GRU, WaveNets combined in a weighted decision-level fusion, GRU-based consolidated versus training models for individual subjects of the OhioT1DM dataset [23], and recurrent self-attention network were explored by Dudukcu et al. [10], Dudukcu et al. [30], and Cui et al. [20], respectively. In 2022, Nemat et al. [21] explored ensembling of simple regression model, Vanilla LSTM, and bidirectional LSTM (BiLSTM). In 2023, Shuvo et al. [22] explored deep multitask learning with stacked LSTM for BG prediction task. Chronologically listed in Table 2 is a comprehensive overview of the prediction methods, preprocessing steps, features utilized, and performance estimation techniques presented in the literature for the OhioT1DM dataset [23].

In general, throughout the literature, the popular metrics for analytical performance estimation on the prediction task were RMSE and MAE, while SEG [29] and Clarkes' error grid (CEG) [53] were the most commonly used metrics for evaluating clinical accuracy. The preprocessing was quite diverse, with some prior work considering interpolation methods [12, 15, 18, 21, 22, 33, 49, 50, 52] and imputation methods [11, 19] to account for missing data. Depending on the research timeline (6 more subjects were incorporated into the OhioT1DM dataset [23] in 2020), the literature considers either 6 [12, 13, 15, 16, 17, 18, 33, 48, 49, 50, 51] or 12 subjects [10, 20, 21, 22, 30]. Some work [12, 18, 19, 48] opted to harness additional datasets like Padova T1DM Simulator [35], TidePool [41], and OAPS [40] dataset to bolster the prediction performance. Additionally, commonly explored preprocessing techniques involve median [18, 19] and Gaussian [22] filtering, as well as data normalization [10, 13, 51]. Regarding features employed in the prediction task, aside from CGM, carbohydrate intake, basal, and bolus insulin are frequently taken into account [12, 13, 15, 16, 18, 19, 48]. While many studies used

Table 2 Summary of related research on type 1 diabetes prediction for OhioT1DM dataset (Adapted from [7])

Reference	Additional dataset	Subjects from OhioT1DM dataset	Features used	Methods	Preprocessing	Performance metrics
[33]	×	559, 563, 570, 575, 588, 591	CGM, insulin, Carbohydrate intake, normalized time index	Casual dilated CNN layers & fast WaveNet	Interpolation, extrapolation combination, filtering	RMSE
[49]	×	540, 544, 552, 567, 584, 596	CGM	Online ARIMA & Residual compensation network	Extrapolation Backward induction	RMSE, MAE
[15]	×	540, 544, 552, 567, 584, 596	CGM, insulin bolus, carbohydrate intake, exercise	Deep multitask networks & convolutional RNN	Interpolation, Extra-polation, Standardization	RMSE, MAE
[12]	UVA/ Padova T1DM Simulator [35]	591, 570, 563, 559, 588, 575	CGM, insulin doses, carbohydrate intake	Dilated RNN	Interpolation, extrapolation, filtering, combination	RMSE
[48]	[41]	567, 544, 552, 596, 540, 584	CGM, bolus insulin, finger stick glucose, carbohydrate intake, sine & cosine of time, CGM missingness indicator	Deep residual Time-series forecasting	Encode time to sine, cosine embedding, variable of missingness, resampling, missing values as zero	RMSE, MAE CEG
[50]	×	540, 544, 552, 567, 584, 596	CGM, basal insulin, bolus insulin, insulin on board	Latent variable-based statistical modeling	Interpolation, statistical modelling of glucose dynamics for missing values	RMSE, MAE CEG

(continued)

Table 2 Summary of related research on type 1 diabetes prediction for OhioT1DM dataset (Adapted from [7])

[51]	×	540, 544, 552, 567, 584, 596	CGM, Insulin on board, carbohydrate on board, slope of CGM	Shallow neural network Error imputation module	Normalization by mean & standard deviation	RMSE, MAE, COD, delay
[13]	×	540, 544, 552, 567, 584, 596	CGM, basal insulin, bolus insulin, carbohydrate intake, timestamp	Multi-scale LSTM with multi-lag structure	Data alignment, outlier detection & reconciliation, extrapolation, data normalization	RMSE, MAE
[16]	×	540, 544, 552, 567, 584, 596	CGM, basal insulin, bolus insulin, carbohydrate intake	CRNN	Data alignment, imputation, interpolation, resampling	RMSE, MAE
[17]	×	540, 544, 552, 567, 584, 596	CGM	Linear model, NN, RNN, RNN + attention mechanisms	Standardization, replace missing values with zero	RMSE, MAE
[18]	[41]	559, 563, 570, 575, 588, 591	CGM, basal insulin, bolus insulin, carbohydrate intake encoder, LSTM	Neural physiological	Interpolation, extrapolation median filtering	RMSE
[11]	×	Testing: 584, 567, 596, 552, 544, 540, Training: 12 subjects	CGM, finger stick, basal rate, galvanic skin response, skin temp., bolus dose	LSTM, BiLSTM, convolutional LSTMs, TCN, & sequence-to-sequence models	Imputation, resampling, parameter merging	RMSE, MAE

[19]	OAPS [40]	Training: all 12 subjects Testing: 540, 544, 552, 567, 584, 596 subjects. Validation: 559, 563, 570, 575, 588, 591 subjects	CGM values, insulin basal rate, bolus amount, carbs intake & difference between consecutive CGM	Knowledge Distillation: RNN	Imputation, median filtering	RMSE, MAE
[52]	×	Pre-train: 559, 563, 570, 575, 588, 591 subjects. Testing: 540, 544, 552, 567, 584, 596 subjects	CGM, insulin, meal, work, sleep, psycho-logical stress, physical exercise	GAN, gated RNN, CNN	Interpolation, extrapolation, Data resolution matching	RMSE, MAE CEG
[10]	×	All 12 subjects	CGM	LSTM, WaveNet, GRU & weighted decision-level fusion	Normalization, missing data intervals dropped	SEG, RMSE, MSA, MAPE, RMSPE
[30]	×	All 12 subjects	CGM	GRU-based consolidated & individual training	Missing data intervals dropped	RMSE, MAE, MAPE, RMSPE
[20]	×	All 12 subjects	Basal rate, bolus intake, carbohydrate intake, CGM	Recurrent self-attention network	Missing data intervals dropped	RMSE, T-test
[21]	×	All 12 subjects	CGM	Regression, Vanilla LSTM, BiLSTM, stacking, Multi-variate, subsequence ensemble	Interpolation, data scaling, reframing for supervised learning task	RMSE, MAE, MCC, SEG, Friedman test, Wilcoxon test, CDD
[22]	×	All 12 subjects	CGM, finger stick glucose, bolus insulin, carbohydrate intake	Deep multitask learning, Stacked LSTM	Interpolation, missing data intervals dropped, Extrapolation, Gaussian filtering	RMSE, MAE, CEG
Presented Work	×	All 12 subjects	CGM	NAS, DRL, pruning, LSTM, GRU	z-score normalization	RMSE, MAE, SEG, CEG

multiple features for the prediction task, we in this chapter consider only past CGM values to predict future BG values and evaluate over the entire OhioT1DM [23] dataset, consistent with [10, 21].

We present a novel method of BO-based NAS in conjunction with DRL, encompassing a dense encoder-decoder network in combination with LSTM or GRU cells. We present z-score normalization on the training set and apply the normalization score obtained from the training set to the test dataset, as the only preprocessing step while considering only CGM values. We evaluate the presented method analytically using RMSE and MAE metrics and clinically via SEG and CEG. The results are compared against the state of the art, which involves weighted decision-level fusion across LSTM, WaveNet, and GRU [10] for the prediction task.

3 Methodology

We present the design of a DRL architecture in conjunction with a NAS framework to automatically derive architectures for predicting BG values for individuals affected with T1DM, as presented in Fig. 2. The input encoding network consists of densely connected layers that harness their representational capabilities to extract informative features for the subsequent LSTM network. This is followed by an

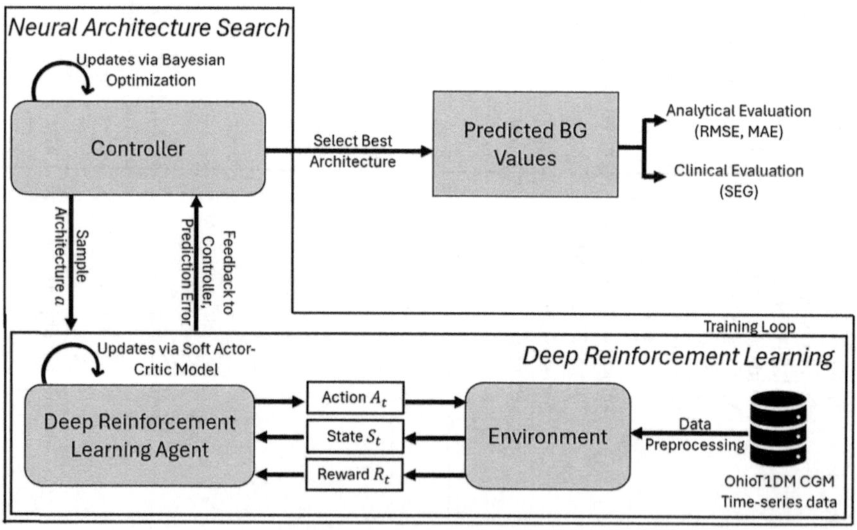

Fig. 2 Overview of the presented work. The neural architecture search samples an architecture from the search space, trains it using the deep reinforcement learning loop, and receives the prediction error on the train dataset as feedback. The deep reinforcement learning agent and neural architecture search are updated using the Soft Actor-Critic model and Bayesian optimization framework, respectively (Adapted from [7])

output encoding network. The output of this encoding network is fed into a projection layer that generates a probability distribution across prediction values, ensuring a nonzero mean and bounded output to improve numerical stability. We formulated BG prediction as a DRL learning task and integrated the Soft Actor-Critic (SAC) to the DRL algorithm. A major drawback of using only DRL-based approach is the need of human experts to manually craft network architectures based on domain knowledge and intuition. This indicates that the architecture for the BG prediction models is built with a standard configuration of densely connected encoding-decoding layers, LSTM layers, and RL parameter settings. However, relying solely on human intuition for this default setup may result in sub-optimal network architecture. In this chapter, the introduction of the NAS framework alongside BO offers a systematic method for (a) crafting tailored networks optimized for predicting future BG levels for individual patients and (b) efficiently navigating through a vast search space to converge on a patient-specific optimal model. The NAS framework starts with the architecture presented in prior work [14] iteratively generating candidate architectures that are optimized in network architecture, prediction performance, and model size. The candidate architecture, generated by the NAS framework, is evaluated on the BG prediction task. The performance of the candidate architecture, or the prediction error, on the BG prediction task for the particular subject is used as a reward signal for the NAS controller. The feedback signal helps the NAS controller decide on how well the generated candidate architecture performed on the prediction task, thereby helping it generate a new candidate architecture to meet the specific objective. We use a BO-based framework called Optuna [28] to find architectures with a low complexity that achieve high performance in the BG prediction task, optimized in a subject-specific manner. In addition to the prior work [14] which only involved DRL with LSTM as the sole variant of the RNN, this chapter includes GRUs as an additional variant of the RNNs, alongside the major contribution to the design of NAS with Bayesian optimization framework on top of the DRL. Figure 3 elucidates the model predictions in a moving window fashion.

Consistent with the state-of-the-art baseline [10], we use only CGM values for the prediction task. Using a patient's past six BG values as inputs, we predict future BG values in the forecasting prediction horizon of 30 and 45 minutes, corresponding to six and nine BG values, respectively. The models are trained and evaluated on all 12 subjects. In the 30-minute prediction horizon, patient ID 570 exhibited the best-case prediction performance, while patient ID 584 represents the worst-case scenario in terms of the analytical performance measured over RMSE on the prediction task using only the DRL method. Therefore, the selection of patient ID 570 and ID 584 for the NAS with DRL experiments is based on their representation of extreme cases. By focusing on these two scenarios, we aim to comprehensively assess the effectiveness of the NAS with DRL approach in improving prediction accuracy across a wide spectrum of patient scenarios. Comprehensive details regarding the network architecture are provided in the following sections.

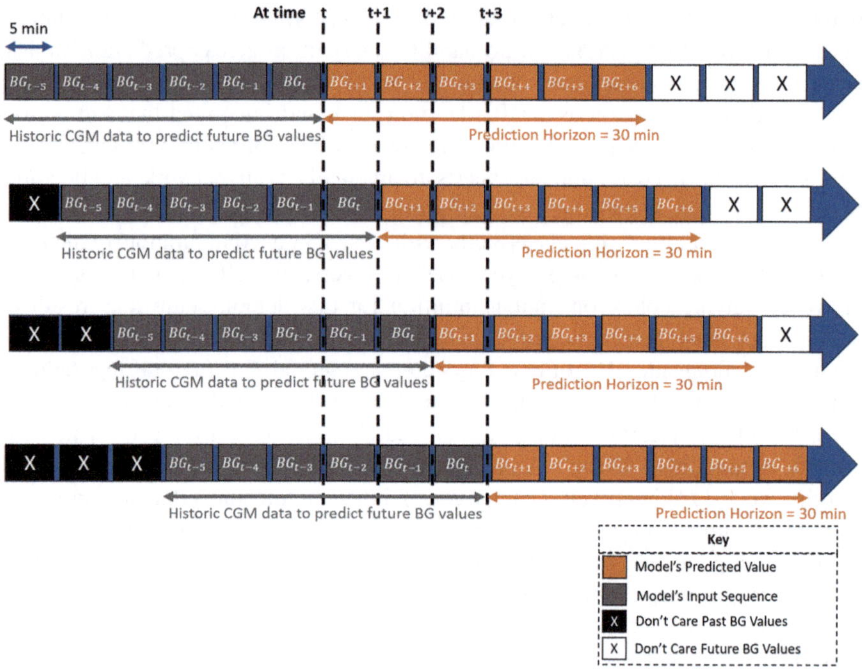

Fig. 3 Using a subject's past six BG values as inputs, our presented method predicts future BG values in the forecasting prediction horizon of 30 and 45 minutes, corresponding to six (as shown in this figure) and nine BG values, respectively, in a moving window fashion (Adapted from [7])

3.1 *Deep Neural Networks*

Prior work identified DNNs as promising models for BG prediction [33, 15] on the OhioT1DM dataset [23]. Variants of the same including CNN [33] and CRNN [16] were also studied. We combine densely connected neural networks with LSTM or GRU, within a DRL framework. Diverging from other studies, we present a DRL and NAS approach over supervised learning, thereby adapting both the architecture search space and strategy to aid the training process.

3.1.1 Densely Connected Encoding Networks

Densely connected networks are a type of neural network architecture where each neuron in one layer is linked to every neuron in the subsequent layer. Reasons for utilizing these networks include their high representational power, flexibility, and effective feature learning, incorporating multiple levels of abstraction. In several applications, they have proven to be effective when integrated with convolutional networks, serving as feature extractors, as exemplified in architectures like DenseNet [54] and similar other models.

Encoding Networks The input encoding network comprises a densely connected neural network with three layers (256, 512, and 256 neurons per layer with ReLU as the activation function). This architecture leverages their representational power to extract informative features for the subsequent LSTM or GRU network. Likewise, the output encoding networks capture features from the LSTM or GRU for the final BG prediction through the projection layer.

Projection Layer It performs the prediction of the BG values. The size of the (densely connected) layer aligns with the number of time steps in the prediction horizon. Employing a tanh normal projection, the projection layer transforms the output to create a normal distribution with a mean and standard deviation, which represents the probability distribution across prediction values (referred to as actions) in the DRL framework, as discussed in Sect. 3.2. The use of tanh normal projection ensures an efficient distribution with a nonzero mean, promoting bounded output and enhancing numerical stability.

3.1.2 Long Short-Term Memory

An RNN variant specifically crafted for handling sequential data, such as time-series data, employs a feedback mechanism (referred to as memory) to incorporate prior network outputs for current predictions. Illustrated in Fig. 1, the LSTM structure features multiple gates (input, forget, and output gates) regulating information flow through time within the memory [55]. The LSTM cell processes the input x_t at time step t, the previous cell memory c_{t-1}, and the previous output h_{t-1} of the cell to produce the outputs c_t and h_t, respectively. The input gate in (1) and the forget gate in (2) govern the cell state c_t, as seen in (4), wherein the operator $\odot$ denotes the Hadamard product. Along with the outcome of the output gate in (3), the hidden state h_t of the cell is determined in (5):

$$i_t = \text{sigmoid}\left(W_i x_t + U_i h_{t-1} + b_i\right) \tag{1}$$

$$f_t = \text{sigmoid}\left(W_f x_t + U_f h_{t-1} + b_f\right) \tag{2}$$

$$o_t = \text{sigmoid}\left(W_o x_t + U_o h_{t-1} + b_o\right) \tag{3}$$

$$c_t = f_t \odot c_{t-1} + i_t \odot \tanh\left(W_c x_t + U_c h_{t-1} + b_c\right) \tag{4}$$

$$h_t = o_t \odot \tanh\left(c_t\right) \tag{5}$$

The presence of these gates and the recurrent structure within the network enable us to mitigate the vanishing gradient problem and enhance the learning of long-term dependencies, as discussed in [56].

3.1.3 Gated Recurrent Unit

A GRU is similar in structure to LSTMs but without a forget gate, thus consisting only of a reset and an update gate [57]. Therefore, GRUs have fewer trainable parameters than LSTMs and are widely used in tasks involving natural language processing, speech recognition, and time-series prediction. The general structure of a GRU cell is given in Fig. 1. As shown, the input at a given time t is x_t, and the output of the previous time step is h_{t-1}. The resulting hidden state h_t is calculated as given in Eqs. (8) and (9), including the calculations of the reset and update gates in Eqs. (6) and (7). In the following equations, W_z, U_z, and b_z are (trainable) parameter matrices or vectors, respectively:

$$z_t = \text{sigmoid}\left(W_z x_t + U_z h_{t-1} + b_z\right) \tag{6}$$

$$r_t = \text{sigmoid}\left(W_r x_t + U_r h_{t-1} + b_r\right) \tag{7}$$

$$\tilde{h}_t = \tanh\left(W_h x_t + U_h\left(r_t \odot h_{t-1}\right) + b_h\right) \tag{8}$$

$$h_t = \left(1 - z_t\right) \odot h_{t-1} + z_t \odot \tilde{h}_t \tag{9}$$

Despite the widespread adoption of GRUs, the effectiveness of GRUs and LSTMs depends on the specific application [58].

3.2 *Deep Reinforcement Learning*

DRL merges the representational capabilities of DL with the decision-making proficiency of RL. It constitutes a subset of ML tailored for complex, high-dimensional decision-making tasks, where an agent learns optimal behavior by interacting with its environment, receiving rewards or penalties based on the outcomes of its actions [59]. DRL tasks are conceptualized as Markov decision processes, defining the spaces of states $\mathcal{S}$, agent actions $\mathcal{A}$, and a scalar training signal $\mathcal{R}$ (reward). At each time step t, the agent gets a state $s_t \in \mathcal{S}$ and chooses an action $a_t \in \mathcal{A}$ following a policy $\pi\left(a_t|s_t\right)$. Consequently, the agent receives a reward $r_t \in \mathcal{R}$ and transitions to the next state $s_{t+1} \in \mathcal{S}$. We parameterize the policy π using a DNN and aim to maximize the cumulative reward (return) during training. Analogous to the approach in [60], we adopt Actor-Critic training algorithms. These algorithms employ a value function to predict future rewards, where the value function in Eq. (10) denotes the anticipated total (discounted) reward originating from state s, with γ representing the discount factor:

$$V(s) = \mathrm{E}_\pi\left[\sum_{k=0}^{\infty} \gamma^k r_{t+k+1} | s_t = s\right] \tag{10}$$

Hereafter, our emphasis is on the Soft Actor-Critic (SAC) algorithm [61], identified as the most effective among the algorithms studied in [60].

Soft Actor-Critic The DRL algorithm designed for continuous control tasks adheres to the Actor-Critic framework [62]. This algorithm incorporates two networks—the actor and the critic network—to learn the policy and estimate a soft value function akin to Eq. (10). The soft value function is employed to update the policy using temporal difference (TD) error. SAC's value function and entropy regularization enhance exploration and prevent the policy from becoming overly deterministic. Additionally, the automatic temperature tuning simplifies the hyperparameter adjustment process, promoting stability and robustness in training [61]. Nevertheless, the training process may demand more computational resources, potentially extending the training time compared to conventional supervised approaches.

In the subsequent experiments, we employ an actor and a critic network with a structure illustrated in Fig. 1. To frame the BG prediction as a DRL task, we define the state $\mathcal{S}$ at time t as a vector:

$$\mathcal{S}(t) = [BG_{t-5}, BG_{t-4}, BG_{t-3}, BG_{t-2}, BG_{t-1}, BG_t]^T \tag{11}$$

The reward is determined as described in Eq. (12), and actions are represented by continuous BG values within the range of 35–500 mg/dL for a given prediction horizon. Throughout the training process, the agent generates continuous BG values from 35 to 500 based on the observed state of the environment:

$$R_{t+1} = -|BG_t - a_t| \tag{12}$$

The reward is formulated in a way that it approaches zero when the output action a_t closely aligns with the ground truth BG value B_t and deviates significantly from zero (in the negative direction) otherwise.

3.3 Automated Machine Learning

Automated machine learning (AutoML) is a subfield of ML that focuses on algorithms, techniques, and systems to automate and optimize ML pipelines [63]. AutoML encompasses various areas and aspects of ML pipelines to make AI more accessible, efficient, and accountable in different applications, e.g., healthcare [64]. We focus on hyperparameter optimization, NAS, and resource efficiency, e.g., for resource-constrained environments like edge devices. In the following experiments, we use model-based Bayesian optimization for NAS to optimize resource efficiency.

3.3.1 Neural Architecture Search

NAS is used to automatically design optimal or highly efficient neural network structures for a given task. It aims to optimize the network parameters and topology, such as the number of layers or units per layer, to maximize the performance on the given task. The process of NAS involves multiple components where key considerations are the definition of the search space, the scoring metric to evaluate the quality of different architectures, and the search algorithm to explore the search space. Whereas the search space encodes a set of task-specific architectural hyperparameters and the scoring metric a task-specific quality measure, different search algorithms exist [65]. We opt for BO, which is a probabilistic, model-based optimization technique that uses a surrogate model (usually a Gaussian process) of the objective function and a corresponding acquisition function iteratively to guide the selection of new samples in the search space. After each iteration, the surrogate model and acquisition function are updated to select the next configuration. BO is commonly used in AutoML frameworks and other applications, e.g., drug discovery. Thus, many software packages for BO are available. We use Optuna [28] with a tree-structured Parzen estimator (TPE) to perform NAS. Among other methods such as RL or genetic algorithms, BO is especially appealing to us as follows:

- BO is particularly efficient in scenarios where the evaluation of the objective function is time-consuming or expensive. This is true in our case because the DRL training needs to be performed in every iteration.
- BO is designed to find global optimum, making it less likely to get stuck in local optima.
- BO can adaptively balance exploration and exploitation in hyperparameter optimization. Additionally, BO provides room for iterative refinement.
- BO can also handle black box and noisy objective functions.

4 Experimental Setup

4.1 Dataset

This chapter utilizes the widely used OhioT1DM dataset [23], a key resource in blood glucose level prediction (BGLP) research. After the 2020 BGLP Challenge, the dataset now includes data from 12 subjects managing T1DM with insulin pump therapy. Collected over 8 weeks per individual, the dataset incorporates CGM, insulin, physiological sensors, and self-reported life event data. Participants used Medtronic 530G or 630G insulin pumps and Medtronic Enlite CGM sensors, ensuring complete anonymization with randomly assigned ID numbers. Six subjects used Empatica sensor band, while the other six used Basis peak sensor to monitor real-time physiological signals. The dataset comprises 19 features, including CGM values monitored at 5-minute intervals. Notably among those features are BG values from periodic self-monitoring of BG via finger sticks, insulin doses involving both

Table 3 Description of the OhioT1DM dataset properties [23] (Adapted from [7])

Patient ID	Gender	Age range	Pump model	Sensor band	Training samples	Test samples
540	Male	20–40	630G	Empatica	11,947	2884
544	Male	40–60	630G	Empatica	10,623	2704
552	Male	20–40	630G	Empatica	9080	2352
567	Female	20–40	630G	Empatica	10,858	2377
584	Male	40–60	530G	Empatica	12,150	2653
596	Male	60–80	530G	Empatica	10,877	2731
559	Female	40–60	530G	Basis	10,796	2514
563	Male	40–60	530G	Basis	12,124	2570
570	Male	40–60	530G	Basis	10,982	2745
575	Female	40–60	530G	Basis	11,866	2590
588	Female	40–60	530G	Basis	12,640	2791
591	Female	40–60	530G	Basis	18,847	2760

bolus and basal separately, self-reported meal times with estimated carbohydrate intake, and self-reported times of exercise, sleep, work, stress, and illness. Demographically, it involves seven male and five female subjects, aged between 20 and 80 years. Table 3 provides a detailed overview of the dataset characteristics. Consistent with the state-of-the-art baseline [10], we use only CGM values for the prediction task.

4.2 Preprocessing

To standardize the data, we employ z-score normalization on the training set and apply the normalization score obtained from the training set to the test dataset. This method rescales the values of each feature within the data to achieve zero mean and unit variance. The computation of z-score normalization on the considered CGM feature is outlined in Eq. (13). It requires the determination of the distribution mean (μ_x) and standard deviation (σ_x) for the specific feature (x) on the training dataset:

$$x' = \frac{x - \mu_x}{\sigma_x} \tag{13}$$

The rationale behind employing z-score normalization lies in the correlation between average BG values and key indicators such as HbA1C (A1C), a metric to measure the amount of blood glucose attached to hemoglobin and conditions like hyperglycemia. In this chapter, the z-score normalization is done on the training set, and the normalization score obtained from the training set is applied to the test dataset, as the only preprocessing step while considering CGM values. Z-score normalization centrally revolves around the mean, and clinically speaking, differences in A1C and mean glucose levels have been noted among different racial groups, and these variations are even more pronounced among individuals belonging to the same racial group [66, 67]. Additionally, this normalization method prioritizes identifying potentially risky fluctuations in BG values while disregarding statistical aspects in the data that do not contribute meaningful information for accurate CGM prediction [68].

4.3 Evaluation Metrics

Evaluation metrics serve as tools for quantifying the error of a system. Numerous metrics are available to assess the error of BG prediction tasks. Analytical error evaluation involves quantitative approaches to depict the proximity of predictions to the ground truth. In contrast, clinical error serves as a qualitative measure, assessing the clinical outcome of the prediction results. As such, its definition encompasses statistical metrics and the expertise of clinicians.

Analytical Error The primary numerical metric employed is the RMSE, as outlined in Eq. (14). Furthermore, for assessing the analytical error of BG prediction in this study, the MAE, detailed out in Eq. (15), is also used:

$$\text{RMSE} = \sqrt{\frac{1}{n}\sum_{i=1}^{n}\left(y_{\text{predicted}} - y_{\text{measured}}\right)^2} \tag{14}$$

$$\text{MAE} = \frac{1}{n}\sum_{i=1}^{n}\left|y_{\text{predicted}} - y_{\text{measured}}\right| \tag{15}$$

Clinical Error The SEG [29] is a recently adopted grid-based visualization method for assessing clinical error. It serves as a metric for both error and clinical risk evaluations in BG measurements [29]. Generally, the error grid displays a series of risk zones with assigned scores (risk levels) reflecting clinical impact, ranging from 0 (none) to 4 (extreme). In Fig. 4a, a simplified SEG [29] is presented, discretely structured with limits spanning from 0 to 600 mg/dL and risk zones separated by 120 mg/dL intervals. The predictions are overlaid on a continuously color-coded SEG. The color indicates the average risk rating determined by clinician respondents in a survey [29]. It represents the collective decision of experts, reflecting the mean group consensus. Furthermore, CEG [53], which was developed in 1987 to assess the clinical risk, is used to assess the clinical risk for the predicted BG values to the ground truth in the test set. The visualization of the CEG risk zones is presented in Fig. 4b.

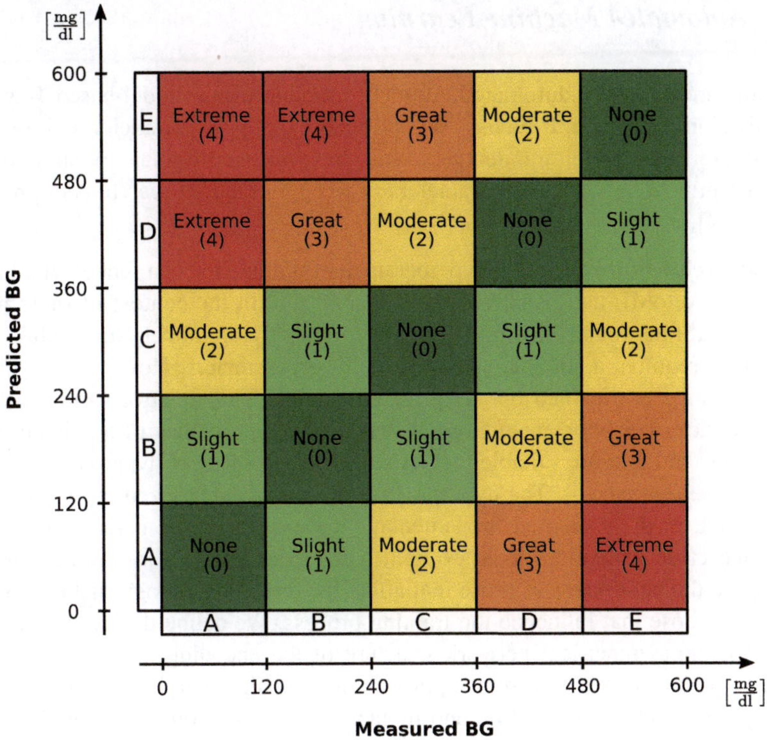

(a)

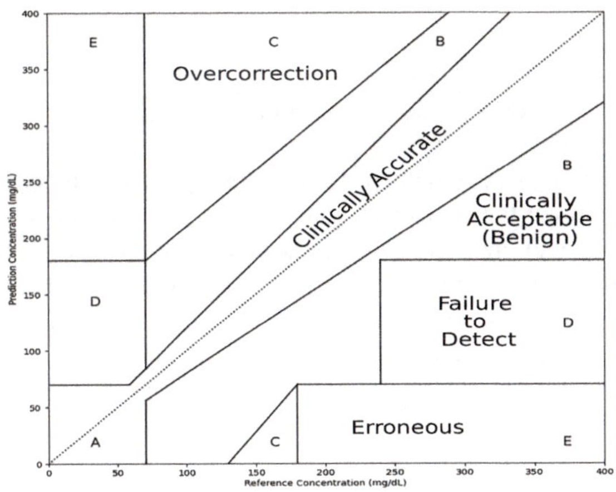

(b)

Fig. 4 Visualization of risk zones for two different metrics to evaluate the performance of the predicted BG values in clinical settings. (**a**) Visualization of SEG risk zones. (**b**) Visualization of CEG risk zones (Adapted from [7])

4.4 Automated Machine Learning

We implement a fully automated AutoML pipeline using model-based Bayesian optimization in Optuna. It focuses on optimizing the hyperparameter and topological structures of model architectures. Still, the presented framework supports the manual optimization of hyperparameters or to apply random search as in previous works [14].

Neural Architecture Search Hyperparameter optimization is a fundamental component of AutoML and usually plays a crucial role in the context of (automatic) training and fine-tuning ML models. Therefore, applying models successfully to a given task requires a suitable selection of hyperparameters. However, the hyperparameters are not learned but set prior to training. They control various aspects of the model's behavior, including the structure and properties of architectures or algorithms, and thus are suitable for NAS. The search space of hyperparameters for NAS is given in Table 4. The hyperparameters can be categorized in two different ways: First, in the presented implementation, we differ in the parameters for the actor and critic network, which are both part of the SAC algorithm. Second, we categorize the parameters as those that affect the model's topology (network structures) and those that influence the training process. We define the hyperparameter such that the symmetrical network structure of the encoding networks shown in Fig. 1 is preserved. We extend the approach of manually applying a random search for hyperparameter optimization and replace it with an automatized, model-based Bayesian methodology using Optuna. Therefore, we define the objective metric as given in Eq. (16):

$$f_{\text{NAS}} = \frac{\text{RMSE}_{\text{train, NAS}}}{\text{RMSE}_{\text{train, initial}}} + \frac{\text{model size}}{\text{initial model size}} \tag{16}$$

The objective function f_{NAS} is defined in a multi-objective manner where the first term aims to minimize the RMSE on the training data and the second aims to reduce the overall model size. To balance the optimization of both terms and thus find the best models with low error and small model size, we normalize the terms with the initial values of the baseline. Initially, the presented architecture (actor network) has a size of around 1.39 M trainable parameters and a test RMSE error of approximately 20, depending on the patient. We chose the value ranges of individual hyperparameters in alignment with existing literature [14]. In the following, if not mentioned otherwise, we generate 100 candidate architectures for every NAS iteration, considering the time and computing expenses associated with conducting experiments. The individual candidate architectures are evaluated concerning the performance of the training data and model size to determine the best hyperparameters.

Table 4 Description of hyperparameter search space for NAS (Adapted from [7])

Parameter		Category	Values
Cell type		Actor/critic structure	{LSTM, GRU}
Cell size		Actor/critic structure	{4, 8, 16, 32, 64, 128, 256}
2*Observation FCN	Layers	Critic structure	{1, 2, 3}
	Neurons	Critic structure	{4, 8, 16, 32, 64, 128, 256, 512}
2*Input FCN	Layers	Actor structure	{1, 2, 3}
	Neurons	Actor structure	{4, 8, 16, 32, 64, 128, 256, 512}
Action FCN		Critic structure	{4, 8, 16, 32, 64, 128, 256, 512}
2*Joint FCN	Layers	Critic structure	{1, 2, 3}
	Neurons	Critic structure	{4, 8, 16, 32, 64, 128, 256, 512}
2*Output FCN	Layers	Actor/critic structure	{1, 2, 3}
	Neurons	Actor/critic structure	{4, 8, 16, 32, 64, 128, 256, 512}
Activation function		Actor/critic structure	{ReLU, Tanh, Sigmoid}
Target update period		Training	[1, 100]
Target update τ		Training	[0.001, 1.0]

5 Results

5.1 Model Performance

The presented DRL and NAS algorithms are trained on the OhioT1DM dataset [23], which includes continuous BG level monitoring at 5-minute intervals for 12 patients. During the training phase, we utilize only the BG recordings of each patient from the OhioT1DM dataset in an 80–20 train-test split ratio. For each patient, training is conducted over 30-minute intervals (equivalent to six consecutive BG level recordings), followed by a multistep prediction for the BG level in the subsequent 30 or 45 minutes. The obtained results are clinically evaluated using SEG and CEG and compared analytically to the baseline [10] in terms of RMSE and MAE. The rest of the section analyzes the prediction results for the entire dataset spanning 250 hours, focusing on multistep predictions.

Multistep Predictions for 30 Minutes Using DRL Framework The SAC model presented in this chapter predicts the BG level for 6 consecutive steps using a batch size of 1024 and undergoes 50,000 training steps, considering input data over a 30-minute duration. The evaluation of RMSE and MAE for all 12 patients is presented in Table 5 along with the standard deviation, for predicting all values throughout the 30-minute prediction interval, corresponding to six data points. Comparative results with prior work which uses the same number of features, and all subjects for train and test cases, including LSTM, WaveNet, GRU, ensemble methods, and a potential combination of these approaches, are highlighted in Table 6. It is evident that our presented method consistently demonstrates significant improvement over the baseline [10] across all patients, as indicated by both metrics. On average, for all

Table 5 Evaluation of the presented SAC-based DRL approach for all 12 subjects while predicting BG values throughout the 30-minute prediction horizon, with respect to RMSE and MAE metrics (Adapted from [7])

	Root mean square error			Mean absolute error		
Patient ID	Our presented method	Baseline [10] (LSTM + WaveNet + GRU)	% Improvement over baseline	Our presented method	Baseline [10] (LSTM + WaveNet + GRU)	% Improvement over baseline
540	19.25 ± 10.49	25.28	23.83	13.80 ± 9.58	18.77	12.99
544	15.28 ± 7.33	19.76	22.63	11.34 ± 6.45	14.36	12.99
552	14.49 ± 7.14	19.43	25.37	10.58 ± 6.42	14.66	27.83
559	19.37 ± 11.02	21.78	19.77	13.47 ± 9.83	15.35	12.22
563	17.77 ± 10.09	20.43	12.97	12.30 ± 9.34	14.39	14.46
567	22.20 ± 14.17	23.96	7.32	14.68 ± 12.74	17.41	12.62
570	15.80 ± 7.69	18.06	12.49	11.71 ± 6.97	12.85	8.83
575	20.37 ± 12.21	25.02	18.58	14.01 ± 11.16	16.77	16.45
584	24.30 ± 14.33	24.84	3.23	16.60 ± 12.76	18.57	10.57
588	15.02 ± 7.55	21.26	29.33	11.04 ± 6.77	15.55	29.00
591	18.09 ± 9.74	23.76	23.83	12.99 ± 8.82	18.61	28.43
596	17.31 ± 9.28	19.23	9.93	12.62 ± 8.47	13.58	7.03

Table 6 Comparison of the presented SAC-based DRL approach with prior work which uses only historical CGM values to predict future BG values for all 12 subjects while predicting all values throughout the 30-minute prediction horizon. (Adapted from [7])

Method	Avg. root mean square error	Avg. mean absolute error
LSTM [10]	22.13	16.02
BiLSTM [21]	20.05	14.29
Vanilla LSTM [21]	19.83	14.09
Ensemble via stacking [21]	19.63	13.88
Ensemble via multivariate [21]	19.64	13.92
Ensemble via subsequences [21]	19.62	13.88
GRU-NN [30]	21.54	15.39
WaveNet [10]	22.49	16.47
GRU [10]	22.00	15.91
WaveNet + LSTM [10]	22.35	16.29
WaveNet + GRU [10]	22.21	16.15
LSTM + GRU [10]	21.98	15.86
LSTM + WaveNet + GRU [10]	21.90	15.87
Deep RL (**presented method**)	**18.32**	**12.93**

Table 7 Classification results of the presented SAC-based DRL approach in terms of accuracy, precision, recall, and F1 score for hypoglycemic, normoglycemic, and hyperglycemic range for all 12 subjects, corresponding to all predicted BG values throughout the 30-minute prediction horizon (Adapted from [7])

Patient ID	Accuracy	Hypoglycemia Precision	Recall	F1 score	Normoglycemia Precision	Recall	F1 score	Hyperglycemia Precision	Recall	F1 score
540	0.885	0.546	0.528	0.545	0.915	0.896	0.906	0.875	0.916	0.895
544	0.918	0.243	0.273	0.257	0.920	0.963	0.941	0.946	0.845	0.893
552	0.923	0.581	0.507	0.541	0.950	0.951	0.951	0.861	0.872	0.867
559	0.909	0.671	0.662	0.667	0.921	0.922	0.922	0.909	0.908	0.908
563	0.885	0.300	0.167	0.214	0.912	0.891	0.902	0.851	0.887	0.869
567	0.869	0.879	0.437	0.584	0.870	0.946	0.906	0.864	0.807	0.834
570	0.947	0.333	0.091	0.143	0.894	0.928	0.911	0.971	0.959	0.965
575	0.896	0.704	0.496	0.582	0.914	0.922	0.918	0.884	0.913	0.898
584	0.888	0.667	0.074	0.133	0.903	0.917	0.910	0.864	0.864	0.864
588	0.895	1.000	0.250	0.400	0.876	0.936	0.905	0.920	0.849	0.883
591	0.881	0.575	0.615	0.595	0.894	0.931	0.912	0.912	0.812	0.859
596	0.900	0.635	0.446	0.524	0.897	0.976	0.935	0.942	0.718	0.815
Average	0.899	0.594	0.378	0.432	0.905	0.931	0.918	0.899	0.862	0.879

12 patients, our presented method enhances RMSE by 16.34% and MAE by 18.4%. Additionally, the BG prediction regression task was translated to a classification task using data labeling approach [69] for hypoglycemia, normoglycemia, and hyperglycemia for BG values BG ≤ 70 mg/dL, 70 mg/dL $<$ CGM $<$ 180 mg/dL, and BG ≥ 180 mg/dL, respectively. For the predictions of all the BG values throughout the 30-minute prediction interval, corresponding to six data points, the

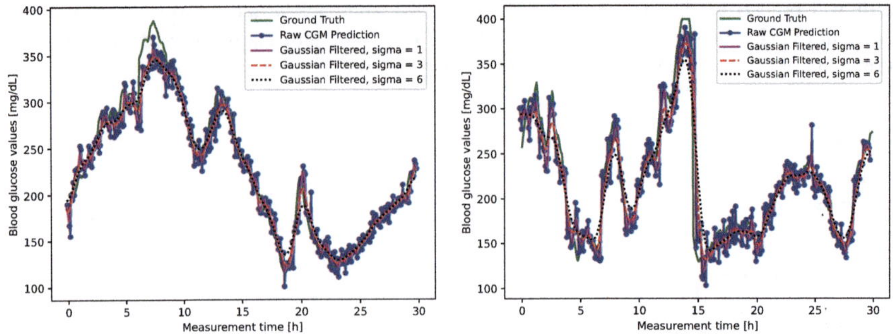

(a) (b)

Fig. 5 BG ground truth in comparison to the predictions and Gaussian smoothing for a representative sample of the testing dataset of 30 hours for patient ID 570 and patient ID 584. (**a**) Patient ID 570. (**b**) Patient ID 584 (Adapted from [7])

classification results are presented in Table 7. Table 8 highlights our prediction results in terms of RMSE and MAE for all patients for predicting only BG values at 30-minute prediction interval.

Multistep Predictions for 45 Minutes Using DRL Framework The presented SAC model predicts the BG level for 9 consecutive steps, utilizing a batch size of 1024 and undergoing 50,000 training steps, considering input data over a 30-minute interval. The evaluation of RMSE and MAE for all 12 patients is detailed in Table 9. Table 10 provides a comparative analysis of the presented method against the baseline [10], which is a combination of LSTM, WaveNet, and GRU. Notably, the presented method exhibits significant improvement over the baseline [10], as reflected in both metrics. On average, across all 12 patients, the presented method enhances RMSE by 19.53% and MAE by 22.5%. For a visual representation of the presented model's fitting to the data, Fig. 5 depicts the ground truth compared to the prediction results for a representative test dataset spanning 30 hours for patient IDs 570 and 584. Gaussian filtering with varying σ is used to reduce the fluctuations arising from sensor faults, connectivity problems, electrical interference, and manufacturing variability [22, 70, 71], see Fig. 5. Additionally, the BG prediction regression task was translated to a classification task using data labeling approach [69], for hypoglycemia, normoglycemia, and hyperglycemia for BG values BG ≤ 70 mg/dL, 70 mg/dL $<$ CGM $<$ 180 mg/dL, and BG $\geq$ 180 mg/dL respectively. For the predictions of all the BG values throughout the 45-minute prediction interval, corresponding to nine data points, the classification results are presented in Table 11. Also, a comparative analysis of the presented method against prior work is highlighted in Table 12, many of which use additional datasets to train the model, perform testing on six out of the twelve OhioT1DM subjects, and harness more than one feature to predict the future BG values.

Table 8 Evaluation of the presented SAC-based DRL approach for all 12 subjects while predicting BG values at the 30-minute prediction mark, with respect to RMSE and MAE metrics (Adapted from [7])

Patient ID	Root mean square error Our presented method	Mean absolute error Our presented method
540	27.48	21.19
544	20.23	14.89
552	19.72	14.82
559	25.38	18.61
563	22.61	16.79
567	28.85	20.41
570	20.56	15.64
575	26.03	18.76
584	32.04	22.53
588	21.11	15.84
591	24.41	17.69
596	23.06	17.23

Table 9 Evaluation of the presented SAC-based DRL approach for all 12 subjects while predicting BG values throughout the 45-minute prediction horizon, with respect to RMSE and MAE metrics (Adapted from [7])

Patient ID	Root mean square error Our presented method	Mean absolute error Our presented method
540	24.99 ± 13.21	18.02 ± 11.67
544	20.52 ± 10.90	14.84 ± 9.87
552	19.82 ± 10.29	14.42 ± 9.28
559	24.83 ± 13.72	17.60 ± 12.23
563	20.88 ± 12.40	14.35 ± 11.29
567	27.87 ± 17.54	18.49 ± 15.72
570	19.32 ± 9.50	14.15 ± 8.53
575	25.15 ± 14.97	17.24 ± 13.66
584	31.51 ± 16.10	22.68 ± 14.70
588	20.45 ± 10.92	14.70 ± 9.91
591	23.08 ± 12.78	16.38 ± 11.50
596	22.72 ± 10.35	16.94 ± 9.35
Average	23.43 ± 12.72	16.65 ± 11.47

Table 10 Comparison of the presented SAC-based DRL approach with prior work which uses only historical CGM values to predict future BG values for all 12 subjects while predicting all values throughout the 45-minute prediction horizon (Adapted from [7])

Method	Root mean square error	Mean absolute error
LSTM [10]	29.28	21.61
WaveNet [10]	29.68	22.19
GRU [10]	29.22	21.50
WaveNet + LSTM [10]	29.46	21.87
WaveNet + GRU [10]	29.44	21.83
LSTM + GRU [10]	29.26	21.56
LSTM + WaveNet + GRU [10]	29.12	21.52
Deep RL (**presented method**)	**23.43**	**16.66**

Table 11 Classification results of the presented SAC-based DRL approach in terms of accuracy, precision, recall, and F1 score for hypoglycemic, normoglycemic, and hyperglycemic range for all 12 subjects, corresponding to all predicted BG values throughout the 45-minute prediction horizon (Adapted from [7])

Patient		Hypoglycemia			Normoglycemia			Hyperglycemia		
ID	Accuracy	Precision	Recall	F1 score	Precision	Recall	F1 score	Precision	Recall	F1 score
540	0.868	0.614	0.370	0.462	0.850	0.955	0.899	0.940	0.779	0.852
544	0.888	0.222	0.077	0.114	0.879	0.969	0.922	0.929	0.745	0.827
552	0.901	0.411	0.550	0.471	0.940	0.933	0.936	0.850	0.825	0.837
559	0.870	0.522	0.541	0.532	0.896	0.879	0.887	0.859	0.881	0.870
563	0.874	0.342	0.52	0.413	0.913	0.871	0.892	0.834	0.884	0.858
567	0.835	0.798	0.553	0.654	0.861	0.900	0.880	0.767	0.753	0.760
570	0.934	0.182	0.125	0.148	0.883	0.894	0.888	0.959	0.956	0.957
575	0.875	0.669	0.573	0.617	0.872	0.940	0.905	0.918	0.800	0.855
584	0.857	0.294	0.143	0.192	0.889	0.878	0.883	0.814	0.841	0.827
588	0.866	0.125	0.167	0.143	0.849	0.911	0.879	0.891	0.815	0.852
591	0.851	0.615	0.417	0.497	0.866	0.918	0.891	0.845	0.776	0.809
596	0.877	0.525	0.294	0.376	0.891	0.949	0.919	0.849	0.722	0.781
Average	0.875	0.443	0.361	0.385	0.882	0.916	0.898	0.871	0.815	0.840

Clinical Evaluation for Multistep Prediction Utilizing SEG, we evaluate the clinical error between predicted and ground truth BG levels for all 12 patients. SEG calculates the percentage of predictions within predefined risk zones, as outlined by clinical practices. Table 13 presents the SEG results for all the subjects of the OhioT1DM dataset, followed by the comparison to prior work in Table 14, for the 30-minute PH, wherein six BG values are predicted. Additionally, Fig. 6 reports the clinical error using the CEG metric for patient ID 570 and patient ID 584 corresponding to the best case and worst case, respectively. Table 15 presents the SEG results for all the subjects of the OhioT1DM dataset, followed by the comparison to prior work in Table 16, for the 45-minute PH, wherein nine BG values are predicted. Figure 7 visualizes the clinical error using the SEG metric for patient ID 570 and patient ID 584 corresponding to the best case and worst case, respectively. In general, through SEG analysis, it is evident that a substantial proportion of the prediction results falls within the no risk to slight risk zone, with a few outliers. Furthermore, the predictions consistently either outperform or align with the baseline.

To prove the effectiveness of the presented method over other datasets, we accessed S1 dataset [43], and D1NAMO dataset [44], and used the historical CGM values to predict BG values throughout the prediction horizon of 30 mins, corresponding to six BG values. Both the S1 dataset [43] and the D1NAMO dataset [44] have CGM values recorded from individuals with type 2 diabetes mellitus (T2DM) and T1DM individuals, respectively, at an interval of 5 minutes. Table 17 reports RMSE and MAE for five selected patients from both datasets. A low RMSE and MAE prove the efficacy of the presented method.

Table 12 Comparison for all predictions made throughout the PH of 30 mins and 45 mins over average RMSE and average MAE (Adapted from [7])

Reference	Additional dataset	Subjects from OhioT1DM dataset	Number of features used	PH=30 mins		PH=45 mins	
				Avg. RMSE	Avg. MAE	Avg. RMSE	Avg. MAE
[19]	Yes	Train: 12 subjects Test: 6 subjects	CGM, insulin bolus rate, bolus amount, carbohydrate intake, difference between consecutive glucose values	20.10	13.63	N/A	N/A
[11]	No	Train: 12 subjects Test: 6 subjects	Finger-stick value, basal rate value, galvanic skin response value, skin temperature value, bolus dose value, CGM	20.6	18.3	N/A	N/A
[33]	No	Train: 6 subjects Test: 6 subjects	CGM, bolus amount, carbohydrate intake, time index normalized to the unit for each day	21.72	N/A	N/A	N/A
[21]	No	Train: 12 subjects Test: 6 subjects	CGM	19.63	13.88	N/A	N/A
[10]	No	Train: 12 subjects Test: 12 subjects	CGM	21.90	15.86	29.12	21.50
[30]	No	Train: 12 subjects Test: 12 subjects	CGM	21.54	15.39	N/A	N/A
[18]	Yes	Train: 6 subjects Test: 6 subjects	CGM, basal insulin dose, bolus insulin dose, carbohydrate intake	17.80	N/A	N/A	N/A
[20]	No	Train: 12 subjects Test: 6 subjects	Interstitial fluid glucose concentration from CGM recordings, basal insulin infusion rate, insulin bolus delivery, carbohydrate oral intake	18.90	N/A	N/A	N/A
[52]	No	Train: 6 subjects Test: 6 subjects	CGM, carbohydrate intake, bolus insulin	18.34	13.37	N/A	N/A
[22]	No	Train: 12 subjects Test: 12 subjects	CGM, carbohydrate intake, bolus insulin, finger-stick glucose	16.06	10.64	N/A	N/A
[49]	No	Train: 6 subjects Test: 6 subjects	CGM	20.03	14.52	N/A	N/A
[15]	No	Train: 6 subjects Test: 6 subjects	CGM, insulin, carbohydrate intake, exercise	19.79	13.62	N/A	N/A

(continued)

Table 12 Comparison for all predictions made throughout the PH of 30 mins and 45 mins over average RMSE and average MAE (Adapted from [7])

[12]	Yes	Train: 6 subjects Test: 6 subjects	Sampling time, CGM values, meal intake, and insulin dose	18.90	N/A	N/A	N/A
[48]	Yes	Train: 6 subjects Test: 6 subjects	CGM, finger stick, bolus dose, carbohydrate input, time of day encoded using sine and cosine embeddings	18.22	12.83	N/A	N/A
[50]	No	Train: 6 subjects Test: 6 subjects	CGM, basal insulin, bolus insulin, insulin on board	19.37	13.76	N/A	N/A
[51]	No	Train: 6 subjects Test: 6 subjects	CGM, insulin on board, carbohydrates on board, exercise on board, slope of CGM, time of day	18.69	10.08	N/A	N/A
[13]	No	Train: 6 subjects Test: 6 subjects	CGM, basal insulin dosage, bolus insulin dosage, carbohydrate intake, time stamp	19.05	13.50	N/A	N/A
[16]	No	Train: 6 subjects Test: 6 subjects	Basal insulin, bolus insulin, carbohydrates, CGM	17.45	23.25	N/A	N/A
[17]	No	Train: 6 subjects Test: 6 subjects	CGM	18.23	14.37	N/A	N/A
Presented work	No	Train: 12 subjects Test: 12 subjects	CGM	18.32	12.93	23.43	16.66

Table 13 Evaluation of the presented SAC-based DRL approach for all 12 subjects while predicting BG values throughout the 30-minute prediction horizon, with respect to SEG metric (Adapted from [7])

Patient ID	None (0)	Slight (1)	Moderate (2)	Great (3)	Extreme (4)
540	85.104	14.896	0.000	0.000	0.000
544	90.815	9.185	0.000	0.000	0.000
552	89.541	10.459	0.000	0.000	0.000
559	89.928	9.952	0.120	0.000	0.000
563	89.085	10.915	0.000	0.000	0.000
567	86.869	12.963	0.168	0.000	0.000
570	90.367	9.560	0.073	0.000	0.000
575	89.161	10.801	0.039	0.000	0.000
584	85.332	14.517	0.075	0.075	0.000
588	91.361	8.531	0.108	0.000	0.000
591	89.592	10.408	0.000	0.000	0.000
596	89.011	10.989	0.000	0.000	0.000
Average	88.847	11.098	0.048	0.000	0.000

Table 14 Comparison of the presented SAC-based DRL approach with prior work which uses only historical CGM values to predict future BG values for all 12 subjects while predicting all values throughout the 30-minute prediction horizon, with respect to the SEG metric (Adapted from [7])

Method	None (0)	Slight (1)	Moderate (2)	Great (3)	Extreme (4)
LSTM [10]	86.42	13.56	0.02	0.00	0.00
WaveNet [10]	85.91	14.06	0.03	0.00	0.00
GRU [10]	86.41	13.55	0.04	0.00	0.00
WaveNet + LSTM [10]	86.43	13.54	0.03	0.00	0.00
WaveNet + GRU [10]	86.39	13.57	0.04	0.00	0.00
LSTM + GRU [10]	86.52	13.44	0.04	0.00	0.00
LSTM + WaveNet + GRU [10]	86.53	13.45	**0.02**	0.00	0.00
Deep RL (**Presented method**)	**88.85**	**11.10**	0.05	**0.00**	**0.00**

5.2 *Performance Optimization*

In all prior work [10, 14], hyperparameters and network architectures were manually defined which often necessitated a significant depth of domain knowledge. Moreover, these models were not optimized for performance on analytical and clinical accuracy, network architecture, and model size in a subject-specific manner. This chapter presents a NAS approach using sequential, model-based BO. We implement Optuna as a BO framework and optimize the model obtained from the DRL framework with respect to performance and model size, respectively. We perform 100 trials using the defined hyperparameter search space in 4.4. In each trial, the NAS generates candidate architectures, which are trained using the DRL framework. The parameters *Target Update Period* and *Target Update* τ are used as the feedback to the NAS controller in each trial since they strongly affect the training

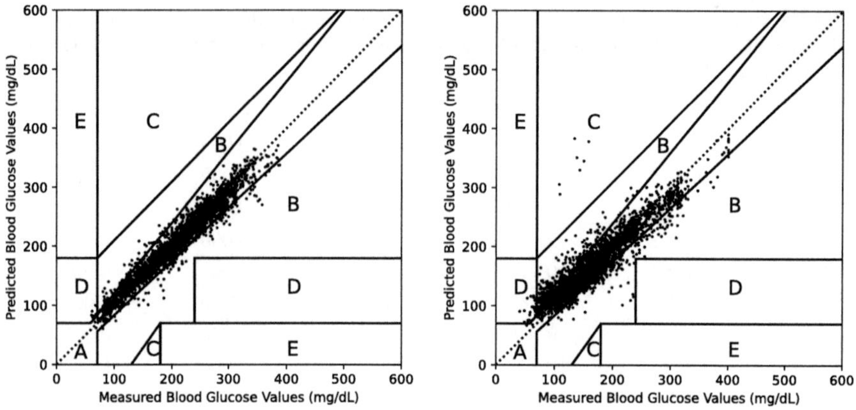

(a) (b)

Fig. 6 CEG for BG predictions in comparison to measurements for a representative sample of the testing dataset of 30 hours for patient ID 570 and patient ID 584. (**a**) CEG for patient ID 570. (**b**) CEG for patient ID 584 (Adapted from [7])

Table 15 Evaluation of the presented SAC-based DRL approach for all 12 subjects while predicting BG values throughout the 45-minute prediction horizon, with respect to SEG metric (Adapted from [7])

Patient ID	None (0)	Slight (1)	Moderate (2)	Great (3)	Extreme (4)
540	86.343	13.542	0.116	0.000	0.000
544	89.605	10.395	0.000	0.000	0.000
552	89.344	10.656	0.000	0.000	0.000
559	89.022	10.951	0.027	0.000	0.000
563	91.111	8.863	0.0261	0.000	0.000
567	85.035	14.515	0.366	0.084	0.000
570	92.015	7.790	0.195	0.000	0.000
575	86.817	13.157	0.0259	0.000	0.000
584	82.212	17.611	0.151	0.025	0.000
586	90.380	9.620	0.000	0.000	0.000
591	86.847	13.153	0.000	0.000	0.000
596	87.151	12.849	0.000	0.000	0.000

of SAC. The remaining hyperparameters are used to optimize the model's structure and performance.

5.2.1 Neural Architecture Search

Figure 8 shows the performance and model size of the individual candidate architectures presented by the NAS framework. We investigate the trade-off between RMSE on the test data and the size (number of trainable parameters) of the

Table 16 Comparison of the presented SAC-based DRL approach with prior work which uses only historical CGM values to predict future BG values for all 12 subjects while predicting all values throughout the 45-minute prediction horizon, with respect to the SEG metric averaged for all 12 subjects (Adapted from [7])

Method	None (0)	Slight (1)	Moderate (2)	Great (3)	Extreme (4)
LSTM [10]	80.95	18.98	0.07	0.00	0.00
WaveNet [10]	79.88	20.07	0.05	0.00	0.00
GRU [10]	81.08	18.85	0.07	0.00	0.00
WaveNet + LSTM [10]	81.04	18.90	0.06	0.00	0.00
WaveNet + GRU [10]	81.01	18.92	0.07	0.00	0.00
LSTM + GRU [10]	81.10	18.83	0.07	0.00	0.00
LSTM + WaveNet+ GRU [10]	81.14	18.80	**0.06**	**0.00**	0.00
Deep RL (**Presented method**)	**87.99**	**11.93**	0.08	0.01	**0.00**

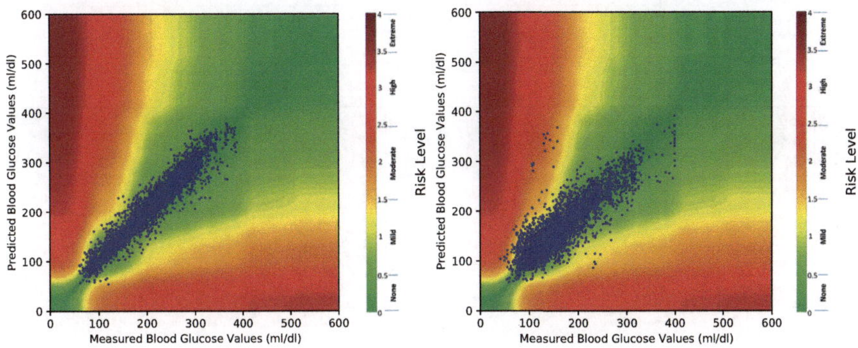

(a) (b)

Fig. 7 SEG for BG predictions in comparison to measurements for a representative sample of the testing dataset of 30 hours for patient ID 570 and patient ID 584 while predicting all values throughout the 45-minute prediction horizon. (**a**) SEG patient ID 570. (**b**) SEG patient ID 584 (Adapted from [7])

actor network (critic network is only necessary for training). For both patients, comparative results of LSTM-based and GRU-based candidate architectures with prior work in [14] are indicated by the red dashed lines. Both architecture types show a high sensitivity to different hyperparameter settings, including outliers with respect to test performance with an RMSE up to 140 (training is no longer converging). However, Fig. 8 visualizes that the BO-based approach can identify higher amounts of suitable hyperparameter sets, as shown by the clusters in the scatterplots. With these hyperparameter sets, we achieve competitive results but using a fraction of network parameters. Depending on the specific patient and type of architecture, the presented approach reduces the model size by a factor between 20 and 150 without sacrificing performance w.r.t. RMSE. A more detailed analysis with respect to test performance and model size is given in Tables 18 and 19, respectively. In Table 18, we show the top five candidate architectures with

Table 17 Root mean square error (RMSE) and mean absolute error (MAE) evaluation of the presented method for BG predictions in T2DM individuals [43] and BG predictions on T1DM for D1NAMO dataset [44] for a prediction horizon of 30 mins (Adapted from [7])

Dataset	Patient ID	RMSE Our presented method	MAE Our presented method
S1 dataset [43]	1636 - 69 - 001	11.01	7.98
	1636 - 69 - 026	12.62	8.57
	1636 - 69 - 028	17.41	12.62
	1636 - 69 - 048	8.88	6.37
	1636 - 69 - 053	18.57	14.37
D1NAMO [44]	001	22.71	17.31
	002	27.54	21.21
	003	14.91	12.00
	005	18.86	14.59
	007	15.93	11.710

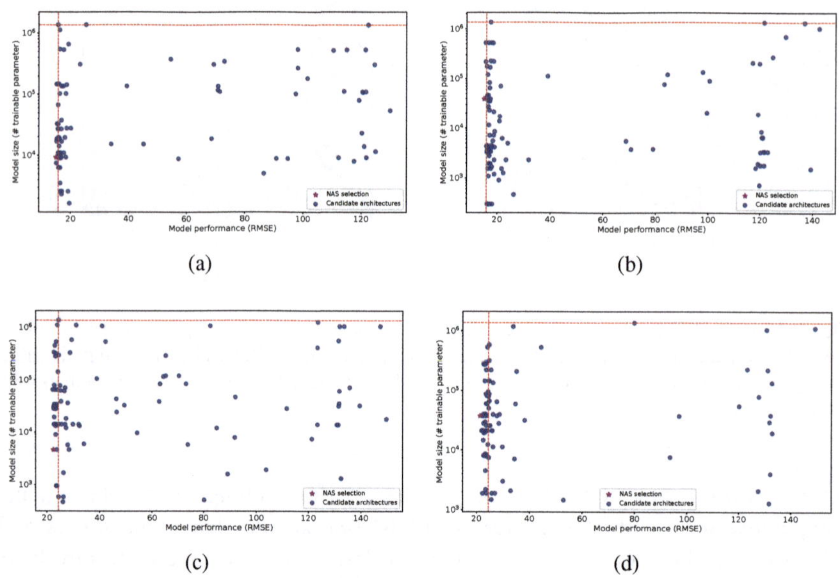

Fig. 8 Model size vs. test performance (RMSE) for candidate architectures presented by our NAS framework. The scatterplots 8a and 8b show the result on patient ID 570 for different architecture types. Similarly, the plots 8c and 8d show results on patient ID 584, respectively. (**a**) LSTM-based NAS candidate architectures for patient ID 570. (**b**) GRU-based NAS candidate architectures for patient ID 570. (**c**) LSTM-based NAS candidate architectures for patient ID 584. (**d**) GRU-based NAS candidate architectures for patient ID 584 (Adapted from [7])

Table 18 Overview of top five candidate architectures sorted by test performance (here, RMSE). Additionally, we specify the number of parameters to show the different scaling behavior of the architecture types (Adapted from [7])

Top five	LSTM-based architecture		GRU-based architecture	
architectures (RMSE)	Parameters	Error (RMSE)	Parameters	Error (RMSE)
5*Patient ID 570	**9428**	**15.050**	**19,212**	**14.898**
	7552	15.175	1,192,460	15.594
	148,980	15.267	279,436	15.625
	17,076	15.291	101,516	15.909
	18,716	15.335	224,524	15.926
5*Patient ID 584	**65,884**	**21.876**	**37,920**	**21.246**
	4668	22.384	21,428	21.809
	30,556	22.479	37,920	21.839
	28,096	22.503	21,716	22.220
	338,316	22.580	21,408	22.224

Table 19 Overview of top five candidate architectures sorted the number of parameters. Additionally, we specify the test RMSE to show the different scaling behavior of the architecture types (Adapted from [7])

Top five	LSTM-based architecture		GRU-based architecture	
architectures (complexity)	Parameters	Error (RMSE)	Parameters	Error (RMSE)
5*Patient ID 570	1672	19.557	308	16.176
	2348	16.922	308	16.179
	2620	19.141	308	17.257
	2620	16.556	308	16.680
	2620	17.060	308	16.517
5*Patient ID 584	472	25.940	1320	131.642
	520	80.117	1492	52.776
	588	24.336	1508	25.458
	588	26.125	1920	22.226
	956	23.756	1920	26.814

respect to their performance on the test data. In the given experimental setting, our framework prefers network architectures with high performance over a small model size. Therefore, we select architectures that improve previous results, e.g., in [14], between 5% and 12%, while sacrificing a possible reduction in model size by a factor of up to 4500. Depending on the patient, GRU-based architectures outperform LSTM-based structures but do not always show better scaling behavior with respect to test performance and small model sizes; see Fig. 8 and Table 19.

Additionally, the BG prediction regression task was translated to a classification task using data labeling approach [69] for hypoglycemia, normoglycemia, and hyperglycemia for BG values BG ≤ 70 mg/dL, 70 mg/dL $<$ CGM $<$ 180 mg/dL, and BG $\geq$ 180 mg/dL, respectively. For the predictions of all the BG values

Table 20 Classification results on patient ID 570 and patient ID 584 for the presented NAS in conjunction to DRL approach in terms of accuracy, precision, recall, and F1 score for hypoglycemic, normoglycemic, and hyperglycemic range for all 12 subjects, corresponding to all predicted BG values throughout the 30-minute prediction horizon (Adapted from [7])

Patient		Hypoglycemia			Normoglycemia			Hyperglycemia		
ID	Accuracy	Precision	Recall	F1 score	Precision	Recall	F1 score	Precision	Recall	F1 score
570 (LSTM architecture)	0.949	0.235	0.364	0.286	0.925	0.898	0.912	0.965	0.973	0.969
570 (GRU architecture)	0.949	0.364	0.364	0.364	0.912	0.913	0.913	0.968	0.967	0.967
584 (LSTM architecture)	0.891	0.5	0.074	0.129	0.912	0.911	0.911	0.86	0.881	0.87
584 (GRU architecture)	0.922	0.583	0.259	0.359	0.931	0.942	0.937	0.91	0.906	0.908

throughout the 30-minute prediction interval, corresponding to six data points, the classification results are presented in Table 20.

Figure 9 visualizes the distribution of the model parameters across the individual architectural components. Interestingly, the input encoding seems less relevant and thus typically requires less trainable parameters. The remaining parameters are almost equally distributed across the output encoder and the RNN, with a slight preference for one of the two depending on the network type and patient in the dataset.

Multistep Predictions for 30 Minutes In the following, we predict BG values for six consecutive time steps while considering input data with a time frame of 30 minutes using the best model of the NAS run for LSTM-based and GRU-based architectures and patient ID 570 as well as 584. For patient ID 570, the presented method achieved improvements of 4.8% (RMSE: 15.050) using LSTM-based architectures and 5.7% (RMSE: 14.898) with GRU-based architectures, respectively. Furthermore, increasing improvement for patient ID 584 with 10.0% (RMSE: 21.876) and 12.6% (RMSE: 21.246), respectively, was observed. Figure 10 shows how well the selected candidate architectures fit the test data. Therefore, it visualizes the ground truth BG values compared to the model predictions for a representative test data of 30 hours for both patients. Similar to Fig. 5, we observe that both architecture types fit the data well but show small imprecision predictions at sudden changes in BG values. Gaussian filtering with varying σ is used to reduce the fluctuations arising from sensor faults, connectivity problems, electrical interference, and manufacturing variability [22, 70, 71]; see Fig. 10.

SEG for Multistep Prediction We assess the risk between using the prediction of the selected candidate architectures and the ground truth BG values. The SEG visualizations of patient ID 570 and ID 584 for the different architecture types are shown in Fig. 11. Using the SEG metric to evaluate clinical risks, we conclude that most predictions are in no risk to slight risk zones. Moreover, the results remain

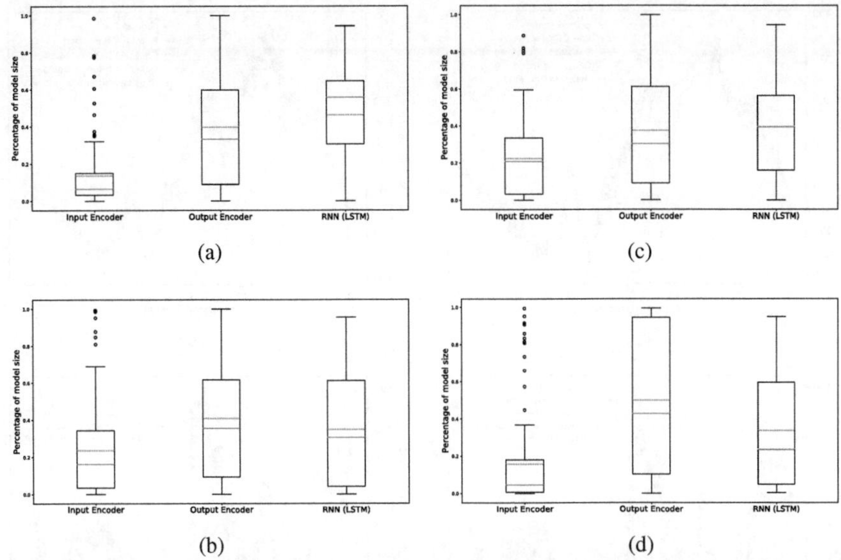

Fig. 9 The box plots show the distribution of (trainable) model parameters across the individual model components: the input encoder, the output encoder, and the RNN. (**a**) Distribution of model size for LSTM-based candidates, patient ID 570 (**b**) Distribution of model size for LSTM-based candidates, patient ID 584. (**c**) Distribution of model size for GRU-based candidates, patient ID 570. (**d**) Distribution of model size for GRU-based candidates, patient ID 584 (Adapted from [7])

consistent with state of the art [10, 14]. Thus, the presented architectures effectively reduce the model size without increasing clinical risks severely (Fig. 12).

In Table 21, we present the average SEG values in multistep prediction of 30 minutes using the presented architectures of the NAS framework. Compared with the latest results in [14], we encounter only small changes in the none and slight risk zones in a range of ±3% (see Table 21).

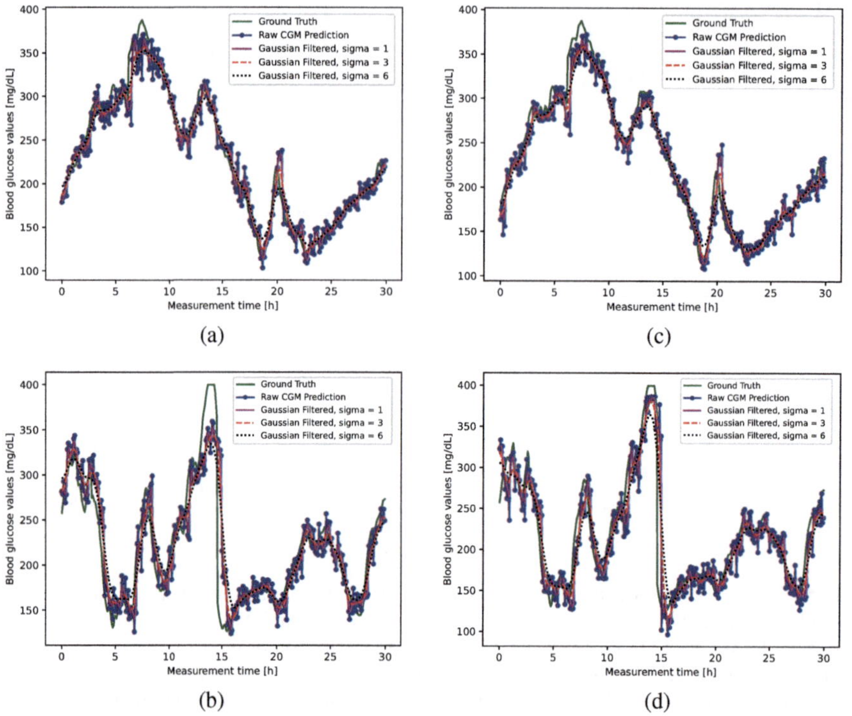

Fig. 10 BG ground truth in comparison to the predictions and Gaussian smoothing for a representative sample of the testing dataset of 30 hours for patient ID 570 and patient ID 584 for our presented method of DRL in conjunction with NAS. (**a**) Patient ID 570 (LSTM architecture). (**b**) Patient ID 584 (LSTM architecture). (**c**) Patient ID 570 (GRU architecture). (**d**) Patient ID 584 (GRU architecture) (Adapted from [7])

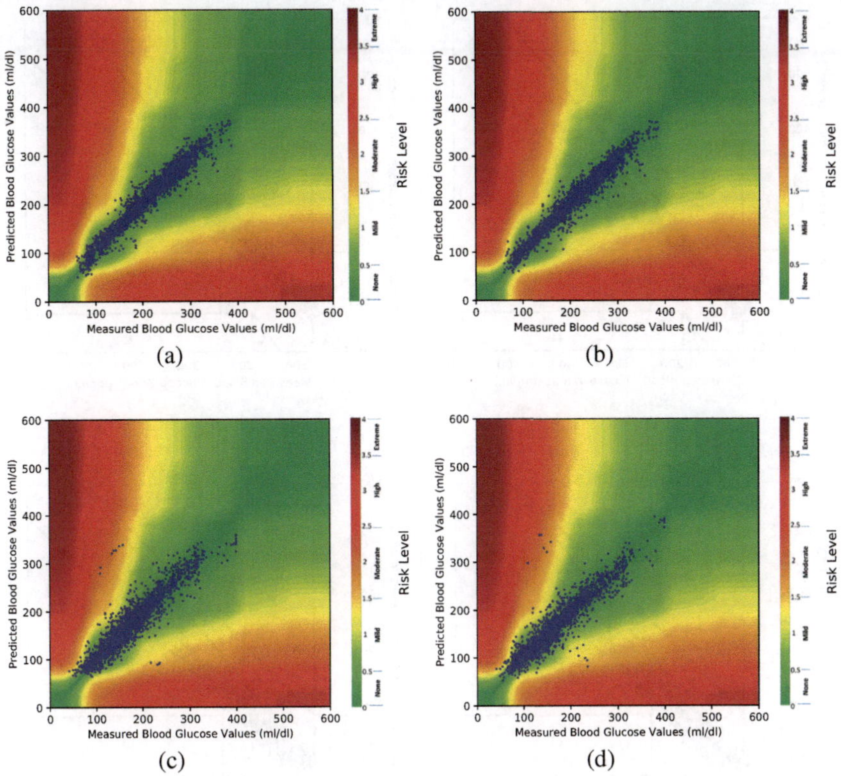

Fig. 11 SEG for NAS-based BG predictions in comparison to the measurements for a representative sample of the testing dataset of 30 hours for patient ID 570 (11a, 11b) and patient ID 584 (11c, 11d). (**a**) SEG for LSTM-based architectures on patient ID 570. (**b**) SEG for GRU-based architectures on patient ID 570. (**c**) SEG for LSTM-based architectures on patient ID 584. (**d**) SEG for GRU-based architectures on patient ID 584 (Adapted from [7])

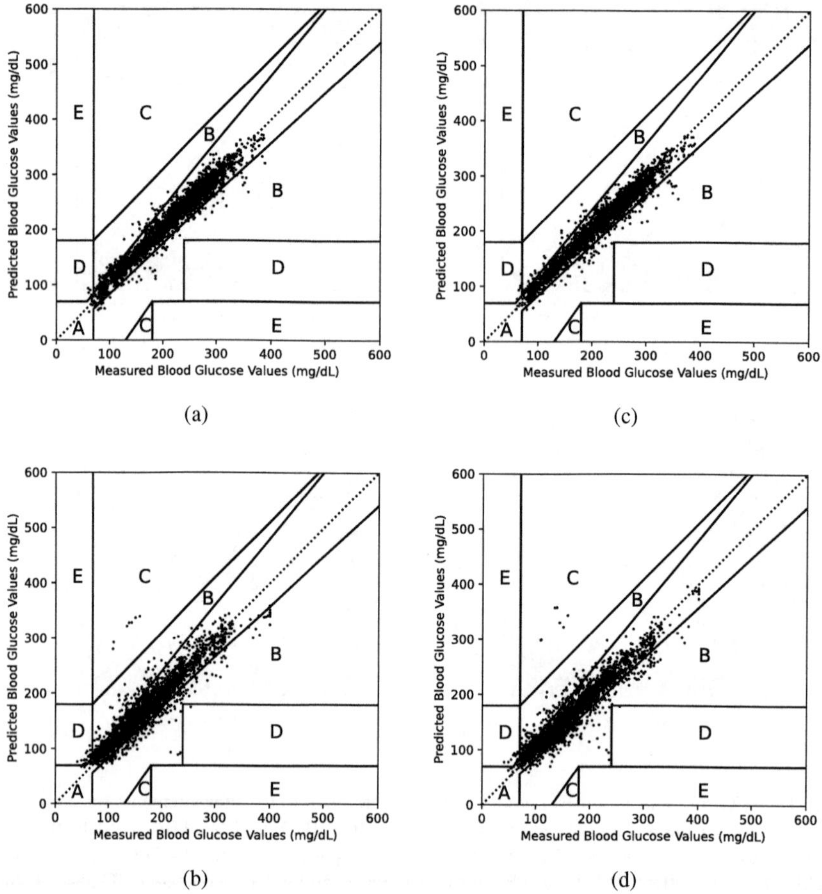

Fig. 12 SEG for NAS-based BG predictions in comparison to the measurements for a representative sample of the testing dataset of 30 hours for patient ID 570 (12a, 12c) and patient ID 584 (12b, 12d). (**a**) CEG for LSTM-based architectures on patient ID 570. (**b**) CEG for LSTM-based architectures on patient ID 584. (**c**) CEG for GRU-based architectures on patient ID 570. (**d**) CEG for GRU-based architectures on patient ID 584 (Adapted from [7])

Table 21 Average SEG in NAS-based multistep prediction results (PH=30 mins) (Adapted from [7])

	None (0)			Slight (1)			Moderate (2)			Great (3)			Extreme (4)		
		NAS			NAS			NAS			NAS			NAS	
Patient ID	[14]	LSTM	GRU	[14]	LSTM	GRU	[14]	LSTM	GRU	[14]	LSTM	GRU	[14]	LSTM	GRU
570	90.367	89.267	91.575	9.560	10.659	8.425	0.073	0.073	0.000	0.000	0.000	0.000	0.000	0.000	0.000
584	85.332	88.311	86.953	14.517	11.614	12.971	0.075	0.075	0.075	0.075	0.000	0.000	0.000	0.000	0.000

6 Conclusion

We presented a novel methodology for time-series modeling of BG levels in T1DM using a densely connected encoder-decoder network and an LSTM or GRU formulated as a DRL problem. We have evaluated the results for the OhioT1DM dataset benchmark. Furthermore, we presented an automatized optimization workflow, effectively introducing NAS in BG prediction tasks. Compared to the prior work that achieved the best prediction error, on average and for the MAE, the presented method has improved by 18.4% and 22.5% in 30-minute and 45-minute prediction horizons without applying additional optimization of the network's architectural structure. The prediction results on the best- (ID 570) and worst-case (ID 584) patients using only the DRL method were further enhanced by integrating the BO-based NAS to the DRL framework. The presented method achieved additional improvements of 4.8% using LSTM-based architectures and 5.7% with GRU-based architectures for patient ID 570 by integrating NAS. The patient with the lowest performance (ID 584) on the DRL method shows an even greater performance boost, with improvements of 10.0% and 12.6% observed for the LSTM or GRU, respectively. Moreover, for risk assessment in our predictions, we have visualized the error and evaluated clinical risk through an SEG and a CEG approach. We have shown that we outperform or remain consistent with the results of the previous risk assessments and that NAS can reduce the model size by a significant factor without increasing clinical risks to unacceptable levels.

For future work, the successful compression of the model sizes paves the way for applying selected candidate architectures on resource-constrained devices, e.g., edge or smart devices. Therefore, deploying the model efficiently on edge devices will be another goal of the future work.

Acknowledgments The authors acknowledge the use of GenAI tools for editing this chapter.

References

1. Domanski P, et al. Advancing blood glucose prediction with neural architecture search and deep reinforcement learning for type 1 diabetics. Biocybernetics and biomedical engineering, biocybernetics and biomedical engineering 44(3):481–500.
2. American Diabetes Association. Diagnosis and classification of diabetes mellitus. Diabetes care. 2014 Jan 1;37(Supplement_1):S81–90.
3. Mahmoudi, G. Farahani, P. Domanski, B. Farahani, F. Firouzi, K. Chakrabarty, "DiabLLM: An LLM-Based Framework for Blood Glucose Prediction in Type 1 Diabetes", *IEEE Journal of Biomedical and Health Informatics*, 2026. IEEE
4. Deshpande AD, Harris-Hayes M, Schootman M. Epidemiology of diabetes and diabetes-related complications. Physical therapy. 2008 Nov 1;88(11):1254–64.
5. Ogurtsova K, Guariguata L, Barengo NC, Ruiz PL, Sacre JW, Karuranga S, Sun H, Boyko EJ, Magliano DJ. IDF diabetes Atlas: Global estimates of undiagnosed diabetes in adults for 2021. Diabetes research and clinical practice. 2022 Jan 1;183:109118.

6. Elsayed N, ElSayed Z, Ozer M. Early Stage Diabetes Prediction via Extreme Learning Machine. In: SoutheastCon 2022 2022 Mar 26 (pp. 374–379). IEEE.
7. Domanski, Peter, Aritra Ray, Kyle Lafata, Farshad Firouzi, Krishnendu Chakrabarty, and Dirk Pflüger. "Advancing blood glucose prediction with neural architecture search and deep reinforcement learning for type 1 diabetics." *Biocybernetics and Biomedical Engineering*, vol. 44, no. 3, 2024, pp. 481–500.
8. Domanski, Peter, Aritra Ray, Farshad Firouzi, Kyle Lafata, Krishnendu Chakrabarty, and Dirk Pflüger. "Blood Glucose Prediction for Type-1 Diabetics using Deep Reinforcement Learning." In *2023 IEEE International Conference on Digital Health (ICDH)*, pp. 339–347. IEEE, 2023.
9. Liardo, Anthony, Aritra Ray, Farshad Firouzi, Kyle Lafata, and Krishnendu Chakrabarty. "Neural Architecture Search for Blood Glucose Prediction in Type-1 Diabetics." In *2024 IEEE 20th International Conference on Body Sensor Networks (BSN)*, pp. 1–4. IEEE, 2024.
10. Dudukcu HV, Taskiran M, Yildirim T. Blood glucose prediction with deep neural networks using weighted decision level fusion. Biocybernetics and Biomedical Engineering. 2021 Jul 1;41(3):1208–23.
11. Bhimireddy A, Sinha P, Oluwalade B, Gichoya JW, Purkayastha S. Blood glucose level prediction as time-series modeling using sequence-to-sequence neural networks. CEUR Workshop Proceedings.
12. Zhu T, Li K, Chen J, Herrero P, Georgiou P. Dilated recurrent neural networks for glucose forecasting in type 1 diabetes. Journal of Healthcare Informatics Research. 2020 Sep;4:308–24.
13. Yang T, Wu R, Tao R, Wen S, Ma N, Zhao Y, Yu X, Li H. Multi-Scale Long Short-Term Memory Network with Multi-Lag Structure for Blood Glucose Prediction. KDH@ ECAI. 2020 Aug 29;45:136–40.
14. Domanski P, Ray A, Firouzi F, Lafata K, Chakrabarty K, Pflüger D. Blood Glucose Prediction for Type-1 Diabetics using Deep Reinforcement Learning. In: 2023 IEEE International Conference on Digital Health (ICDH) 2023 Jul 2 (pp. 339–347). IEEE.
15. Daniels J, Herrero P, Georgiou P. Personalised Glucose Prediction via Deep Multitask Networks. In: KDH@ ECAI 2020 Aug 29 (pp. 110–114).
16. Freiburghaus J, Rizzotti A, Albertetti F. A deep learning approach for blood glucose prediction of type 1 diabetes. In: Proceedings of the Proceedings of the 5th International Workshop on Knowledge Discovery in Healthcare Data co-located with 24th European Conference on Artificial Intelligence (ECAI 2020), 29–30 August 2020, Santiago de Compostela, Spain 2020. 29–30 August 2020.
17. Bevan R, Coenen F. Experiments in non-personalized future blood glucose level prediction. In: CEUR Workshop Proceedings 2020 Jan 1 (Vol. 2675, pp. 100–104).
18. Gu K, Dang R, Prioleau T. Neural physiological model: A simple module for blood glucose prediction. In: 2020 42nd annual international conference of the IEEE Engineering in Medicine & Biology Society (EMBC) 2020 Jul 20 (pp. 5476–5481). IEEE.
19. Hameed H, Kleinberg S. Investigating potentials and pitfalls of knowledge distillation across datasets for blood glucose forecasting. InProceedings of the 5th Annual Workshop on Knowledge Discovery in Healthcare Data 2020 Jan.
20. Cui R, Hettiarachchi C, Nolan CJ, Daskalaki E, Suominen H. Personalised short-term glucose prediction via recurrent self-attention network. In: 2021 IEEE 34th International Symposium on Computer-Based Medical Systems (CBMS) 2021 Jun 7 (pp. 154–159). IEEE.
21. Nemat H, Khadem H, Eissa MR, Elliott J, Benaissa M. Blood glucose level prediction: advanced deep-ensemble learning approach. IEEE Journal of Biomedical and Health Informatics. 2022 Jan 25;26(6):2758–69.
22. Shuvo MM, Islam SK. Deep Multitask Learning by Stacked Long Short-Term Memory for Predicting Personalized Blood Glucose Concentration. IEEE Journal of Biomedical and Health Informatics. 2023 Jan 2;27(3):1612–23.
23. Marling C, Bunescu R. The OhioT1DM dataset for blood glucose level prediction: Update 2020. In: CEUR workshop proceedings 2020 Sep (Vol. 2675, p. 71). NIH Public Access.

24. Benidis K, Rangapuram SS, Flunkert V, Wang Y, Maddix D, Turkmen C, Gasthaus J, Bohlke-Schneider M, Salinas D, Stella L, Aubet FX. Deep learning for time series forecasting: Tutorial and literature survey. ACM Computing Surveys. 2022 Dec 7;55(6):1–36.
25. Tang S, Pan Y. Feature Extraction via Recurrent Random Deep Ensembles and its Application in Gruop-level Happiness Estimation. arXiv preprint arXiv:1707.09871. 2017 Jul 24.
26. Keren G, Schuller B. Convolutional RNN: an enhanced model for extracting features from sequential data. In2016 International Joint Conference on Neural Networks (IJCNN) 2016 Jul 24 (pp. 3412–3419). IEEE.
27. Fu Q, Han Z, Chen J, Lu Y, Wu H, Wang Y. Applications of reinforcement learning for building energy efficiency control: A review. Journal of Building Engineering. 2022 Jun 1;50:104165.
28. Akiba T, Sano S, Yanase T, Ohta T, Koyama M. Optuna: A next-generation hyperparameter optimization framework. In: Proceedings of the 25th ACM SIGKDD international conference on knowledge discovery & data mining 2019 Jul 25 (pp. 2623–2631).
29. Klonoff DC, Lias C, Vigersky R, Clarke W, Parkes JL, Sacks DB, Kirkman MS, Kovatchev B, Error Grid Panel. The surveillance error grid. Journal of diabetes science and technology. 2014 Jul;8(4):658–72.
30. Dudukcu HV, Taskiran M, Yildirim T. Consolidated or individual training: Which one is better for blood glucose prediction?. In: 2021 International Conference on INnovations in Intelligent SysTems and Applications (INISTA) 2021 Aug 25 (pp. 1–6). IEEE.
31. Francescato MP, Geat M, Fusi S, Stupar G, Noacco C, Cattin L. Carbohydrate requirement and insulin concentration during moderate exercise in type 1 diabetic patients. Metabolism. 2004 Sep 1;53(9):1126–30.
32. Zhu T, Li K, Kuang L, Herrero P, Georgiou P. An insulin bolus advisor for type 1 diabetes using deep reinforcement learning. Sensors. 2020 Sep 6;20(18):5058.
33. Zhu T, Li K, Herrero P, Chen J, Georgiou P. A Deep Learning Algorithm for Personalized Blood Glucose Prediction. InKHD@ IJCAI 2018 Jul 13 (pp. 64–78).
34. Ubl M, Koutny T, Della Cioppa A, De Falco I, Tarantino E, Scafuri U. Distributed Assessment of Virtual Insulin-Pump Settings Using SmartCGMS and DMMS. R for Diabetes Treatment. Sensors. 2022 Dec 2;22(23):9445.
35. Man CD, Micheletto F, Lv D, Breton M, Kovatchev B, Cobelli C. The UVA/PADOVA type 1 diabetes simulator: new features. Journal of diabetes science and technology. 2014 Jan;8(1):26–34.
36. Kovatchev BP, Breton M, Dalla Man C, Cobelli C. In silico preclinical trials: a proof of concept in closed-loop control of type 1 diabetes.
37. Colmegna P, Bisio A, McFadden R, Wakeman C, Oliveri MC, Nass R, Breton M. Evaluation of a web-based simulation tool for self-management support in type 1 diabetes: A pilot study. IEEE Journal of Biomedical and Health Informatics. 2022 Sep 23;27(1):515–25.
38. [dataset] Xie J. jxx123/simglucose. GitHub. 2021. Available from: https://github.com/jxx123/simglucose
39. [dataset] Hettiarachchi C. chirathyh/GluCoEnv. GitHub. 2024 [cited 2024 Feb 1]. Available from: https://github.com/chirathyh/GluCoEnv
40. [dataset] What is OpenAPS? – OpenAPS.org. Openaps.org. 2018. Available from: https://openaps.org/what-is-openaps/
41. [dataset] Tidepool. www.tidepool.org. Available from: https://www.tidepool.org/
42. Broll S, Urbanek J, Buchanan D, Chun E, Muschelli J, Punjabi NM, Gaynanova I. Interpreting blood glucose data with R package iglu. PLoS One. 2021 Apr 1;16(4):e0248560.
43. Colás A, Vigil L, Vargas B, Cuesta–Frau D, Varela M. Detrended Fluctuation Analysis in the prediction of type 2 diabetes mellitus in patients at risk: Model optimization and comparison with other metrics. PloS one. 2019 Dec 18;14(12):e0225817.
44. Dubosson F, Ranvier JE, Bromuri S, Calbimonte JP, Ruiz J, Schumacher M. The open D1NAMO dataset: A multi-modal dataset for research on non-invasive type 1 diabetes management. Informatics in Medicine Unlocked. 2018 Jan 1;13:92–100.
45. Hall H, Perelman D, Breschi A, Limcaoco P, Kellogg R, McLaughlin T, Snyder M. Glucotypes reveal new patterns of glucose dysregulation. PLoS biology. 2018 Jul 24;16(7):e2005143.

46. Weinstock RS, DuBose SN, Bergenstal RM, Chaytor NS, Peterson C, Olson BA, Munshi MN, Perrin AJ, Miller KM, Beck RW, Liljenquist DR. Risk factors associated with severe hypoglycemia in older adults with type 1 diabetes. Diabetes Care. 2016 Apr 1;39(4):603–10.
47. Neinstein A, Wong J, Look H, Arbiter B, Quirk K, McCanne S, Sun Y, Blum M, Adi S. A case study in open source innovation: developing the Tidepool Platform for interoperability in type 1 diabetes management. Journal of the American Medical Informatics Association. 2016 Mar 1;23(2):324–32.
48. Rubin-Falcone H, Fox I, Wiens J. Deep Residual Time-Series Forecasting: Application to Blood Glucose Prediction. InKDH@ ECAI 2020 Jun (pp. 105–109).
49. Ma N, Zhao Y, Wen S, Yang T, Wu R, Tao R, Yu X, Li H. Online Blood Glucose Prediction Using Autoregressive Moving Average Model with Residual Compensation Network. In: KDH@ ECAI 2020 (pp. 151–155).
50. Sun X, Rashid MM, Sevil M, Hobbs N, Brandt R, Askari MR, Shahidehpour A, Cinar A. Prediction of Blood Glucose Levels for People with Type 1 Diabetes using Latent-Variable-based Model. In: KDH@ ECAI 2020 (pp. 115–119).
51. Pavan J, Prendin F, Meneghetti L, Cappon G, Sparacino G, Facchinetti A, Del Favero S. Personalized Machine Learning Algorithm based on Shallow Network and Error Imputation Module for an Improved Blood Glucose Prediction. In: KDH@ ECAI 2020 (pp. 95–99).
52. Zhu T, Yao X, Li K, Herrero P, Georgiou P. Blood glucose prediction for type 1 diabetes using generative adversarial networks. In: CEUR Workshop Proceedings 2020 Jan 1 (Vol. 2675, pp. 90–94).
53. Clarke WL. The original Clarke error grid analysis (EGA). Diabetes technology & therapeutics. 2005 Oct 1;7(5):776–9.
54. Zhou T, Ye X, Lu H, Zheng X, Qiu S, Liu Y. Dense convolutional network and its application in medical image analysis. BioMed Research International. 2022 Apr 25;2022.
55. Hochreiter S, Schmidhuber J. Long short-term memory. Neural computation. 1997 Nov 15;9(8):1735–80.
56. Hochreiter S, Bengio Y, Frasconi P, Schmidhuber J. Gradient flow in recurrent nets: the difficulty of learning long-term dependencies.
57. Cho K, Van Merriënboer B, Gulcehre C, Bahdanau D, Bougares F, Schwenk H, Bengio Y. Learning phrase representations using RNN encoder-decoder for statistical machine translation. arXiv preprint arXiv:1406.1078. 2014 Jun 3.
58. Chung J, Gulcehre C, Cho K, Bengio Y. Empirical evaluation of gated recurrent neural networks on sequence modeling. arXiv preprint arXiv:1412.3555. 2014 Dec 11.
59. Sutton RS, Barto AG. Reinforcement learning: An introduction. MIT press; 2018 Nov 13.
60. Liu T, Tan Z, Xu C, Chen H, Li Z. Study on deep reinforcement learning techniques for building energy consumption forecasting. Energy and Buildings. 2020 Feb 1;208:109675.
61. Haarnoja T, Zhou A, Abbeel P, Levine S. Soft actor-critic: Off-policy maximum entropy deep reinforcement learning with a stochastic actor. In: International conference on machine learning 2018 Jul 3 (pp. 1861–1870). PMLR.
62. Konda V, Tsitsiklis J. Actor-critic algorithms. Advances in neural information processing systems. 1999;12.
63. He X, Zhao K, Chu X. AutoML: A survey of the state-of-the-art. Knowledge-Based Systems. 2021 Jan 5;212:106622.
64. Waring J, Lindvall C, Umeton R. Automated machine learning: Review of the state-of-the-art and opportunities for healthcare. Artificial intelligence in medicine. 2020 Apr 1;104:101822.
65. Elsken T, Metzen JH, Hutter F. Neural architecture search: A survey. The Journal of Machine Learning Research. 2019 Jan 1;20(1):1997–2017.
66. Bergenstal RM, Gal RL, Connor CG, Gubitosi-Klug R, Kruger D, Olson BA, Willi SM, Aleppo G, Weinstock RS, Wood J, Rickels M. Racial differences in the relationship of glucose concentrations and hemoglobin A1c levels. Annals of internal medicine. 2017 Jul 18;167(2):95–102.

67. Beck RW, Connor CG, Mullen DM, Wesley DM, Bergenstal RM. The fallacy of average: how using HbA1c alone to assess glycemic control can be misleading. Diabetes care. 2017 Aug 1;40(8):994–9.
68. Peters AL. Role of Continuous Glucose Monitoring in Diabetes Treatment. Arlington (VA): American Diabetes Association. 2018 Aug:3–7.
69. D'Antoni F, Petrosino L, Marchetti A, Bacco L, Pieralice S, Vollero L, Pozzilli P, Piemonte V, Merone M. Layered meta-learning algorithm for predicting adverse events in type 1 diabetes. IEEE Access. 2023 Jan 18;11:9074–94.
70. Staal OM, Sælid S, Fougner A, Stavdahl Ø. Kalman smoothing for objective and automatic preprocessing of glucose data. IEEE journal of biomedical and health informatics. 2018 Mar 1;23(1):218–26.
71. Facchinetti A, Del Favero S, Sparacino G, Cobelli C. Modeling transient disconnections and compression artifacts of continuous glucose sensors. Diabetes technology & therapeutics. 2016 Apr 1;18(4):264–72.

GPSR Compliance

The European Union's (EU) General Product Safety Regulation (GPSR) is a set of rules that requires consumer products to be safe and our obligations to ensure this.

If you have any concerns about our products, you can contact us on ProductSafety@springernature.com

In case Publisher is established outside the EU, the EU authorized representative is:

Springer Nature Customer Service Center GmbH
Europaplatz 3
69115 Heidelberg, Germany

Batch number: 10542183

Printed by Printforce, the Netherlands